AF353030

Modern Technologies and Their Influence in Fermentation Quality

Modern Technologies and Their Influence in Fermentation Quality

Special Issue Editor

Santiago Benito

MDPI • Basel • Beijing • Wuhan • Barcelona • Belgrade • Manchester • Tokyo • Cluj • Tianjin

Special Issue Editor
Santiago Benito
Department of Chemistry and
Food Technology, Polytechnic
University of Madrid
Spain

Editorial Office
MDPI
St. Alban-Anlage 66
4052 Basel, Switzerland

This is a reprint of articles from the Special Issue published online in the open access journal *Fermentation* (ISSN 2311-5637) (available at: https://www.mdpi.com/journal/fermentation/special_issues/technologies_fermentation).

For citation purposes, cite each article independently as indicated on the article page online and as indicated below:

LastName, A.A.; LastName, B.B.; LastName, C.C. Article Title. *Journal Name* **Year**, *Article Number*, Page Range.

ISBN 978-3-03928-947-9 (Pbk)
ISBN 978-3-03928-948-6 (PDF)

Cover image courtesy of Santiago Benito.

Contents

About the Special Issue Editor

Santiago Benito is Director of the Polytechnic Madrid University Experimental Winery. He has written more than 50 scientific/indexed publications in renowned international journals, mostly on the topic of wine microbiology and winemaking. The author studied at the Polytechnic University of Madrid, where he was awarded his Food Engineering degree in 2004 and PhD degree in 2008. During 2004 to 2009, he worked as Technical Director of Bodegas Urbina Winery and was involved in numerous engineering projects involved in the construction of new different wineries in La Rioja wine region. He has been teaching Winemaking, Food Safety, and Refrigeration Engineering at the Polytechnic University of Madrid since his appointments as University Professor in 2009. Additionally, he has been teaching Wine Chemistry and Winemaking as part of the international Master Erasmus Mundus Vinifera at the Montpellier SubAgro University since 2006. He has enjoyed a stay as guest Researcher at the Accredited Wine laboratory "Estación Enológica de Haro" in 2015 and has taught as guest Professor at Geisenheim University on Microbiology and Oenology in Vinifera Euromaster as well as in the Bachelor course in International Wine Business during the 2013–2014 and 2017–2018 courses of the Professor's Erasmus Mundus program.

 *fermentation*

Editorial

Modern Technologies and Their Influence in Fermentation Quality

Santiago Benito

Departamento de Química y Tecnología de Alimentos, Universidad Politécnica de Madrid, Ciudad Universitaria S/N, 28040 Madrid, Spain; santiago.benito@upm.es; Tel.: +34-913363710 or +34-913363984

Received: 13 January 2020; Accepted: 16 January 2020; Published: 19 January 2020

Keywords: *Lachancea thermotolerans*; non-*Saccharomyces*; *Saccharomyces*; acidity; food safety; HACCP; wine quality; color; human health-promoting compounds; biocontrol; wine flavor; low ethanol wine; Vineyard Microbiota; wine color; wine aroma; climate change

---

Since the beginning of enology and fermentation research, wine quality has been parametrized from a chemical and sensory point of view. The main chemical compounds employed nowadays to parameterize the quality of wine or other fermented beverages are acids, polyphenols, volatile particles, and polysaccharide compounds [1]. All these chemical compounds directly influence sensory parameters commonly perceived by consumers such as general acidity, variety character, aroma quality, structure, and overall impression [1].

Before starting to study technologies that enhance alcoholic fermentation quality parameters, there is a need to reduce the incidence of spoilage microorganisms such as *Brettanomyces/Dekkera* or *Zygosaccharomyces rouxii* able to produce undesirable molecules such as ethyl phenols or acetic acid [2,3] that mask the influence of positive molecules. Traditionally additives such as SO_2 were used to inhibit these undesirable microorganisms. However, modern legislation started to regulate their use due to allergenic food safety problems [4]. A new technology that reduces the incidence of spoilage microorganisms without generating any health collateral effects for specific consumers, is the use of bio controller technologies [3]. Selected strains of yeast species such as *Wickerhamomyces anomalus* and *Metschnikowia pulcherrima* have been proven to be especially efficient against undesirable spoilage microorganisms [3].

Color is the first perception that a wine consumer appreciate in a sensory analysis. This quality parameter depends mainly on the anthocyanin concentration. Modern enology has studied ways to increase the extraction and to increase the stability of these molecules during the winemaking process. Recent technologies such as must replacement and hot pre-fermentative maceration increase the phenolic content and enhance the chromatic characteristics of wine while inactivating polyphenol oxidases enzymes able to degrade colored molecules and promoting condensation between anthocyanins and tannins [5]. Other modern technologies to increase wine color from a microbiological point of view are related to the production of highly stable forms of anthocyanins during alcoholic fermentation. Specific yeasts are able to produce high levels of pyruvic acid that increases the formation of high stable anthocyanins such as vitisin A [1,6] or allow to avoid the malolactic fermentation process [7,8] where color intensity usually gets reduced.

The modern food safety standards demanded by most popular food distributors require wines free of hazards compounds. Additionally, most countries start to stablish legal limits for some hazardous molecules. This fact oblige winemakers to control these undesirable compounds form a winemaking point of view. The main parameters to control are ochratoxin A, biogenic amines [9], ethyl carbamate, sulfur dioxide, allergens, pesticides, genetically modified organisms, physical hazards and phthalates [4].

Modern wine consumers usually prefer wines with moderate ethanol levels. This fact promoted the development of new strategies to reduce the high ethanol levels, especially in warm viticulture areas. One interesting strategy is the use of less efficient yeasts than *S. cerevisiae* in the conversion of sugar into ethanol. Sequential fermentation inoculations involving *Hanseniaspora uvarum* show interesting results in ethanol reduction while also increase wine quality parameters such as fruity aroma or color intensity [10]. Additionally, climate change is making it difficult in some countries/regions to control some quality parameters during alcoholic fermentation such as the presence of undesirable microorganisms, excessive sugar, lack of acidity, high pH, imbalanced color, undesirable flavors or food safety problems. Modern wine microbiology management offers interesting alternatives to mitigate these problems [11].

Although traditionally some non-*Saccharomyces* species have been considered spoilage microorganisms [2]. The use of some specific non-*Saccharomyces* species allow to control and to improve several wine quality parameters [1,12]. The most popular ones are *Torulaspora delbrueckii* [13], *Lachancea thermotolerans* [14–16], *Metschnikowia pulcherrima* [12,17], *Schizosaccharomyces pombe* [18], *Hanseniaspora uvarum* [10] and *Pichia kluyveri* [12]. Some groups are studying the microbiota of vineyards and soils to look for other microorganism different from *S. cerevisiae* able to enhance quality parameters of alcoholic and malolactic fermentation [19].

Modern biotechnologies based on the use of some conventional and non-conventional yeasts allow to produce wine or beer with functional properties for human health [20]. The last studies show interesting results to improve the content of specific neuroprotectives and neurotrasmitters such as serotonin or melatonin [20].

Most studies involving fermentative industries are focused on alcoholic fermentation. However, during the last decade the knowledge regarding malolactic fermentation has increased due to the industrial difficulties that this process shows in some occasions. The use of lactic bacteria species different from *Oenococus oeni* and the use of combinations of non-*Saccharomyces* and lactic bacteria are of current interest [21]. Combinations between *Hanseniaspora uvarum*, *S. cerevisiae* and *Lactobacilus plantarum* show improvements in malolactic fermentation time, wine body and aroma [21].

Other new alcoholic beverages different from wine and beer start to be developed and optimized. One of those modern alternatives to grape wine is cashew apple fermentation. This alcoholic beverage show interesting properties such as low ethanol content and significant amounts of antioxidants such as ascorbic acid or polyphenols. The fermentation process of cashew apple has been optimized using *Hanseniospora guillermondii* that increases phenyl ethanol and acetate ester [22]. Additionally, the fermentation industry is being optimized in industries different from wine, beer or other alcoholic industries. One interesting example of this is the optimization of itaconic acid production using *Aspergillus terrus* [23].

Saccharomyces cerevisiae remains the main option to perform alcoholic fermentation due to its high fermentation reliability. Nevertheless, the genome of *S. cerevisiae* is huge and there is a high variability depending on the selected strain. The use of commercial strains can produce standardized wines without personal differentiations. For that reason, some researchers are developing *S. cerevisiae* selection processes applied to specific regions and grape varieties to enhance their typicity, a good example is Narince wines [24]. Specific selected autochthonous *S. cerevisae* strains are able to enhance specific esters and terpenes that increase the sensory quality parameters such as floral and fruity characters. Selections of *S. cerevisiae* strains from "Vinos de Madrid" viticultural region (D.O.) show a way to preserve regional sensory properties different from those of commercial strains that promote biodiversity while improve the personality of wine in parameters such as fruity or floral characters [25]. Recent studies for Bombino bianc wine show how it is possible to select specific *S. cerevisiae* strains able to enhance arbutin splitting (β-glucosidase) and with moderate pectolytic activity that improves the quality of wine [26].

References

1. Benito, Á.; Calderón, F.; Benito, S. The Influence of Non-*Saccharomyces* Species on Wine Fermentation Quality Parameters. *Fermentation* **2019**, *5*, 54. [CrossRef]
2. Benito, S.; Palomero, F.; Morata, A.; Calderón, F.; Suárez-Lepe, J.A. A method for estimating *Dekkera/Brettanomyces* populations in wines. *J. Appl. Microbiol.* **2009**, *106*, 1743–1751. [CrossRef] [PubMed]
3. Kuchen, B.; Maturano, Y.P.; Mestre, M.V.; Combina, M.; Toro, M.E.; Vazquez, F. Selection of native non-*Saccharomyces* yeasts with biocontrol activity against spoilage yeasts in order to produce healthy regional wines. *Fermentation* **2019**, *5*, 60. [CrossRef]
4. Benito, S. The Management of Compounds that Influence Human Health in Modern Winemaking from an HACCP Point of View. *Fermentation* **2019**, *5*, 33. [CrossRef]
5. Piccardo, D.; González-Neves, G.; Favre, G.; Pascual, O.; Canals, J.M.; Zamora, F. Impact of Must Replacement and Hot Pre-Fermentative Maceration on the Color of Uruguayan Tannat Red Wines. *Fermentation* **2019**, *5*, 80. [CrossRef]
6. Benito, S.; Palomero, F.; Gálvez, L.; Morata, A.; Calderón, F.; Palmero, D.; Suárez-Lepe, J.A. Quality and composition of red wine fermented with *Schizosaccharomyces pombe* as sole fermentative yeast, and in mixed and sequential fermentations with *Saccharomyces cerevisiae*. *Food Technol. Biotechnol.* **2014**, *52*, 376.
7. Benito, Á.; Calderón, F.; Benito, S. Combined use of *S. pombe* and *L. thermotolerans* in winemaking. Beneficial effects determined through the study of wines' analytical characteristics. *Molecules* **2016**, *21*, 1744. [CrossRef]
8. Benito, A.; Calderón, F.; Benito, S. The combined use of *Schizosaccharomyces pombe* and *Lachancea thermotolerans*—Effect on the anthocyanin wine composition. *Molecules* **2017**, *22*, 739. [CrossRef]
9. Mylona, A.E.; Del Fresno, J.M.; Palomero, F.; Loira, I.; Bañuelos, M.A.; Morata, A.; Calderón, F.; Benito, S.; Suárez-Lepe, J.A. Use of *Schizosaccharomyces* strains for wine fermentation—Effect on the wine composition and food safety. *Int. J. Food Microbiol.* **2016**, *232*, 63–72. [CrossRef]
10. Mestre, M.V.; Maturano, Y.P.; Gallardo, C.; Combina, M.; Mercado, L.; Toro, M.E.; Carrau, F.; Vazquez, F.; Dellacassa, E. Impact on sensory and aromatic profile of low ethanol malbec wines fermented by sequential culture of *Hanseniaspora uvarum* and *Saccharomyces* cerevisiae native yeasts. *Fermentation* **2019**, *5*, 65. [CrossRef]
11. Berbegal, C.; Fragasso, M.; Russo, P.; Bimbo, F.; Grieco, F.; Spano, G.; Capozzi, V. Climate changes and food quality: The potential of microbial activities as mitigating strategies in the wine sector. *Fermentation* **2019**, *5*, 85. [CrossRef]
12. Dutraive, O.; Benito, S.; Fritsch, S.; Beisert, B.; Patz, C.-D.; Rauhut, D. Effect of Sequential Inoculation with Non-*Saccharomyces* and *Saccharomyces* Yeasts on Riesling Wine Chemical Composition. *Fermentation* **2019**, *5*, 79. [CrossRef]
13. Benito, S. The impact of *Torulaspora delbrueckii* yeast in winemaking. *Appl. Microbiol. Biotechnol.* **2018**, *102*, 3081–3094. [CrossRef] [PubMed]
14. Vilela, A. *Lachancea thermotolerans*, the Non-*Saccharomyces* Yeast that Reduces the Volatile Acidity of Wines. *Fermentation* **2018**, *4*, 56. [CrossRef]
15. Benito, S. The impacts of *Lachancea thermotolerans* yeast strains on winemaking. *Appl. Microbiol. Biotechnol.* **2018**, *102*, 6775–6790. [CrossRef]
16. Porter, T.J.; Divol, B.; Setati, M.E. *Lachancea* yeast species: Origin, biochemical characteristics and oenological significance. *Food Res. Int.* **2019**, *119*, 378–389. [CrossRef]
17. Ruiz, J.; Belda, I.; Beisert, B.; Navascués, E.; Marquina, D.; Calderón, F.; Rauhut, D.; Santos, A.; Benito, S. Analytical impact of *Metschnikowia pulcherrima* in the volatile profile of Verdejo white wines. *Appl. Microbiol. Biotechnol.* **2018**, *102*, 8501–8509. [CrossRef]
18. Benito, S. The impacts of *Schizosaccharomyces* on winemaking. *Appl. Microbiol. Biotechnol.* **2019**, *103*, 4291–4312. [CrossRef]
19. Alonso, A.; De Celis, M.; Ruiz, J.; Vicente, J.; Navascués, E.; Acedo, A.; Ortiz-Álvarez, R.; Belda, I.; Santos, A.; Gómez-Flechoso, M.Á.; et al. Looking at the origin: Some insights into the general and fermentative microbiota of vineyard soils. *Fermentation* **2019**, *5*, 78. [CrossRef]
20. Vilela, A. The importance of yeasts on fermentation quality and human health-promoting compounds. *Fermentation* **2019**, *5*, 46. [CrossRef]

21. Du Plessis, H.; Du Toit, M.; Nieuwoudt, H.; Van der Rijst, M.; Hoff, J.; Jolly, N. Modulation of wine flavor using *Hanseniaspora uvarum* in combination with different *Saccharomyces cerevisiae*, lactic acid bacteria strains and malolactic fermentation strategies. *Fermentation* **2019**, *5*, 64. [CrossRef]

22. Gamero, A.; Ren, X.; Lamboni, Y.; de Jong, C.; Smid, E.J.; Linnemann, A.R. Development of a low-alcoholic fermented beverage employing cashew apple juice and non-conventional yeasts. *Fermentation* **2019**, *5*, 71. [CrossRef]

23. Komáromy, P.; Bakonyi, P.; Kucska, A.; Tóth, G.; Gubicza, L.; Bélafi-Bakó, K.; Nemestóthy, N. Optimized pH and its control strategy lead to enhanced itaconic acid fermentation by *Aspergillus terreus* on glucose substrate. *Fermentation* **2019**, *5*, 31. [CrossRef]

24. Çelik, Z.D.; Erten, H.; Cabaroglu, T. The influence of selected autochthonous *Saccharomyces cerevisiae* strains on the physicochemical and sensory properties of narince wines. *Fermentation* **2019**, *5*, 70. [CrossRef]

25. García, M.; Esteve-Zarzoso, B.; Crespo, J.; Cabellos, J.M.; Arroyo, T. Influence of Native *Saccharomyces cerevisiae* Strains from D.O. "Vinos de Madrid" in the Volatile Profile of White Wines. *Fermentation* **2019**, *5*, 94. [CrossRef]

26. Speranza, B.; Campaniello, D.; Petruzzi, L.; Sinigaglia, M.; Corbo, M.R.; Bevilacqua, A. Preliminary Characterization of Yeasts from Bombino Bianco, a Grape Variety of Apulian Region, and Selection of an Isolate as a Potential Starter. *Fermentation* **2019**, *5*, 102. [CrossRef]

 *fermentation*

Review

The Management of Compounds that Influence Human Health in Modern Winemaking from an HACCP Point of View

Santiago Benito

Departamento de Química y Tecnología de Alimentos, Universidad Politécnica de Madrid, Ciudad Universitaria S/N, 28040 Madrid, Spain; santiago.benito@upm.es; Tel.: +34-913363984

Received: 25 January 2019; Accepted: 31 March 2019; Published: 10 April 2019

Abstract: The undesirable effects of some hazardous compounds involved in the different steps of the winemaking process may pose health risks to consumers; hence, the importance of compliance with recent international food safety standards, including the Hazard Analysis and Critical Control Point (HACCP) standards. In recent years, there has been a rise in the development of new technologies in response to the hazardous effects of chemical compounds detected during the winemaking process, whether naturally produced or added during different winemaking processes. The main purpose was to reduce the levels of some compounds, such as biogenic amines, ethyl carbamate, ochratoxin A, and sulfur dioxide. These technological advances are currently considered a necessity, because they produce wines free of health-hazardous compounds and, most importantly, help in the management and prevention of health risks. This review shows how to prevent and control the most common potential health risks of wine using a HACCP methodology.

Keywords: biogenic amines; ethyl carbamate; ochratoxin A; sulfur dioxide; phthalates; HACCP

---

1. Introduction

During the last few decades, grape fermentation products have shown positive health effects when consumed responsibly. Wine is common in the diet of many countries whose populations have high life expectancies, such as Spain. However, there are several health risks related to alcoholic beverages and specifically to wine. Those risks are usually related to specific groups of consumers, such as people suffering from allergies, pregnant women, or alcoholics. In this work, we focus on those health risks that can be avoided by a responsible consumer.

The Hazard Analysis and Critical Control Point (HACCP) theories emerged during the 1970s. Implementation of HACCP is now compulsory for food industries in most countries in order to protect consumers [1]. This article discusses the hazards associated with wine consumption, following the principles of the HACCP, in order to make it easy to understand and applicable for those who work in the wine industry. The HACCP theory is a preventive measure rather than a reactive policy. For this reason, this work shows that most of the known ways to prevent the appearance of human health hazards in wine begin with vineyard management. The first goal of HACCP is to control micro-organisms that could potentially harm regular consumers. From this perspective, wine is a simple food product to control, as no dangerous pathogens (such as *Clostridium botulinum*, *Salmonella enteritidis*, *Escherichia coli*, *Listieria monocytogenes*, *Bacillus cereus*, *Staphylococcus aureus*, *Campylobacter jejuni*, or *Aeromonas hydrophila*) can develop in a medium that contains an ethanol level of approximately 10–14%, high acidity, phenols, and sulfide. Indeed, in big cities (before chloride made water safer to drink) alcoholic beverages were consumed instead of water in order to avoid water pathogens that develop under unhygienic conditions. Nowadays, all developed countries and most developing countries have high-quality public water from a food safety point of view. This fact makes the situation completely different, and although no pathogenic micro-organisms can easily develop in wine, new food safety problems (unknown until recent years) have begun to appear.

The old approaches of HACCP were based on the belief that no pathogenic micro-organisms could reach the consumer through wine and were focused on other food safety hazards, such as chemical or physical risks [2,3]. However, recent research has shown that some potentially indirect pathogenic micro-organisms that are not able to colonize a human body, such as lactic bacteria or grape fungi, can generate dangerous metabolites under specific circumstances. These compounds, in fixed concentrations, can be put into danger-specific groups of the population, or even regular consumers. Some of these compounds are biogenic amines, ethyl carbamate, or ochratoxin A (OTA).

Food safety controls were originally based on testing analyses of final products. The main problem of this approach was the impossibility of analyzing entire productions. In the case of winemaking, it would mean analyzing each bottle. Another specific problem of the enology industry is the price of specific analyses related to food safety, which can easily reach 100 € per unit and analysis, depending on the studied hazard. For these reasons, HACCP theories are based on preventive principles, such as routine control measures during manufacturing, in order to keep production under controlled conditions. In the past, the HACCP focused on pathogenic micro-organisms; however, today it also seeks to control physical and chemical hazards [4]. Such hazards are of great importance in the wine industry. For that reason, we discuss chemical hazards, such as pesticides, commonly used in vineyards or common additives, such as sulfites or fining agents. Physical hazards common in wine industries, such as glass, are also studied. These problems are generally easier to avoid than microbiological hazards, as they are more predictable than micro-organisms.

Because the HACCP is considered to be the most international system for preventing food hazards, we will discuss in detail how to follow a structure based on the seven principles that constitute this theory. This methodology easily allows the reader to identify where potential hazards appear in the winemaking process, their dangerous levels, their origins, and how to prevent them through systematic controls. It also shows how to verify from time to time that the whole system is under control by using more complex and expensive methodologies.

The Codex Alimentarius Guidelines [5] show seven principles to guide the implementation of a HACCP system, as follows.

1.1. Principle 1: To Conduct a Hazard Analysis

All hazards relating to a food product that can negatively influence the health of any consumer must be identified at their source. Possible preventive measures should also be described. Hazards must be divided into three groups: microbiological, chemical, and physical. As we explained before, from a microbiological point of view, no human pathogenic bacteria, fungi, or virus can successfully develop in wine due to its ethanol content. However, some micro-organisms that commonly appear in wine or grapes, such as lactic bacteria or fungi, are able to produce some potentially dangerous compounds, such as biogenic amines, ochratoxin A, or ethyl carbamate. There is generally low awareness of these problems of microbial origin in the wine industry, and there is some controversy about which preventive measures are most effective. These compounds constitute the main health hazards of microbiological origin in the wine industry. The main chemical hazards are the pesticides used in the vineyard to protect the plant and grapes from diseases produced by fungi. Migrations emanating from the packaging or containers where the wine is stored or manipulated are also chemical problems. Some fining agents that, on occasion, can be potential allergen compounds for specific groups of people are used to fine the wine in order to reduce the initial turbidity. Additives that can stabilize wine against micro-organism spoilage or against spoilage processes, such as oxidation, in over dosage can also produce health risks. The main physical hazards in the winemaking process are remains of machinery particles that can end up in the wine and glass particles from deteriorated bottles in which the final wine is stored.

1.2. Principle 2: To Determine the Critical Control Points

After conducting a study of all the possible hazards and their potential detriment to health and the probability of occurrence, we must establish how to control these risks. Critical control points (CCPs) are phases in the food process where it is essential to control some parameter that can prevent or eliminate the potential food safety hazard or reduce it to an acceptable level. For example, if a hazard comes from the grapes row material, the best moment to control it is before processing so as to make it easier to isolate the source. Therefore, it would not make any sense to control it at the last stage of the process. Thus, efforts must be made to identify the problem as soon as possible.

1.3. Principle 3: To Establish Critical Limits

Once it has been established where a hazard is going to be controlled, we must establish a criterion that allows for differentiating between what is acceptable and what is not. That criterion is defined according to a critical limit. Most of the time, critical limits are established according to the legal limits defined by legislation, such as that pertaining to histamine, ochratoxin A, ethyl carbamate, or legalized additives.

1.4. Principle 4: To Establish a Monitoring System

Once the stage where we have to control a hazard and its critical limit have been established, we must establish the kind of control to use, its frequency, and the qualified responsible person to use it. These controls are usually analyses that are fast and economical but allow for very quick decision-making. It is very common to use semi-quantitative methodologies that are not the official methods and are usually expensive and require specific equipment not commonly available from every winery. The official methods are commonly used in HACCP Principle 6.

1.5. Principle 5: To Establish Corrective Actions

When a deviation from the established critical limits occurs, a corrective action must be performed in order to restore the control and avoid potentially dangerous wine reaching the consumer. The most drastically corrective action is to eliminate the product. Nevertheless, several other options permit removing the hazard or procuring a secondary product less valuable but with a residual economical value. The principle also proposes to review the cause of the mistake or the imprecise action that generated the deviation in order to correct the procedure.

1.6. Principle 6: To Establish Verification Procedures

Hazard Analysis and Critical Control Point (HACCP) verification is defined as those activities, other than monitoring, that establish the validity of the HACCP plan and ensure that the HACCP system is operating according to the plan. Verification is done to determine whether the HACCP plan is being implemented properly, whether practices used are consistent with the HACCP plan, whether the HACCP system is working to control significant hazards, and whether modifications of the HACCP plan are required to reduce the risk of recurrence of deviations [6]. In winemaking, to verify the success and correct implementation of control measures, which are in most cases based on fast and semi-quantitative analyses, the most common procedure is to perform periodic checks using the official methodology. For that reason, it is very common to perform the verification analyses in accredited laboratories that possess advanced equipment, such as HPLC or GC/MS, and qualified professionals to run them.

1.7. Principle 7: To Establish Documentation Concerning All Procedures and Records That Are Appropriate to These Principles and Their Applications

A HACCP manual must be written. It describes the methodologies to follow in the HACCP system and how to apply them to this specific industry. It also describes potential hazards and their effect on human health, critical control points, critical limits, corrective actions, control measures, and verification measures. The manual also keeps records of all performed operations in order to help produce safe products.

The main purpose of this review is to show wine manufacturers the main hazards in the wine industry and how to manage them according to HACCP theories (Table 1).

2. Ochratoxin A

2.1. Toxicity

Mycotoxins are toxic compounds of fungal origin that, when ingested, absorbed, or inhaled, can cause illness or death in humans. Ochratoxin A is a common compound in wines. It is considered hazardous to human health because of its nephrotoxic, neurotoxic, immunotoxic, mutagenic, and teratogenic properties [6–8]. Recently, the International Agency for Research of Cancer classified OTA as a carcinogenic compound [9]. The tolerable daily intake of OTA ranges from 0.3 to 0.89 µg/day for a person weighing 60 kg. It can cause instant poisoning in doses between 12 and 3000 mg for a person of that weight [10]. The Food and Agricultural Organization (FAO) and the World Health Organization set the daily upper limit intake to 14 ng/kg and the weekly intake to 100 ng/kg of body weight [10].

2.2. Origin

The origin of mycotoxins in enology are several fungi species in rotten grapes that are able to produce them. The OTA formation is related to the raw grapes, and it is not possible for OTA-producing fungi to develop in liquid juice or wine, as all fungi responsible for its formation are strictly aerobic, such as *Aspergillus carbonarius* [11,12]. The main species able to produce OTA in grapes, must and wine are *A. carbonarius* [13], *Aspergillus fumigatus* [14], *Aspergillus niger* [15], *Aspergillus tubingensis* [16], *Aspergillus japonicus*, and *Penicillium tubingensis* [10].

2.3. Critical Limit

Nowadays, OTA concentration in wine is regulated in certain European Union countries. We propose a critical limit that corresponds with the European legal limit of 2 µg/kg (available online: http://europa.eu/rapid/press-release_IP-04-1215_en.htm). The average value of OTA in European wines is about 0.19 µg/L [10]. According to some research, Spanish wines show an incidence of 1% of being over the legal limit [17].

2.4. Preventive Measures

Preventive measures mainly involve vineyard management being used to avoid the development of undesirable fungi capable of rotting the grapes. Some of those species are powdery mildew [18], *Rhizopus stolonifera*, or *Botrytis cinerea* [19] that favor berry colonization by the *Aspergillus* genus. Those vineyard diseases are well-known by viticulturists and in most cases are easily treated through phytosanitary controls. The insect known as *Lobesia botrana* usually produces small injuries in grapes that favor the latter's colonization by the former fungi. The insect plays an important role in OTA formation as fungi, such as *A. carbonarius*, are not able to attack the grape skin and invade the pulp by themselves [20]. Thus, previous skin damage is needed for colonization [12]. This insect management is also well-known by viticulturists. Nowadays, there is a trend to use a methodology based on sexual confusion through hormones in order to avoid the use of dangerous chemical compounds. Some alternative options for avoiding undesirable fungal developments and the use of pesticides are the biocontrol agents, such as *Aureobasidium pullulans* [21], *Kluyveromyces thermotolerans* [22], and *Lanchacea thermotolerans* [23,24]. The biocontrol strategy consists of colonizing plant surfaces or wounds for long periods under dry conditions before fungal attacks take place under wet conditions. Another trend is to use vineyard management that exposes the grapes to the sun and allows for higher air-stream circulation. In such microclimates, the development of fungi is more limited.

Table 1. The main wine-industry hazards and their management from a Hazard Analysis and Critical Control Point (HACCP) point of view.

Hazard	Toxicity	Origin	Critical Limit	Preventive Measures	Control Measures	Corrective Measures	Verification
Ochratoxin A	Nephrotoxic, neurotoxic, immunotoxic, mutagenic, teratogenic, carcinogenic	Fungus *Aspergillus Penicillium*	2 µg/kg	Vineyard management, phytosanitary controls, yeast biocontrol agents	Fungi visual control, gluconic acid, immunoaffinity	Maceration, Finning agents, selected yeast, amicrobic filtration	HPLC with fluorescent detector, 80 €
Biogenic amines	Several allergenic disorders	*Lactic bacteria, Pediococcus, Oenococcus, Lactobacilluas, Leuconostoc*	2 mg/L	Antibacterial agents, sulfur dioxide, lysozyme, chitosan, yeast inoculation	Selected lactic bacteria, *Schizosaccharomyces pombe/Lachancea thermotolerans,* semi-quantitative control	Unknown	fluorescence HPLC, 40 €
Ethyl carbamate	Carcinogenic and genotoxic	Urea evolution and lactic bacteria metabolism	15 µg/L	Nitrogen management, alternatives to malolactic fermentation	Urease enzyme, selected yeasts or bacteria.	Unknown	GC/MS, 40 €
Sulfur Dioxide	Irritation, bronchospasm, pulmonary edema, pneumonitis, and acute airway obstruction	Common additive in wine (antimicrobial and antioxidant)	150 mg/L	Sulfur dioxide alternatives: sorbic acid, lysozyme, chitosan, ascorbic acid, thermovinification, high hygiene	Calculation, sulfur dioxide stock control, Ripper method	To oxygenate, dilution	Paul method, 5 €
Wine Food Allergens	Allergic reactions	Present in fining agents (egg white, caseinates, or fish gelatin)	Traces	Fining agent dose evaluation before adding. Stabilization tests. No animal origin fining agents.	Label control	Unknown	ELISA tests, 50 €
Pesticides	Dermatological, gastrointestinal, neurological, carcinogenic, respiratory, reproductive, and endocrine negative effects	Vineyard protection against fungi and insects	Ditianon 5 mg/kg	Vineyard management	Field practice notebook registration (residual period control)	Unknown	ECD gas chromatography, 21 €
Genetically Modified Organisms (GMOs)	Precaution measures until total harmlessness is proved	Better performance of GMO yeasts	Residual presence, ML01 and ECMo01 0.005% (mass/mass)	Spontaneous alcoholic and malolactic fermentations	Yeast labeling control before use (free GMO product)	Unknown	Polymerase Chain Reaction, 100 €
Physical hazard	Cuts, bleeding, infection, and choking	Installations, raw materials, bad manipulation	2 mm	Raw material inspection, preventative equipment maintenance, good practice guidelines	Filtration	To refiltration	Random control
Phthalates	Endocrine disrupting, estrogenic, carcinogenic, and mutagenic	Equipment, pipes, plastic boxes, or epoxy resin surfaces	DBP 0.3 mg/kg, DEHP 1.5 mg/kg and DINP 9 mg/kg	Food quality material free of phthalates	Food quality material inspection	Unknown	ECD gas chromatography, 80 €

2.5. Possible Control Measures

The main strategy to avoid possible contaminations by OTA is to control the sanitary status of the grapes by visual control prior to processing them. This methodology allows the producers not to accept any spoiled grapes or to remove the affected grape bunches in a selective process before fermentation. Some authors report a success of about 98% using this methodology [25], when they establish a critical limit of tolerating just 1% of infected grapes.

However, in some cases, early contaminations by fungi cannot be detected by human eyes. Therefore, the control of fungi chemical indicator parameters, such as gluconic acid in grape reception, allows one to eliminate possible human subjectivities. A fast enzyme test able to analyze that compound is commonly used as a HACCP control measure, as it is relatively cheap: about 1 €/sample [26]. There are also commercial kits based on immunoaffinity [27] that offer good accuracy and a rapid solution. The official detection method of OTA is commonly used as a verification measure in accredited laboratories due to its price, which is around 30–40 € [28,29] currently. Early detection allows for the removal of traceability lots and allows one to apply corrective measures that are quite effective in this specific case.

2.6. Corrective Measures

Once the presence of OTA concentration over the critical limit is detected, several corrective measures can be applied before eliminating the lot. Some methodologies, such as reducing maceration in the case of contaminated grapes, fining activated carbon [30], or fermenting with selected yeast, can reduce OTA concentration in final wine from 70 to 32% [31]. Non-*Saccharomyces*, such as *Schizosacchromyces*, look to be very promising in reducing the content by about 70% during fermentation [31,32]. A regular amicrobic filtration before bottling about 0.45 µm of wine can easily reduce the final concentration in OTA by about 80% [33].

All these options make it easy to manage OTA when it is detected. For that reason, affected lots are usually not disqualified due to the high number of possibilities of corrective measures.

2.7. Verification

The official methodology in Europe is HPLC with a fluorescent detector. The detection price in an accredited laboratory varies from 30 to 40 € currently [28,29].

3. Biogenic Amines

3.1. Toxicity

Biogenic amines are over-specific concentrations able to produce undesirable effects, such as headaches, respiratory distress, blushing, heart palpitation, hyper or hypotension, tachycardia, itching, skin irritation, vomiting, and several allergenic disorders [34,35]. The levels found in wines are far from being able to produce such harmful effects in regular consumers. There are some specific groups of people, such as those who are allergic to histamine, for whom the effects could be especially dangerous. The most toxic biogenic amine that can appear in wine is histamine [36,37]. Human metabolism possesses several enzymes, such as monoamine oxidase and diamine oxidase, that degrade the toxic compound histamine for regular cases. However, specific groups of people have gradually inhibited those enzymatic activities. Another specific parameter of wine is the presence of ethanol, which can also inhibit those enzymes or alternative medication [36].

3.2. Origin

Biogenic amines production is mainly related to bacteria metabolism [38–40]. The main bacteria genera involved in the process are *Pediococcus, Oenococcus, Lactobacillus*, and *Leuconostoc*. Histamine formation depends on the genes of histamine decarboxylase activity. Lactic acid bacteria produce biogenic amines during the malolactic fermentation that takes place in almost every red wine after alcoholic fermentation [41], although other micro-organisms, such as yeasts, are able to produce biogenic amines in smaller amounts [42].

3.3. Critical Limit

Although there are no specific laws, several countries have established rules for the specific biogenic amine histamine, whereas other biogenic amines remain free of control. Some recommended limits are 10 mg/L in Australia and Switzerland, 8 mg L in France, 3.5 mg L in the Netherlands, 6 mg L in Belgium, and 2 mg/L in Germany [37,43]. These recommended levels could become compulsory in the near future. According to these data, we can establish an industry critical limit of 2 mg/L, which is the most restrictive reported concentration.

3.4. Preventive Measures

All the preventive measures are designed to avoid uncontrolled bacteria developments in the grapes or during alcoholic fermentation. The use of sulfur dioxide is the traditional way of inhibiting lactic bacteria during alcoholic fermentation. The conventional enology sulfur dioxide doses allow yeasts to develop and ferment while bacteria are inhibited. An alternative is the inoculation of a high number of commercial yeasts that makes the development of other competitor micro-organisms impossible. Nevertheless, other modern products, such as lysozyme or chitosan, also effectively inhibit lactic acid bacteria development, consequently reducing the incidence of biogenic amines in wine. These products can also be used if an undesirable lactic bacteria development takes place during alcoholic fermentation, in order to stop it as soon as it is detected. Nevertheless, there are other types of management than additive provision, such as high levels of hygiene, that reduce the initial population of undesirable micro-organisms, such as wild lactic bacteria, in any installation that is in contact with wine [44]. Biofilm techniques can considerably reduce the risk of bacteria able to produce biogenic amines. Biofilm techniques consist of directly colonizing the wine and preventing the development of spoilage micro-organisms. For that purpose, species such as *Torulaspora delbrueckii* are used to minimize the use of additives, such as sulfites [45].

3.5. Possible Control Measures

The management of these risky compounds at the industry level is commonly based on the use of selected lactic bacteria that do not possess histamine decarboxylase enzymatic activity [40]. Approximately 20% of bacteria do not possess that undesirable enzymatic activity [40]. Nowadays, it is relatively easy to detect which bacteria are able to decarboxylase amino acids precursors to the unhealthy biogenic amine forms [46]. All lactic bacteria available in the market underwent selection processes in order not to develop such enzymatic activities, a part of classic selection parameters, such as malic acid degradation, performance at low temperatures, and sulfur dioxide tolerance. Therefore, the inoculation of those strains, instead of performing a spontaneous process, and the control of a proper devolving of malolactic fermentation through the monitoring of bacteria implantation through microscopic observation or more advanced techniques or malic acid degradation and evolution after bacteria inoculation are some of the most common ways to control enzymatic activity.

Nevertheless, during the last several years, new biotechnologies based on the use of yeasts able to remove the malic acid from wine while avoiding any possible bacteria activity are becoming popular, especially in those regions where performing malolactic fermentation can mean a drop in quality [41]. The use of *Schizosaccharomyces pombe* is the best option, although in grape juices that are not very acidic it must be combined with *L. thermotolerans* to avoid excessive deacidification [41]. These new biotechnologies are usually combined with other technologies, such as lysozyme or chitosan, to avoid any undesirable bacteria development that could generate detrimental biogenic amines production. Therefore, in those cases, the production of biogenic amines is not possible, and the final concentration is null.

Another control measure is direct analysis of biogenic amines, such as histamine. This control is highly recommended in wineries that perform spontaneous malolactic fermentations. Some affordable options are the use of rapid techniques, such as enzymatic analysis [26], which is fast and relatively cheap: about 1 € per sample. The official methodology is usually performed for verification purposes, as it is much more expensive and requires specific instrumental equipment.

3.6. Corrective Measures

Even though some yeasts can remove small amounts of biogenic amines during alcoholic fermentation or during lees contact processes, there is no effective way of removing biogenic amines when they appear in finished wine. For that reason, all efforts must be focused in order to avoid their undesirable formation, as to date it has not had any effective corrective solution.

3.7. Verification

After using protocols that reduce the incidence of biogenic amines, in most cases the verification measure is performed by fluorescence HPLC chromatography in accredited laboratories [29]. The price in the market varies from 40 to 70 € [29].

4. Ethyl Carbamate

4.1. Toxicity

Ethyl carbamate (EC) is a known carcinogen compound present in a variety of fermented foods [47]. Since the 1940s, the literature has considered it a toxic compound. In 1943, it was proven to be carcinogenic [48,49]. A common use of ethyl carbamate was as a sedative and anesthetic for animals. Ethyl carbamate is carcinogenic and genotoxic for several species, including hamsters, rats, mice, and monkeys, which suggests a high potential carcinogenic risk for humans [50,51]. Ethyl-carbamate absorption implicates three pathways: N-hydroxylation or C-hydroxylation, hydrolysis, and side-chain oxidation. The main pathway is Ethyl carbamate (EC) hydrolysis through liver microsomal esterases to carbon dioxide, ammonia, and ethanol. The International Agency for Research on Cancer (IARC) classifies ethyl carbamate as a group 2A carcinogen (i.e., probably carcinogenic to humans) [52].

4.2. Origin

Ethyl carbamate is mainly produced in wines due to the evolution of urea. Urea is a regular metabolite produced by most yeasts and bacteria during their regular metabolisms. Urea is slowly combined with ethanol, producing ethyl carbamate. This is why incidence is higher in old aged wines. Other secondary production pathways can be created by the action of lactic bacteria and specific amino acids metabolism. Citrulline is an intermediate of arginine degradation by wine lactic acid bacteria during malolactic fermentation. Citrulline is the second precursor in the formation of ethyl carbamate after urea [53]. A high percentage of heterofermentative wine lactic bacteria, such as *Oenococus oeni*, are able to degrade arginine. The enzyme arginine deiminase produces that phenomenon.

4.3. Critical Limit

European legislation does not specify any legal limit regarding ethyl carbamate. Nevertheless, some specific countries possess a legal limit or a recommended level. Some examples are Canada (30 μg/L), Czech Republic (30 μg/L), South Korea (30 μg/L), and the United States (15 μg/L) [54]. We propose the lowest referenced level of 15 μg/L as the critical limit to be considered, especially for companies from countries where ethyl carbamate is not legislated but with possibilities of exporting to countries with legal limits.

4.4. Preventive Measures

Possible preventive measures are to reduce nitrogen fertilization in vineyards, especially the direct use of urea. Another measure is to use only the necessary nutrient supplementation before and during fermentation, as increases in nitrogen composition will increase the final production of urea [55,56]. Any alternative to malolactic fermentation is an effective way to avoid ethyl-carbamate formation from that bacteria metabolism or their urea formation [45].

4.5. Control Management

Current strategies are based on the use of urease enzyme [57], which can reduce levels of urea down to 0 mg/L or non-detectable levels. Some companies commercialize the enzyme, and its use is common in companies that export to countries with legal limits. Another more recent alternative is the use of yeast species that naturally possess urease activity. Some of them are able to complete an alcoholic fermentation by themselves, such as *S. pombe*, whereas others can be used in combined fermentations with a more powerful fermenter, such as *Saccharomyces cerevisiae* [41]. Some experiences demonstrate that the final urea values in these cases are close to 0 mg/L. Another option is the use of selected malolactic bacteria that cannot excrete citrulline from arginine degradation [53].

4.6. Corrective Measures

Once it is produced, there is no corrective methodology that can effectively reduce the final concentration to regular levels.

4.7. Verification

Accredited laboratories offer GC/MS as a detection technique. The price in the market varies from 40 to 100 € [29].

5. Sulfur Dioxide

5.1. Toxicity

According to the Agency for Toxic Substances and Disease Registry (ATSDR) [58], sulfur dioxide may cause irritation, and is especially dangerous when exposed to the eyes, mucous membranes, skin, and respiratory tract. Direct exposure can cause such problems as bronchospasm, pulmonary edema, pneumonitis, and acute airway obstruction.

Nevertheless, for the regular levels that can appear in wine, the main issue is people who suffer from chronic pulmonary diseases, such as asthma [59], that can easily evolve to bronchospasm. For that reason, in some markets, it must be indicated in the labeling that the wine contains sulfites in order to protect that specific high-risk group, as they can easily identify any risks by reading the label before consumption.

5.2. Origin

Although some toxicological properties are attributed to sulfur dioxide, its use caused a revolution in winemaking, as it is a common additive that possesses several interesting properties from a technological point of view, such as antioxidant, antimicrobial, and inactivator of oxidase enzymes, such as laccase or tyrosinase, properties. Therefore, such properties notably increased the quality of wines once its use became generalized in most wines. The management of rotten grapes is especially difficult without sulfur dioxide if a good-quality wine is the objective of vinification. Nowadays, there is no other single additive that provides a solution to all the former properties.

Sulfur dioxide is commonly used in different phases of the winemaking process, such as reception, grape crushing, alcoholic fermentation, and barrel aging or storage. The main point about using it in winemaking is to inhibit possible bacteria development during alcoholic fermentation or storage, while protecting against oxidation, which can spoil a wine's aroma and color.

5.3. Critical Limit

The legal limit in Europe varies, depending on the content of sugar and the type of wine, from 150 mg/L to 350 mg/L total sulfur dioxide content [60]. A recent trend is to reduce the legal limit gradually due to its toxicity. Sweet wines and wines produced from rotten grapes are those that have been allowed to reach the highest limits due to their more difficult management from a microbiological and technological point of view. The most consumed wines in Europe—dry red and dry white—have a legal limit of 150 and 200 mg/L, respectively [60,61]. The higher permitted levels for white wines are justified due to their lower protection in antioxidant compounds, such as anthocyanins and tannins, that must be compensated for with higher additions of sulfur dioxide.

5.4. Preventive Measures

Although there are not at this moment any additives that can totally replace sulfur dioxide, many can replace some of its technological properties. The best examples are the ones that possess antimicrobial activity, such as sorbic acid [62], lysozyme [63], and chitosan [64]. Physical methods, such as high-pressure processing, allow one to greatly reduce the need for preservatives due to their capacity for undesirable micro-organism inactivation [65]. Ascorbic acid [66] is effective against oxidations; products that combine sulfur dioxide and ascorbic acid have started to become common in the market. Another option is the removal of oxygen that can react with oxygen before bottling [67]. Therefore, theoretically, it is possible to replace sulfur dioxide with a combination of several additives with different properties. The selection of yeasts with a low production of compounds able to bound to sulfur dioxide, such as acetaldehyde, which decrease the efficiency of sulfur dioxide additions, is an alternative to reducing initial doses [45].

Another alternative is the use of thermovinification, which inhibits most micro-organisms and inactivates such enzymes as tyrosinase or lacassa so that high doses of sulfur dioxide are no longer required. The sanitary initial state of grapes and the hygienic state of winery conditions influence the initial state of microbiota and can contribute to reducing the initial sulfur dioxide doses in winemaking.

5.5. Control Management

Most problems are mistakes in calculations before addition. A regular control measure is to calculate the proper dose and to obtain approval from the enologist before physical addition. Then, the added sulfur dioxide amount is registered and contrasted to the stocked sulfur dioxide in order to detect a possible mistake.

There are several techniques for analyzing sulfur dioxide. It is very common to use, after additions, the cheap and fast analytical method named Ripper, which, though not as accurate as the official method, is accurate enough to detect additions that are excessively high. The official method, which takes 30 min and is named the Paul method, is usually reserved for verification purposes.

5.6. Corrective Measures

One alternative to reducing the concentration is oxygenating [67] the wine through rankings. However, the reduction of high concentrations is very slow and requires large investments of energy to pump. The most common solution is to dilute the wine with another wine whose concentration is below the legal limit. An unrecognized International Organisation of Vine and Wine (OIV) practice is the use of hydrogen peroxide.

5.7. Verification

Verification is performed using the official Paul method. Several companies usually contract the service out to accredited laboratories to verify their internal analyses. The price varies from 2 to 5 € [31].

6. Wine/Food Allergens

6.1. Toxicity

Wines that have been fined using some potentially allergenic products, such as proteins or non-grape tannins, can produce clinical allergic reactions, especially in people who suffer from an allergy to food allergenic proteins [68].

6.2. Origin

During alcoholic and malolactic fermentation, turbidity is an inevitable effect. Potentially allergenic food proteins are used in most wines to achieve specifications related to low turbidity units. Most consumers and wine distributors demand a lack of turbidity in the end product. Although turbidity by itself is not a food safety problem, it is a common reason to refuse lots. Therefore, winemakers commonly use fining agents to produce bottled wines free of any turbidity that could lead to refusal in the market. Several of those fining agents possess allergenic properties, such as egg white, caseinates, or fish gelatin. Residual traces of those compounds could occasion allergenic reactions in allergic individuals [68]. Other modern additives, such as lysozyme (egg allergen), have started to become an interesting option for reducing sulfur dioxide additions in the control of undesirable spoilage bacteria during alcoholic fermentation.

6.3. Critical Limit

Although there is no prohibition of the use of fining agents, there are some that are considered targeted food proteins, and they must be indicated on the labels. European legislation obliges winemakers to label any wine treated with allergenic additives or processing aids if their presence can be detected in the final product [69,70]. We propose as a critical limit to label the product where there is a presence of traces.

6.4. Preventive Measures

The most common preventive measure is performing stabilization tests to determine whether a fining process is needed and which minimum proper dose is possible to achieve the desired effect. Currently, there are several alternatives to fining processes, e.g., cold stabilization and subsequent filtration. Nevertheless, those processes require specific installations, energy, and more time to be performed. A more recent alternative is the use of fining agents whose proteins are from plants, such as wheat or lupine [71–73]. However, we must take into account that, although it is possible to reduce the turbidity in a similar way to those of an animal nature, some of them can also generate risks for specific groups of people, such as those with celiac disease. Nevertheless, peas and potatoes are nowadays not included in the list of main allergens, and they do not need to be included in labeling [72,73].

6.5. Control Management

One control-management measure is to use alternative fine agents, being always aware of their nature in order to avoid other allergenic reactions. Nevertheless, the most common measure is to label the products according to the specific legislation [74] so as to make it easy for allergic people to identify potentially risky products and avoid accidents. It is common for the quality control manager to check the label before proceeding to bottle any lot. Another option is chemical control [74–76], although it is more commonly used for verification purposes.

6.6. Corrective Measures

If those agents are used, it is difficult to remove them completely from the wine as traces will be detected. Nevertheless, some techniques, such as filtration, or secondary finning treatments can achieve final concentrations that make it impossible to detect their presence in the final product.

6.7. Verification

The official detection methods are based on ELISA tests [29], and the price offered by accredited laboratories is about 50 €. Some recent alternative methodologies are based on mass spectrometry [74,75].

7. Pesticides

7.1. Toxicity

Although there are numerous families of pesticides with different negative health effects, the most common harmful health effects are associated with dermatological, gastrointestinal, neurological, carcinogenic, respiratory, reproductive, and endocrine negative effects [77]. Furthermore, high occupational, accidental, or intentional exposure to pesticides can result in hospitalization and death [78].

7.2. Origin

Several fungi or insects can attack the grapes in the vineyards. On occasion, the fungi attacks can seriously decrease the quality of wine and its final value. Therefore, it is normal to protect the vineyard against some common plagues that appear quite often depending on the year and climatic conditions. In some cases, high residual values can help avoid the proper development of yeast that is also fungi. Some of the most common plagues are mildew, *Lobesia botrana*, and *Botrytis cinerea*. These are much-studied, and most agriculturalists possess some notions about how to treat them in a successful way.

7.3. Legal Limit

Most pesticides can cause serious food poisoning. Thus, the pesticides that can be used in viticulture are legislated and need authorization. Those that are authorized possess legal residual limits in final wine of about µg/L [79,80]. The legal limits vary between countries. Nevertheless, the Codex Alimentarius establishes international standards for grapes [81] of 99 pesticides and their limits, and a limit of 5 mg/kg for dithianon for wine grapes [82].

7.4. Preventive Measures

The main preventive measures are based on the use of vineyard management that avoids propitious microclimates for fungi development. Some examples are the removal of leaves and exposure of grape clusters to sunlight and air currents that avoid high moisture conditions. The development of resistant plants is also of great interest [83,84]. Cultivation in dry areas also significantly reduces the risk of fungal attack and makes it easy to produce organic wines free of pesticides.

7.5. Control Management

Although some pesticides are authorized, their use must be justified and registered in the field practice notebook. The aim of registers is to respect the authorized periods of residuality in order to avoid possible residual values over the limit in final wines. When wineries do not control vineyard management, they sell the grapes to viticulturists. They usually request the field practice notebook in order to register performed treatments before accepting the grapes.

7.6. Corrective Measures

Once contamination takes place in wine, it is not possible to remove it.

7.7. Verification

The most common way to verify whether the system is properly controlling the risk of pesticides in our industry is by contracting for the analysis of the official method in an accredited laboratory. The official methodology includes the analyses of the legislated pesticides by ECD gas chromatography [29], which costs about 21 €.

8. Genetically Modified Organisms

8.1. Toxicity

Several scientific studies confirm that there is no proof of recombinant proteins contained in foods produced by genetically modified organisms (GMOs) being more harmful than regular proteins [85]. Nevertheless, as there is also no proof of their being totally innocuous, they are considered a potential food safety hazard in some countries.

8.2. Origin

Laboratory experimentation with genetically modified yeasts shows remarkable improvements in several quality fermentation parameters, e.g., acidity management [86,87], that led to producing better wines or reducing industrial risks under difficult situations, such as high levels of sugar or a lack of nutrients. These scientific findings tempt winemakers to use those GMO yeasts in real industry.

8.3. Critical Limit

Although there is no scientific proof that GMO wine is more dangerous to human health than non-GMO wine, there is very strict preventive legislation in several countries [88]. They will remain valid for several years until new results prove this biotechnology to be totally safe.

8.4. Preventive Measures

The safest preventive measure derived from HACCP to avoid GMO problems would be to perform spontaneous alcoholic and malolactic fermentations. Nevertheless, some controlled fermentations can originate other problems that can influence wine quality or even food safety from other hazard nature food safety problems. For that reason, another preventive measure is to buy products, such as dried commercial yeasts, from registered companies that comply with the legislation, produce safe products, and have a sanitary registration that certifies their conduct.

8.5. Control Measures

Control measures in wineries are commonly based on supplier controls at the moment when the materials are received, when the sanitary registration of the company is identified before product acceptance. The enologist or qualified assistant will check the labeling of the microbiological dried product, identifying the food quality and the GMO-free indication. If the product does not have food quality certification, the company does not have sanitary registration, or the product is not properly labeled, the use will be voided.

8.6. Corrective Measures

Once the product is identified as GMO wine, there is no known solution. Nevertheless, there are some food industries where the use of modified organisms is not legislated, as the product will not be directly consumed by humans, e.g., bioethanol for cars or for disinfection purposes. In those cases, wine could be sold to those industries, obtaining a residual value.

8.7. Verification

The detection of GMOs in food produce is mostly based on Polymerase Chain Reaction (PCR) methodologies that offer enough sensitivity, accuracy, and precision to use on wine [89]. The price of the analysis is about 100 €. Therefore, this instrumental technique is commonly used only for verification purposes or for inspection reasons for governmental controls.

9. Physical Hazard

9.1. Toxicity

According to the Institute of Agriculture and Natural Resources [90], small physical hazards, such as pieces of glass, can cause potential cuts and bleeding and may require surgery to find and remove. Other foreign objects, such as metal from installations or personal effects from employees, can also generate cuts, infection, and choking.

9.2. Origin

Physical hazards are foreign objects, distinct from the food products, that are able to cause injury or illness to the final consumers. The most common extraneous matters found in food products are bones, metal fragments, pieces of product packaging, stones, wood fragments, insects, or other personal items. The most common physical hazard described in wine is glass, because it is the main material of the final vessel that usually contains the wine when it is sold to the final consumer. The main origin is improper raw materials, such as deteriorated bottles, from the beginning of or during their manipulation. Other origins are improper facilities or equipment or a lack of maintenance that enables strange objects to pass into the wine. Another significant origin is improper performance by employees due to human error, which leads to the dropping of personal effects that can end up in the wine.

9.3. Critical Limit

Several countries' legislation does not indicate quantifiable limits or recommended ones. According to the Division of Compliance Management and Operations (HFC-210) [91], a hard or sharp foreign object larger than 7 mm can produce a serious health hazard for humans. However, other countries consider more restrictive sizes down to 2 mm. Health Canada evaluates injurious extraneous material in food, and it considers 2 mm or greater as the threshold size for consideration as a health risk [92]. Besides size, the risk associated with extraneous material is further evaluated through an assessment of shape, hardness, material, source, target consumer groups, etc.

9.4. Preventive Measures

The main preventive measures in winemaking consist of raw material inspection based on specifications, seller quality certifications, preventative equipment maintenance, and employee training based on good practice guidelines.

9.5. Control Management

The most common way to guarantee that no foreign objects are in a commercial wine is to perform a prior filtration with a proper pore nominal diameter below 2 mm before bottling. The use of filtration materials with a pore diameter down to 0.22 μm is common in some winemaking protocols [93].

9.6. Corrective Measures

In case some deviation takes place during the filtration process, the most common corrective measure is to filter the lot again when the filtration equipment has been optimized.

9.7. Verification

The verification of physical hazard critical control point management is commonly based on the inspection of random bottles of each traceability unit.

10. Phthalates

10.1. Toxicity

Phthalates have been scientifically proven to be endocrine-disrupting [94], estrogenic [95], carcinogenic, and mutagenic [96].

10.2. Origin

Equipment that contains phthalates is commonly used in many industries, as these compounds increase the flexibility of plastic [97]. In the winemaking industry, they usually appear in flexible plastic pipes, plastic boxes to collect grapes, or epoxy resin surfaces. Phthalates can migrate to beverages from plastic equipment or packaging materials [98] because they are not covalently bound to plastics [99].

10.3. Critical Limit

According to the European legislation, some phthalates possess a legal limit. Those phthalates are dibutyl phthalate (DBP) 0.3 mg/kg, diethylexyl phthalate (DEHP) 1.5 mg/kg, and diisononyl phthalate (DINP) 9 mg/kg.

10.4. Preventive Measures

Most common preventive measures consist of using food quality material free of phthalates for contact with grapes and wine. Usually, winemakers have to pay special attention to old pipes that were made before the legislation that regulates phthalates. Another option is the use of non-plastic materials, such as stainless steel for pipes.

10.5. Control Management

It is necessary to check that the materials that are going to be in contact with wine possess the proper standard of food quality and are produced by a certified provider. Clearly, pipes designed to move liquids other than food liquids, such as gasoil or water for refrigeration systems, are not suitable for wine management.

10.6. Corrective Measures

When wine contains phthalates over the legal limit, there is no known solution to remove them. Such wines are usually sold as sub-products to companies that make products not aimed at human consumption, such as bioethanol.

10.7. Verification

The most common way to verify if the system is properly controlling the risk of phthalates in winemaking is contracting for the analysis of the official method out to an accredited laboratory. The official methodology includes the analyses of the legislated phthalates by ECD gas chromatography [29], and the cost is about 80 €.

11. Conclusions

Although wine is a food beverage in which no pathogenic micro-organisms can develop, in a similar way to other food products, such risks as physical hazards must be taken into account; these kinds of hazard are easily controlled with such techniques as modern filtrations. Modern research studies have discovered potential toxicological compounds that must be taken into account in order to produce healthier wines and protect specific consumers that can be included in risk groups. Some of these hazards are biogenic amines, ethyl carbamate, and OTA. The management of these modern hazards, such as control measures, corrective measures, and verification methods, is not very well-known yet, which makes them difficult to control. Some new winemaking technologies allow one to effectively control those risks in a successful way, which offers solutions to issues that wine industries face today. All these methodologies can be easily implemented using a HACCP system.

Conflicts of Interest: The author declares no conflict of interest.

References

1. Directive 93/43/EEC on the Hygiene of Food-Stuffs. Available online: https://eur-lex.europa.eu/legal-content/EN/TXT/?uri=CELEX%3A31993L0043 (accessed on 15 August 2018).
2. Hyginov, C. *Elaboración de Vinos, Seguridad, Calidad, Métodos*, 1st ed.; Editorial Acribia: Zaragoza, España, 2000; pp. 14–38, ISBN 978-84-200-0928-5.
3. Briones, A.I.; Ubeda, J.F. Elaboración de un plan APPCC en una bodega. *Tecnología del Vino May/June* **2001**, 89–93.
4. Mortimore, S.; Wallace, C. *HACCP: Enfoque Práctico*, 1st ed.; Editorial Acribia: Zaragoza, España, 1994; pp. 25–79, ISBN 978-84-200-0928-5.
5. Codex Alimentarius International Food Standards. Available online: http://www.fao.org/fao-who-codexalimentarius/codex-texts/guidelines/en/ (accessed on 15 August 2018).
6. Schmidt, R.H.; Newslow, D.L. *Hazard Analysis Critical Control Points (HACCP) Principle 6: Establish Verification Procedures*; Editorial University of Florida IFAS extension: Gainesville, FL, USA, 2007.
7. Petzinger, E.; Ziegler, K. Ochratoxin A from a Toxicological Perspective. *J. Vet. Pharmacol. Ther.* **2000**, *23*, 91–98. [CrossRef]
8. Walker, R. Risk assessment of ochratoxin: Current views of the European Scientific Committee on Food, the JECFA and the Codex Committee on Food Additives and Contaminants. In *Mycotoxins and Food Safety*; Springer: Berlin/Heidelberg, Germany, 2002; pp. 249–255.
9. International Agency for Research on Cancer (IARC)—Summaries & Evaluations Ochratoxin A. Available online: http://www.inchem.org/documents/iarc/vol56/13-ochra.html (accessed on 15 August 2018).
10. Martınez-Rodrıguez, A.J.; Santiago, A.V.C. Application of the Hazard Analysis and Critical Control Point System to Winemaking: Ochratoxin A. *Mol. Wine Microbiol.* **2011**, 319. [CrossRef]
11. Serra, R.; Abrunhosa, L.; Kozakiewicz, Z.; Venâncio, A. Black Aspergillus Species as Ochratoxin A Producers in Portuguese Wine Grapes. *Int. J. Food Microbiol.* **2003**, *88*, 63–68. [CrossRef]
12. Kapetanakou, A.E.; Panagou, E.Z.; Gialitaki, M.; Drosinos, E.H.; Skandamis, P.N. Evaluating the Combined Effect of Water Activity, pH and Temperature on Ochratoxin A Production by Aspergillus Ochraceus and Aspergillus Carbonarius on Culture Medium and Corinth Raisins. *Food Control* **2009**, *20*, 725–732. [CrossRef]
13. Gallo, A.; Perrone, G.; Solfrizzo, M.; Epifani, F.; Abbas, A.; Dobson, A.D.; Mulè, G. Characterisation of a Pks Gene which is Expressed during Ochratoxin A Production by Aspergillus Carbonarius. *Int. J. Food Microbiol.* **2009**, *129*, 8–15. [CrossRef]
14. Battilani, P.; Pietri, A. Ochratoxin A in grapes and wine. In *Mycotoxins in Plant Disease*; Springer: Berlin/Heidelberg, Germany, 2002; pp. 639–643.
15. Bau, M.; Bragulat, M.; Abarca, M.; Minguez, S.; Cabañes, F. Ochratoxigenic Species from Spanish Wine Grapes. *Int. J. Food Microbiol.* **2005**, *98*, 125–130. [CrossRef] [PubMed]
16. Oliveri, C.; Torta, L.; Catara, V. A Polyphasic Approach to the Identification of Ochratoxin A-Producing Black Aspergillus Isolates from Vineyards in Sicily. *Int. J. Food Microbiol.* **2008**, *127*, 147–154. [CrossRef]
17. Gil-Serna, J.; Vázquez, C.; González-Jaén, M.T.; Patiño, B. Wine Contamination with Ochratoxins: A Review. *Beverages* **2018**, *4*, 6. [CrossRef]
18. Cozzi, G.; Paciolla, C.; Haidukowski, M.; De Leonardis, S.; Mulè, G.; Logrieco, A. Increase of fumonisin B2 and ochratoxin A production by black Aspergillus species and oxidative stress in grape berries damaged by powdery mildew. *J. Food Prot.* **2013**, *76*, 2031–2036. [CrossRef]
19. Valero, A.; Sanchis, V.; Ramos, A.J.; Marin, S. Brief in vitro study on Botrytis cinerea and Aspergillus carbonarius regarding growth and ochratoxin A. *Lett. Appl. Microbiol.* **2008**, *47*, 327–332. [CrossRef]

20. Bellí, N.; Marín, S.; Coronas, I.; Sanchis, V.; Ramos, A. Skin Damage, High Temperature and Relative Humidity as Detrimental Factors for Aspergillus Carbonarius Infection and Ochratoxin A Production in Grapes. *Food Control* **2007**, *18*, 1343–1349.

21. Dimakopoulou, M.; Tjamos, S.E.; Antoniou, P.P.; Pietri, A.; Battilani, P.; Avramidis, N.; Tjamos, E.C. Phyllosphere grapevine yeast Aureobasidium pullulans reduces Aspergillus carbonarius (sour rot) incidence in wine-producing vineyards in Greece. *Biol. Control* **2008**, *46*, 158–165. [CrossRef]

22. Ponsone, M.L.; Chiotta, M.L.; Combina, M.; Dalcero, A.; Chulze, S. Biocontrol as a strategy to reduce the impact of ochratoxin A and Aspergillus section Nigri in grapes. *Int. J. Food Microbiol.* **2011**, *151*, 70–77. [CrossRef] [PubMed]

23. Ponsone, M.L.; Nally, M.C.; Chiotta, M.L.; Combina, M.; Köhl, J.; Chulze, S.N. Evaluation of the effectiveness of potential biocontrol yeasts against black sur rot and ochratoxin A occurring under greenhouse and field grape production conditions. *Biol. Control* **2016**, *103*, 78–85. [CrossRef]

24. Benito, S. The Impacts of Lachancea Thermotolerans Yeast Strains on Winemaking. *Appl. Microbiol. Biotechnol.* **2018**, *102*, 6775–6790. [CrossRef] [PubMed]

25. Quintela, S.; Villarán, M.C.; de Armentia, I.L.; Elejalde, E. Ochratoxin A Removal in Wine: A Review. *Food Control* **2013**, *30*, 439–445. [CrossRef]

26. Biosystems Enology Reagents. Available online: https://www.biosystems.es/products/FOODQUALITY/Enology/ENOLOGY%20REAGENTS (accessed on 15 August 2018).

27. Varga, J.; Kozakiewicz, Z. Ochratoxin A in Grapes and Grape-Derived Products. *Trends Food Sci. Technol.* **2006**, *17*, 72–81. [CrossRef]

28. Offer of the VitisLab. Available online: http://agronomiforestalipalermo.it/wp-content/uploads/2014/01/Listino-Analisi-Vini-Terreni-Vitis-Lab-2015.pdf (accessed on 28 March 2019).

29. Offer of the Enological Station of Haro. Available online: http://www.larioja.org/larioja-client/cm/agricultura/images?idMmedia=801034 (accessed on 15 August 2018).

30. Olivares-Marín, M.; Del Prete, V.; Garcia-Moruno, E.; Fernández-González, C.; Macías-García, A.; Gómez-Serrano, V. The Development of an Activated Carbon from Cherry Stones and its use in the Removal of Ochratoxin A from Red Wine. *Food Control* **2009**, *20*, 298–303. [CrossRef]

31. Cecchini, F.; Morassut, M.; García-Moruno, E.; Di Stefano, R. Influence of yeast strain on ochratoxin A content during fermentation of white and red must. *Food Microbiol.* **2006**, *23*, 411–417. [CrossRef]

32. Meca, G.; Blaiotta, G.; Ritieni, A. Reduction of ochratoxin A during the fermentation of Italian red wine Moscato. *Food Control* **2010**, *21*, 579–583.

33. Gambuti, A.; Strollo, D.; Genovese, A.; Ugliano, M.; Ritieni, A.; Moio, L. Influence of Enological Practices on Ochratoxin A Concentration in Wine. *Am. J. Enol. Vitic.* **2005**, *56*, 155–162.

34. Ladero, V.; Calles-Enríquez, M.; Fernández, M.; Alvarez, M.A. Toxicological Effects of Dietary Biogenic Amines. *Curr. Nutr. Food Sci.* **2010**, *6*, 145–156.

35. Martuscelli, M.; Arfelli, G.; Manetta, A.; Suzzi, G. Biogenic Amines Content as a Measure of the Quality of Wines of Abruzzo (Italy). *Food Chem.* **2013**, *140*, 590–597. [CrossRef] [PubMed]

36. Lonvaud-Funel, A. Biogenic Amines in Wines: Role of Lactic Acid Bacteria. *FEMS Microbiol. Lett.* **2001**, *199*, 9–13. [PubMed]

37. Guo, Y.; Yang, Y.; Peng, Q.; Han, Y. Biogenic Amines in Wine: A Review. *Int. J. Food Sci. Technol.* **2015**, *50*, 1523–1532. [CrossRef]

38. Smit, A.Y.; du Toit, M. Evaluating the Influence of Malolactic Fermentation Inoculation Practices and Ageing on Lees on Biogenic Amine Production in Wine. *Food Bioprocess Technol.* **2013**, *6*, 198–206.

39. Moreno-Arribas, M.V.; Polo, M.C.; Jorganes, F.; Muñoz, R. Screening of Biogenic Amine Production by Lactic Acid Bacteria Isolated from Grape must and Wine. *Int. J. Food Microbiol.* **2003**, *84*, 117–123. [CrossRef]

40. Coton, M.; Romano, A.; Spano, G.; Ziegler, K.; Vetrana, C.; Desmarais, C.; Lonvaud-Funel, A.; Lucas, P.; Coton, E. Occurrence of Biogenic Amine-Forming Lactic Acid Bacteria in Wine and Cider. *Food Microbiol.* **2010**, *27*, 1078–1085.

41. Benito, Á.; Calderón, F.; Palomero, F.; Benito, S. Combine use of Selected Schizosaccharomyces Pombe and Lachancea Thermotolerans Yeast Strains as an Alternative to Thetraditional Malolactic Fermentation in Red Wine Production. *Molecules* **2015**, *20*, 9510–9523. [CrossRef]

42. Caruso, M.; Fiore, C.; Contursi, M.; Salzano, G.; Paparella, A.; Romano, P. Formation of Biogenic Amines as Criteria for the Selection of Wine Yeasts. *World J. Microbiol. Biotechnol.* **2002**, *18*, 159–163. [CrossRef]

43. Polo, L.; Ferrer, S.; Peña-Gallego, A.; Hernández-Orte, P.; Pardo, I. Biogenic Amine Synthesis in High Quality Tempranillo Wines. Relationship with Lactic Acid Bacteria and Vinification Conditions. *Ann. Microbiol.* **2011**, *61*, 191–198.

44. Pascual, A.; Llorca, I.; Canut, A. Use of ozone in food industries for reducing the environmental impact of cleaning and disinfection activities. *Trends Food Sci. Technol.* **2007**, *18*, S29–S35. [CrossRef]

45. Benito, S. The Impact of Torulaspora Delbrueckii Yeast in Winemaking. *Appl. Microbiol. Biotechnol.* **2018**, *102*, 3081–3094. [CrossRef]

46. Marcobal, Á.; Martín-Álvarez, P.J.; Polo, M.C.; Muñoz, R.; Moreno-Arribas, M. Formation of Biogenic Amines Throughout the Industrial Manufacture of Red Wine. *J. Food Prot.* **2006**, *69*, 397–404.

47. Xia, Q.; Yuan, H.; Wu, C.; Zheng, J.; Zhang, S.; Shen, C.; Yi, B.; Zhou, R. An Improved and Validated Sample Cleanup Method for Analysis of Ethyl Carbamate in Chinese Liquor. *J. Food Sci.* **2014**, *79*, T1854–T1860.

48. Nettleship, A.; Henshaw, P.S.; Meyer, H.L. Induction of Pulmonary Tumors in Mice with Ethyl Carbamate (Urethane). *J. Natl. Cancer Inst.* **1943**, *4*, 309–319.

49. Haddow, A.; Sexton, W. Influence of Carbamic Esters (Urethanes) on Experimental Animal Tumours. *Nature* **1946**, *157*, 500.

50. Sakano, K.; Oikawa, S.; Hiraku, Y.; Kawanishi, S. Metabolism of Carcinogenic Urethane to Nitric Oxide is Involved in Oxidative DNA Damage. *Free Radic. Biol. Med.* **2002**, *33*, 703–714.

51. Beland, F.A.; Benson, R.W.; Mellick, P.W.; Kovatch, R.M.; Roberts, D.W.; Fang, J.; Doerge, D.R. Effect of Ethanol on the Tumorigenicity of Urethane (Ethyl Carbamate) in B6C3F1 Mice. *Food Chem. Toxicol.* **2005**, *43*, 1–19. [CrossRef]

52. Baan, R.; Straif, K.; Grosse, Y.; Secretan, B.; El Ghissassi, F.; Bouvard, V.; Altieri, A.; Cogliano, V. WHO International Agency for Research on Cancer Monograph Working Group. Carcinogenicity of Alcoholic Beverages. *Lancet Oncol.* **2007**, *8*, 292–293.

53. Mira de Orduña, R.; Liu, S.; Patchett, M.; Pilone, G. Ethyl Carbamate Precursor Citrulline Formation from Arginine Degradation by Malolactic Wine Lactic Acid Bacteria. *FEMS Microbiol. Lett.* **2000**, *183*, 31–35.

54. Ryu, D.; Choi, B.; Kim, E.; Park, S.; Paeng, H.; Kim, C.I.; Lee, J.Y.; Yoon, H.J.; Koh, E. Determination of Ethyl Carbamate in Alcoholic Beverages and Fermented Foods Sold in Korea. *Toxicol. Res.* **2015**, *31*, 289–297.

55. Ough, C.S.; Stevens, D.; Almy, J. Preliminary comments on effects of grape vineyard nitrogen fertilization on the subsequent ethyl carbamate formation in wines. *Am. J. Enol. Vitic.* **1989**, *40*, 219–220.

56. Ough, C.S.; Stevens, D.; Sendovski, T.; Huang, Z.; An, D. Factors contributing to urea formation in commercially fermented wines. *Am. J. Enol. Vitic.* **1990**, *41*, 68–73.

57. Treatments of Wine with Urease. Available online: http://www.oiv.int/public/medias/3542/e-code-ii-3411.pdf (accessed on 15 August 2018).

58. According to the Agency for Toxic Substances and Disease Registry. Available online: https://www.atsdr.cdc.gov/mmg/mmg.asp?id=249&tid=46 (accessed on 15 August 2018).

59. Vally, H.; Thompson, P.J. Role of Sulfite Additives in Wine Induced Asthma: Single Dose and Cumulative Dose Studies. *Thorax* **2001**, *56*, 763–769. [CrossRef] [PubMed]

60. Commission Regulation (EU) No 59/2014 Regarding Sulfur Dioxide. Available online: http://www.fao.org/fao-who-codexalimentarius/codex-texts/guidelines/en/ (accessed on 15 August 2018).

61. Maximum Acceptable Limit. Available online: http://www.oiv.int/public/medias/3741/e-code-annex-maximum-acceptable-limits.pdf (accessed on 15 August 2018).

62. Mota, F.J.; Ferreira, I.M.; Cunha, S.C.; Beatriz, M.; Oliveira, P. Optimisation of Extraction Procedures for Analysis of Benzoic and Sorbic Acids in Foodstuffs. *Food Chem.* **2003**, *82*, 469–473. [CrossRef]

63. Gao, Y.C.; Zhang, G.; Krentz, S.; Darius, S.; Power, J.; Lagarde, G. Inhibition of Spoilage Lactic Acid Bacteria by Lysozyme during Wine Alcoholic Fermentation. *Aust. J. Grape Wine Res.* **2002**, *8*, 76–83.

64. Iriti, M.; Vitalini, S.; Di Tommaso, G.; D'amico, S.; Borgo, M.; Faoro, F. New Chitosan Formulation Prevents Grapevine Powdery Mildew Infection and Improves Polyphenol Content and Free Radical Scavenging Activity of Grape and Wine. *Aust. J. Grape Wine Res.* **2011**, *17*, 263–269.

65. González-Arenzana, L.; Sevenich, R.; Rauh, C.; López, R.; Knorr, D.; López-Alfaro, I. Inactivation of Brettanomyces bruxellensis by high hydrostatic pressure technology. *Food Control* **2016**, *59*, 188–195.

66. Bradshaw, M.P.; Barril, C.; Clark, A.C.; Prenzler, P.D.; Scollary, G.R. Ascorbic Acid: A Review of its Chemistry and Reactivity in Relation to a Wine Environment. *Crit. Rev. Food Sci. Nutr.* **2011**, *51*, 479–498.

67. Du Toit, W.J.; Marais, J.; Pretorius, I.S.; Du Toit, M. Oxygen in must and wine: A review. *S. Afr. J. Enol. Vitic.* **2006**, *27*, 76–94.

68. Peñas, E.; di Lorenzo, C.; Uberti, F.; Restani, P. Allergenic Proteins in Enology: A Review on Technological Applications and Safety Aspects. *Molecules* **2015**, *20*, 13144–13164.

69. Common Organization of the Market in Wine, Amending Regulations. Available online: https://eur-lex.europa.eu/legal-content/EN/TXT/?uri=celex%3A32008R0479 (accessed on 15 August 2018).

70. Potentially Allergenic Residues of Fining Agent Proteins in Wine. Available online: http://www.oiv.int/en/technical-standards-and-documents/resolutions-of-the-oiv/comex-resolutions (accessed on 15 August 2018).

71. Cosme, F.; Ricardo-da-Silva, J.; Laureano, O. Interactions between Protein Fining Agents and Proanthocyanidins in White Wine. *Food Chem.* **2008**, *106*, 536–544. [CrossRef]

72. Cattaneo, A.; Ballabio, C.; Bernardini, R.; Bertelli, A.A.; Novembre, E.; Vierucci, A.; Restani, P. Assessment of Residual Immunoreactivity in Red or White Wines Clarified with Pea or Lupin Extracts. *Int. J. Tissue React.* **2003**, *25*, 159–165.

73. Protein Plant Origin. Available online: http://www.oiv.int/public/medias/5163/e-coei-1-proveg.pdf (accessed on 15 August 2018).

74. Restani, P.; Uberti, F.; Danzi, R.; Ballabio, C.; Pavanello, F.; Tarantino, C. Absence of Allergenic Residues in Experimental and Commercial Wines Fined with Caseinates. *Food Chem.* **2012**, *134*, 1438–1445.

75. Monaci, L.; Losito, I.; De Angelis, E.; Pilolli, R.; Visconti, A. Multi-allergen Quantification of fining-related Egg and Milk Proteins in White Wines by high-resolution Mass Spectrometry. *Rapid Commun. Mass Spectrom.* **2013**, *27*, 2009–2018. [CrossRef]

76. Pilolli, R.; De Angelis, E.; Godula, M.; Visconti, A.; Monaci, L. Orbitrap™ Monostage MS Versus Hybrid Linear Ion Trap MS: Application to multi-allergen Screening in Wine. *J. Mass Spectrom.* **2014**, *49*, 1254–1263.

77. Nicolopoulou-Stamati, P.; Maipas, S.; Kotampasi, C.; Stamatis, P.; Hens, L. Chemical Pesticides and Human Health: The Urgent Need for a New Concept in Agriculture. *Front. Public Health* **2016**, *4*, 148.

78. Gunnell, D.; Eddleston, M.; Phillips, M.R.; Konradsen, F. The Global Distribution of Fatal Pesticide Self-Poisoning: Systematic Review. *BMC Public Health* **2007**, *7*, 357.

79. Regulation (EC) n° 1493/1999. Available online: https://eur-lex.europa.eu/legal-content/ES/ALL/?uri=CELEX%3A31999R1493 (accessed on 15 August 2018).

80. Regulation (EC) n° 423/2008. Available online: https://publications.europa.eu/es/publication-detail/-/publication/640e7ae4-949f-4f80-b23f-b73930d673c1/language-es (accessed on 15 August 2018).

81. Available online: http://www.fao.org/fao-who-codexalimentarius/codex-texts/dbs/pestres/commodities-detail/es/?c_id=113 (accessed on 28 March 2018).

82. Available online: http://www.fao.org/fao-who-codexalimentarius/codex-texts/dbs/pestres/commodities-detail/es/?c_id=406 (accessed on 28 March 2018).

83. Figueiredo, A.; Fortes, A.M.; Ferreira, S.; Sebastiana, M.; Choi, Y.H.; Sousa, L.; Acioli-Santos, B.; Pessoa, F.; Verpoorte, R.; Pais, M.S. Transcriptional and Metabolic Profiling of Grape (Vitis Vinifera L.) Leaves Unravel Possible Innate Resistance Against Pathogenic Fungi. *J. Exp. Bot.* **2008**, *59*, 3371–3381. [CrossRef]

84. Ehrhardt, C.; Arapitsas, P.; Stefanini, M.; Flick, G.; Mattivi, F. Analysis of the Phenolic Composition of fungus-resistant Grape Varieties Cultivated in Italy and Germany using UHPLC-MS/MS. *J. Mass Spectrom.* **2014**, *49*, 860–869.

85. Plahuta, P.; Raspor, P. Comparison of Hazards: Current Vs. GMO Wine. *Food Control* **2007**, *18*, 492–502.

86. Dequin, S. The Potential of Genetic Engineering for Improving Brewing, Wine-Making and Baking Yeasts. *Appl. Microbiol. Biotechnol.* **2001**, *56*, 577–588.

87. Grossmann, M.; Kießling, F.; Singer, J.; Schoeman, H.; Schröder, M.B.; von Wallbrunn, C. Genetically modified wine yeasts and risk assessment studies covering different steps within the wine making process. *Ann. Microbiol.* **2011**, *61*, 103–115.

88. Regulation (EC) No 1829/2003. Available online: https://eur-lex.europa.eu/legal-content/en/ALL/?uri=CELEX%3A32003R1829 (accessed on 15 August 2018).

89. Vaudano, E.; Costantini, A.; Garcia-Moruno, E. An Event-Specific Method for the Detection and Quantification of ML01, a Genetically Modified Saccharomyces Cerevisiae Wine Strain, using Quantitative PCR. *Int. J. Food Microbiol.* **2016**, *234*, 15–23. [CrossRef]

90. Institute of Agriculture and Natural Resources, Physical Hazards. Available online: https://food.unl.edu/physical-hazards (accessed on 15 August 2018).
91. SECTION 555.425-Foods—Adulteration Involving Hard or Sharp Foreign Objects. Available online: https://es.scribd.com/doc/307954392/SECTION-555-425-Peligro-Fisicos-FDA (accessed on 15 August 2018).
92. Canadian Food Inspection Agency. Food Safety Hazards. Available online: http://www.inspection.gc.ca/food/non-federally-registered/product-inspection/inspection-manual/eng/1393949957029/1393950086417?chap=5 (accessed on 15 August 2018).
93. Umiker, N.; Descenzo, R.; Lee, J.; Edwards, C. Removal of Brettanomyces Bruxellensis from Red Wine using Membrane Filtration. *J. Food Process. Preserv.* **2013**, *37*, 799–805.
94. Petrović, M.; Eljarrat, E.; de Alda, M.J.L.; Barceló, D. Analysis and Environmental Levels of Endocrine-Disrupting Compounds in Freshwater Sediments. *TrAC Trends Anal. Chem.* **2001**, *20*, 637–648.
95. Gray, L.E., Jr.; Ostby, J.; Furr, J.; Price, M.; Veeramachaneni, D.R.; Parks, L. Perinatal Exposure to the Phthalates DEHP, BBP, and DINP, but Not DEP, DMP, or DOTP, Alters Sexual Differentiation of the Male Rat. *Toxicol. Sci.* **2000**, *58*, 350–365. [CrossRef]
96. Harrison, P.; Holmes, P.; Humfrey, C. Reproductive Health in Humans and Wildlife: Are Adverse Trends Associated with Environmental Chemical Exposure? *Sci. Total Environ.* **1997**, *205*, 97–106.
97. Del Carlo, M.; Pepe, A.; Sacchetti, G.; Compagnone, D.; Mastrocola, D.; Cichelli, A. Determination of Phthalate Esters in Wine using Solid-Phase Extraction and Gas chromatography–mass Spectrometry. *Food Chem.* **2008**, *111*, 771–777. [CrossRef]
98. Holadová, K.; Prokůpková, G.; Hajšlová, J.; Poustka, J. Headspace Solid-Phase Microextraction of Phthalic Acid Esters from Vegetable Oil Employing Solvent Based Matrix Modification. *Anal. Chim. Acta* **2007**, *582*, 24–33. [CrossRef]
99. Balafas, D.; Shaw, K.; Whitfield, F. Phthalate and Adipate Esters in Australian Packaging Materials. *Food Chem.* **1999**, *65*, 279–287. [CrossRef]

 *fermentation*

Review

The Importance of Yeasts on Fermentation Quality and Human Health-Promoting Compounds [†]

Alice Vilela

CQ-VR, Chemistry Research Centre, University of Trás-os-Montes and Alto Douro (UTAD), School of Life Sciences and Environment, Dep. of Biology and Environment, Enology building, 5000-801 Vila Real, Portugal; avimoura@utad.pt

† I dedicate this small review article to my daughter, Ana Margarida, a 1st-year student of Pharmaceutical Sciences. I would like to encourage her to study these subjects in the search for more natural health-promoting compounds, of natural origin, in detrimental of "laboratory made" ones.

Received: 19 April 2019; Accepted: 29 May 2019; Published: 31 May 2019

Abstract: Non-*Saccharomyces* are important during wine fermentation once they influence wine composition. In the early stages of wine fermentation, and together with indigenous or commercial strains of *Saccharomyces cerevisiae*, non-*Saccharomyces* are able to transform grape-must sugars into ethanol, CO_2, and other important secondary metabolites. A better understanding of yeast biochemistry will allow the selection of yeast strains that have defined specific influences on fermentation efficiency, wine quality, and the production of human health-promoting compounds. Yeast metabolism produces compounds derived from tryptophan, melatonin, and serotonin, which are found in fermented beverages, such as wine and beer. Melatonin is a neurohormone secreted from the pineal gland and has a wide-ranging regulatory and neuroprotective role, while serotonin, as well as being a precursor of melatonin synthesis, is also a neurotransmitter. This review summarizes the importance of some conventional and nonconventional yeast strains' alcoholic fermentations, especially in the production of metabolites that promote human health and thus, attract consumers attention towards fermented beverages. A brief reference is also made on fermented beverages containing probiotics, namely kombucha, also known as kombucha tea, and its interesting microorganism's symbiotic relationships named SCOBY.

Keywords: Yeasts; alcoholic beverages; resveratrol; glutathione; trehalose; tryptophan; melatonin; serotonin; tyrosol; tryptophol; hydroxytyrosol; IAA; probiotics

---

1. Introduction

The term "fermentation" comes from the Latin word *"fermentum"* (meaning, to ferment). The science of fermentation is called "zymology" and the first zymologist was Louis Pasteur who was able to identify and apply yeast in fermentation [1]. Food fermentations date back at least 6000 years. In the 16th century, the beginning of industrialization initiated technological interventions in food and beverage production [2]. However, it was in the last two centuries that significant changes in the world's food system have occurred. In olden days, fermentation of food was meant for food preservation and flavor improvement [3], nowadays, in food and beverages fermentation, various technologies and operations are used with the aim of converting fairly perishable and indigestible raw materials into pleasant foods and drinkable beverages with added value and high stability [4]. The assurance of the quality and safety of the final product is the main goal of the technologies applied [5].

Biotechnology plays a radical role in the production, conservation, nutritional enrichment, and value addition of foods. To be able to understand the science of microbiology in food and beverage applications with identification of new-fermenting species is an advantage to enhance the quality of our food products.

Food and beverage processing using microorganisms is the most suitable technology for the development of innovative fermented food products. Solid state fermentation is used for processing of vinegar, soy sauce, tea, and cheese [6]. Wine, beer, distilled beverages, and yogurt are developed by submerged fermentation. Both methods of fermentation are influenced by numerous factors, including temperature, pH, nature, and composition of the medium, dissolved O_2, dissolved CO_2, operational system, and feeding with precursors, among others. Variation in these factors may affect the rate of fermentation, the product spectrum and yield, the sensory properties of the product (appearances, taste, smell, texture), physic-chemical properties, nutritional quality, and the production of metabolites that promote human health attracting consumers attention towards fermented products, namely beverages.

2. Alcoholic Beverages Consumption and Health-Promoting Compounds

The prevention of diseases by altering lifestyle and dietary conducts may present more benefits than medical care. Up till now, adjusting individual dietary habits is a challenge. Most often, consumers must choose between nutrition, taste, price, convenience, and cost [7]. Nowadays, the nutritional value appears to be the health benefit that has the most impact on a consumer's purchase [8].

In the last 30 years, the effects of wine on human health have been studied by many scientists. In 1992, Renaud and de Lorgeril [9] published a study about the higher wine consumption in the French population, in comparison with other industrialized countries. According to these authors, wine caused a lower incidence of coronary heart disease, notwithstanding the intake of the traditional French diet, rich in saturated fatty acids. This finding constituted the so-called "French paradox". Since then, many studies have been carried out on wine indicating that persons consuming daily moderate amounts of wine display a reduction of cardiovascular mortality and an improvement of antioxidant parameters, when compared with individuals who abstain or who drink alcohol to excess [10,11]. Moreover, Poli et al. [11] also mentioned that in healthy adults, spontaneous consumption of alcoholic beverages (30 g ethanol/day for men and 15 g/day for women) is to be considered acceptable. However, the possible interactions between alcohol and acute or chronic drug use must be discussed with the primary care physician.

Oxidative stress and antioxidant deficiency have been implicated in the pathogenesis of many diseases and conditions, including atherosclerosis, cancer, aging, and respiratory disease. Glutathione (L-g-glutamyl-L-cysteinyl-glycine, GSH) (Figure 1) is a major antioxidant acting as a free radical scavenger that protects the cell from ROS (reactive oxygen species). In addition, GSH is involved in nutrient metabolism and regulation of cellular metabolic functions ranging from DNA and protein synthesis to signal transduction, cell proliferation, and apoptosis [12–14].

Another important molecule is trehalose (Figure 1). This sugar possesses inflammatory properties [15] presenting, also, the ability to protect cellular membranes and labile proteins against denaturation as a result of desiccation and oxidative stress [16].

Yeast metabolism produces compounds derived from tryptophan, which are found in fermented beverages, such as wine and beer. In particular, melatonin and serotonin (Figure 1). Serotonin is a neurohormone that regulates circadian rhythms, and also has an alleged protective effect against neurodegenerative and degenerative diseases (Alzheimer's, Parkinson's and Angiogenesis) [17]. Moreover, serotonin is a neurotransmitter itself, and a precursor of melatonin synthesis.

In humans, melatonin (N-acetyl-5-methoxytryptamine) is a hormone that modulates several physiological processes. This molecule is an indole-amine found in many living organisms like plants, microorganisms, and humans. Melatonin modulates many human physiological processes including the sleep/wake cycle and the reproductive physiology via a receptor-mediated mechanism [18,19] acting, also, as an antioxidant via nonreceptor processes [20]. It is well known that the intake of foods containing melatonin increases its level in plasma and the number of melatonin-derived metabolites [21]. Studies have been carried out to identify melatonin in grapes [22] and beverages such as beer and wine [23,24]. Interesting is the reported concentrations of melatonin in grapes (*Vitis vinifera* L.) and wines: 150 µg/g in Merlot grapes [25]; 130 ng/mL in Tempranillo wine [26].

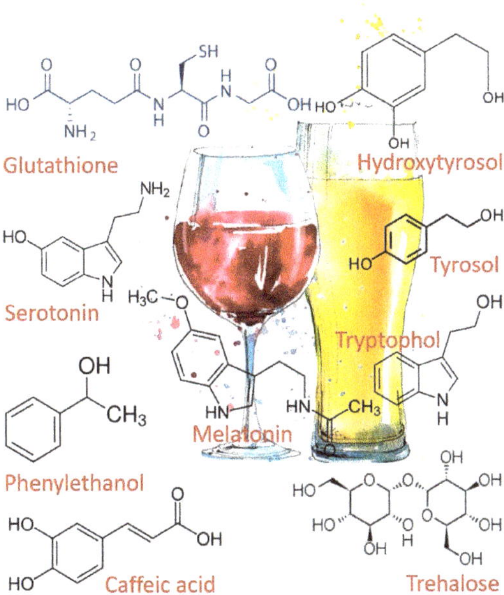

Figure 1. Chemical Structures of the health-promoting compounds mentioned.

Tyrosol and tryptophol (Figure 1) are produced by yeasts during alcoholic fermentation from the catabolism of amino acids tyrosine and tryptophan, respectively, whereas hydroxytyrosol is produced by hydroxylation of its precursor tyrosol. Tyrosol, hydroxytyrosol, tryptophol are reported to possess several health-enhancing activities, deriving from their free radical scavenging, anticarcinogenic, cardioprotective (induces myocardial protection against ischemia-related stress [27]) and antimicrobial properties [28].

It's due to the presence of tyrosol and caffeic acids (Figure 1), that white wine has been reported as having cardioprotective benefits. Tyrosol and caffeic acids are able to activate the cell survival signaling pathway and the *FOXO3a* longevity-associated gene [29,30]. Moreover, tyrosol has been shown to have an important role in the taste of some alcoholic beverages, such as sake [31] and wine [32] by exhibiting a bitter taste above the sensory threshold, but below the recognition threshold.

Tryptophol is also used as a precursor in the synthesis of the drug Indoramin, an α-adrenoreceptor blocking drug used to treat hypertension [33], and in the treatment of benign prostatic hyperplasia [34].

Phenylethanol (Figure 1), also produced by *Candida albicans* as an auto-antibiotic [35] is an aromatic compound that is commonly found in plants, such as roses, possessing pleasant floral rose-like odor. Due to its preservative properties, phenylethanol is often used in soap-based detergents because of its stability in basic conditions. Phenylethanol can also act as a natural preservative in wine and beer to prevent spoilage [35].

3. Mechanisms of Microbial Resistance to Environment Changes that Produce Health-Promoting Compounds

Conservation and commercialization of yeast cultures in fresh liquid or pressed forms are not economically advantageous. Thus, dehydrated yeasts present numerous advantages, such as lower cost, convenient for transport and storage, and ease of handling [36]. However, the drying of the yeasts signifies highly sensitive transformation processes for microorganism's which can lead to cell death or a significant decrease in cell activity potential [37]. The final water volume of the cells, induced by dehydration-rehydration cycles, influence the cells survival [38], and the modification of plasma membrane fluidity during the dehydration-rehydration cycles affects the plasma membrane structure and may induce cell mortality [39].

Increase of contact surface of the cells with air during dehydration also induces accumulation of ROS (reactive oxygen species)—[$O_2^{\bullet-}$ (superoxide anion), $^\bullet OH$ (hydroxyl radical), H_2O_2 (hydrogen peroxide) and ReOOH (hydroperoxides)]—and may contribute to inactivation of several enzymes, leading, also, to cell death [40]. In the presence of these stress conditions yeasts are able to synthesize compounds such as glutathione and trehalose [41].

Glutathione (GSH) is a ubiquitous low molecular weight thiol tripeptide containing glutamate, cysteine, and glycine (Glu-Cys-Gly), it is present in large amounts in yeasts and it can be found in the reduced or oxidized forms (GSH and GSSG, respectively). Glutathione plays a crucial role in redox equilibrium reactions, protecting the cell from oxidative stress, by allowing the formation of native disulfide bonds and by scavenging free radicals present in the cytosol; reactions mediated via glutathione reductase and glutathione peroxidase [12,42].

Hgt1p in yeast *S. cerevisiae* was the first identified high-specificity and high-affinity glutathione transporter (Km 54 mM) [43]. Hgt1p belongs to oligopeptide transporter family which was also found in fungi, plants, and prokaryotes. Genetic and physiological investigations revealed that gene *HGT1* (open reading frame *YJL212c*) as encoding a high-affinity glutathione transporter. Yeast strains deleted in *HGT1* gene did not show any detectable plasma membrane glutathione transport. This transporter is required for the uptake of glutathione from the extracellular medium (Figure 2) [43]. Moreover, mitochondria are a primary source of ROS in cells and mitochondrial thiols are therefore major ROS targets. This phenomenon is exacerbated by the relatively alkaline pH of mitochondria. Therefore, redox regulation is critical for numerous mitochondrial functions, and yeast strains lacking GSH are unable to grow by respiration due to an accumulation of oxidative damage to mitochondrial DNA. Transport of H_2O_2 across yeast cell membranes can be facilitated by transporters such as aquaporins. Hydrogen peroxide causes oxidative stress but also plays important roles as a signaling molecule in the regulation of many biological processes (Figure 2) [44].

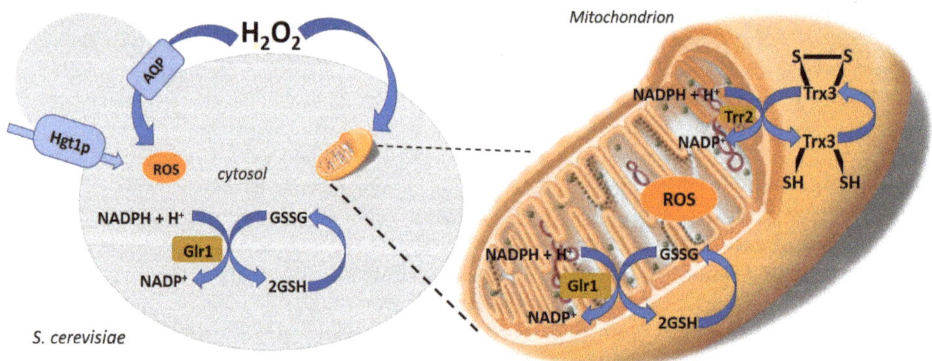

Figure 2. Thiol redox regulation in the response of cells to oxidative stress conditions. Hgt1p (Glutathione transporters in the yeast *S. cerevisiae* [43,46]); AQP (aquaporin-mediated H_2O_2 diffusion transport); Glrl (glutathione reductase); Trr2 (Thioredoxin reductase 2, mitochondrial); Trx3 (Thioredoxin-3, mitochondrial, disulfide oxidoreductase activity) Adapted from Gostimskaya and Grant [45].

Thiol redox regulation plays a role in the response of cells to oxidative stress conditions. Gostimskaya and Grant [45] emphasize the importance of compartmentalized redox regulation when cells are subjected to oxidative stress conditions (Figure 2). At the same time as cytosolic glutathione represents the first major pool of thiols, which would be a target of oxidation in response to exposure to an exogenous oxidant, it is the mitochondrial glutathione pool which is crucial for oxidant tolerance.

Has mentioned before, one interesting disaccharide with potential for medical application, namely in ophthalmology, is trehalose [15]. This sugar, constituted by two glucose residues linked by an α-1-1 glycosidic bond, is widespread in many species of plants, animals and microorganisms, including wine yeasts, where its function is to protect cells against desiccation. However, this sugar is no naturally found in mammals. Nowadays trehalose is used in the biopharmaceutical preservation of labile protein drugs and in the cryopreservation of human cells and is under investigation for the treatment of Huntington's chorea and Alzheimer's disease. It can also be used as a preventive drug to treat dryness in mammalian eyes, a common tear and ocular surface multifactorial disease that can lead to inflammatory reaction [15]. Trehalose also acts as a storage carbohydrate for the cell and it plays a very important role as a protectant against osmotic stress and ethanol stress, in yeast cells [47]. Trehalose acts as a stabilizing effect on the plasma membrane, providing it with increased tolerance to desiccation, dehydration, temperature changes, and high temperature [16]. It can also act as an antioxidant component by reducing oxidation reactions rates while enhancing the viscosity of cell cytoplasm [48].

Câmara et al. [49] studied the effects of glutathione and trehalose biosynthesis in the dehydration stress responses of three non-*Saccharomyces* yeasts strains (*Torulaspora delbrueckii* CBS4865, *Metschnikowia pulcherrima* CBS5833 and *Lachancea thermotolerans* CBS6340). The results obtained will help to better understand certain physiological responses of non-*Saccharomyces* yeasts to dehydration, leading to the promotion and production of new high performance dehydrated non-*Saccharomyces* yeasts strains to be used in food and beverages elaboration. According to the mentioned work [49], yeasts grown in nutrient-rich medium accumulated glutathione leading to a higher resistance to dehydration, whilst the nutrient-poor medium induced the cells to accumulate large amounts of trehalose, which partially protected them from GSSG accumulation.

4. Melatonin and Other Tryptophan Metabolites

In the scientific world, the theme of "wine and health" topics have been focused mainly on polyphenols, once these bioactive compounds are present in plants and are released into fermented products. However, yeast also transforms other molecules into biologically active compounds [19]. Since the pioneering work of Sprenger and co-workers [50] that melatonin molecule, has been reported as being present in wine, and its presence has been related to the activity of the yeast involved in the fermentation process. Originally, seen as a unique product of the pineal gland of vertebrates, called a neurohormone, at the present, it is considered a ubiquitous molecule present in most living organisms [51].

Rodriguez-Naranjo and co-workers [26] studied the capacity of different yeasts to produce melatonin during alcoholic fermentation. Different *Saccharomyces* yeast strains, used for industrial fermentation of beer or as nutritional complements, and non-*Saccharomyces* yeast strains (*Metschnikowia pulcherrima* and *Starmerella bacillaris*) were tested by the referred authors to analyze intracellular and extracellular melatonin production in synthetic grape must. Interestingly, at the beginning of fermentation melatonin was detected, in the intracellular compartment, either in *Saccharomyces* or in non-*Saccharomyces* strains. Production levels differed among strains, being *Starmerella bacillaris* the non-*Saccharomyces* yeast that presented the highest concentration. Nevertheless, extracellular melatonin was detected at different time-points over the fermentation process, depending on the yeast strain. However, the same authors [26] also reported that the presence of tryptophan is essential for melatonin production since it is its principal precursor, it increases final melatonin content and it accelerates its formation. Moreover, the synthesis of melatonin largely depends on the growth phase of the yeast and the concentration of the reducing sugars.

The metabolic pathway for melatonin production in yeast is not completely clarified, nevertheless, the formation of serotonin might be an intermediate metabolite in the pathway [19], Figure 3.

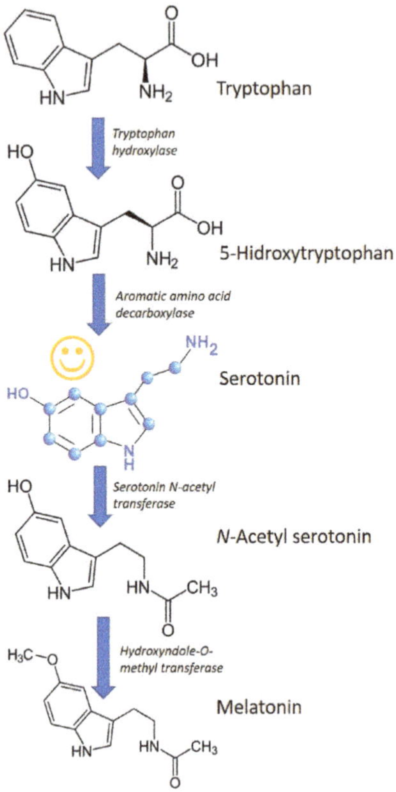

Figure 3. Synthesis of melatonin and serotonin, as an intermediate compound, from tryptophan in yeast. Adapted from Mas et al. [19].

Germann et al. [52], studied *de novo* melatonin biosynthesis from glucose, by genetically modifying *Saccharomyces cerevisiae* strains that comprise heterologous genes encoding one or more variants of L-tryptophan hydroxylase, 5-hydroxy-L-tryptophan decarboxylase, serotonin acetyltransferase, acetylserotonin *O*-methyltransferase, and means for providing the cofactor tetrahydrobiopterin via heterologous biosynthesis and recycling pathways. At the end of the process, yeast strain produced melatonin in concentrations of 14.50 ± 0.57 mg L^{-1} in a 76 hours fermentation, using glucose as sole carbon source.

Although is thought to be an intermediate for melatonin production, there is no evidence for the production of serotonin by *S. cerevisiae*. Serotonin has been found in wines at levels ranging from 2 to 23 mg L^{-1}, mainly as a result of the malolactic fermentation performed by *Lactobacillus plantarum* strains [53,54].

In 2016, Tan and co-workers [55], proposed an alternative pathway for synthesizing melatonin from L-tryptophan via 5-methoxytryptamine with the formation of the intermediate compound N-acetyl-5-hydroxytryptamine [56]. Different *Saccharomyces* strains (Lalvin QA23, Enartis ES488, Lallemand ICV-GRE, and Uvaferm) and non-*Saccharomyces* yeast (*Torulaspora delbrueckii* and *Metschnikowia pulcherrima*) have been reported to produce melatonin during alcoholic fermentation in synthetic must [57,58]. Moreover, *T. delbrueckii* is considered an innovative biotechnological tool, of great commercial interest, to be used in pure culture or in sequential inoculation with *S. cerevisiae*, for bio-modulating wines acidity [59] among other interesting enological features [60].

Tryptophan metabolism includes, in addition to 5-hydroxytryptamine and melatonin, other important metabolites such as indolic compounds like 3-indoleacetic acid (3-IAA) which is the most common plant hormone, of the auxin class, and is known to have many effects including cell proliferation enhancement and antioxidant properties [61]. Kim and co-workers [61] investigated the effects of IAA on H_2O_2 induced oxidative toxicity in human dental pulp stem cells and verified that H_2O_2-induced cytotoxicity was attenuated after IAA treatment. Moreover, according to Fernández-Cruz and co-workers [56], besides IAA, tryptophan metabolites (tryptophol, tryptamine, and l-tryptophan ethyl ester) also present potential as antioxidants and neuroprotective agents. The mentioned authors [56] examined the occurrence of these compounds during the alcoholic fermentation of musts from seven grape cultivars. Fermentations were performed with three *S. cerevisiae* strains and, in two cases, a sequential inoculation with *Torulaspora delbrueckii*. Interestingly, they found that the profile of indolic compounds during alcoholic fermentation depended on the cultivar and not on the yeast strain used. Nonetheless, fermentation time was found to be a more influential factor [56].

5. Fusel Alcohols Formed Via the Ehrlich Pathway

The synthesis of tryptophol by yeast was first described by Felix Ehrlich in 1912 [62,63] as the metabolic conversion of amino acids via the successive steps of transamination, decarboxylation, and reduction [64], Figure 4.

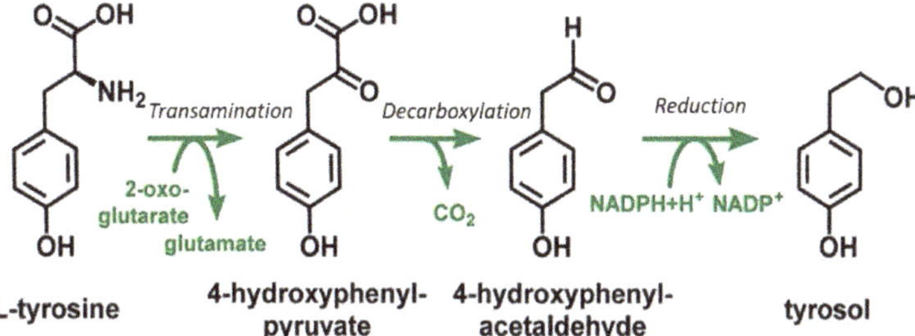

Figure 4. Schematic representation of the Ehrlich pathway [62,63] and the biochemical reactions involved.

Similarly, to tryptophol, phenylethanol, and tyrosol, are phenolic compounds or fusel alcohols formed via the Ehrlich pathway by yeast metabolism. These compounds can yield health benefits as well as contribute to the flavors and aromas of fermented food and beverages [63,65].

Banach and Ooi [65] investigated the possibility of increasing the yield of tyrosol, tryptophol, and phenylethanol in wine (Alexander's Pinot Chardonnay grape juice) and beer [modification of the English Ale recipe composed of chocolate malt barley grain, dried malt extract (DME) and liquid malt extract (LME), supplemented with either the equivalent volume of malt-kiwi purée or with amino acids] using two different yeast strains, and supplementing the substrate with the relevant amino acid precursors or fruits high in these amino acids. At the end of the work, they found that flavor enhancement and enrichment of antioxidants, in wine and beer, could be achieved through supplementing the fermentation (in the case of beer-fruit-supplemented beer) media with precursor amino acids as well as careful choices of the appropriate yeast strain.

6. Fermented Beverages Containing Probiotics

It is common knowledge that most of the fermented milk contains probiotic microorganisms (live microorganisms, which when administered in adequate amounts, confer a health benefit on the host). Yogurt, the most common product of milk lactose fermentation, has on its constitution several lactic acid bacteria. So, the domination of milk-based beverages fermented by LAB, mainly *Leuconostoc*, *lactobacilli*, and *lactococci*, is clear. Milk fermentation in colder climates promotes the growth of mesophilic bacteria such as *Lactococcus* and *Leuconostoc*, whereas beverages produced at higher temperatures usually have greater counts of thermophilic bacteria such as *Lactobacillus* and *Streptococcus* [66,67]. Most often the probiotic bacteria come from *Lactobacillus* or *Bifidobacterium* or a cocktail of both [68].

Another class of fermented beverages is those made from cereals (maize, millet, barley, oats, rye, wheat, rice and sorghum), were the natural microbial component is used to ferment grains. The microbial populations responsible for the fermentation of these beverages are not, yet, well characterized. Of several blends, it has been suggested that fermentation by *S. cerevisiae*, *Leuconostoc mesenteroides*, and *Lactobacillus confusus* produce the most palatable beverages [66].

One example of a known fermented beverage, with probiotic characteristics, is kombucha. It is a fermented sweetened tea, originally from China, but, enjoyed worldwide. *Medusomyces gisevii* Lindau represents a symbiotic microbial community known as "tea fungus" or kombucha tea. During tea fermentation, the added sugar is converted into organic acids and ethanol by yeast and bacteria. The microorganisms co-exist in interdependent symbiotic relationships named SCOBY (symbiotic culture of bacteria and yeast). Organic acids and ethanol formed during tea fermentation protect SCOBY from the colonization of other microorganisms. The yeast present in SCOBY can vary and may include *Brettanomyces/Dekkera*, *Schizosaccharomyces*, *Torulaspora*, *Zygosaccharomyces*, and *Pichia* [69]. The bacteria that we can find in kombucha are species of *Gluconacetobacter xylinus*, *G. kombuchae* sp. nov., *Acetobacter nitrogenifigens* sp. nov, *Acetobacter intermedius*, sp. nov. [70,71].

Nondairy fermented beverage from a blend of cassava and rice based on Brazilian indigenous beverage "cauim" using probiotic lactic acid bacteria and yeast was studied by Freire and co-workers in 2017 [72]. A triple group of microorganisms was used to ferment the blend: *Lactobacillus plantarum* CCMA 0743 (from cauim), *Torulaspora delbrueckii* CCMA 0235 (from tarubá), and the commercial probiotic *L. acidophilus* LAC-04. According to the authors [72], the bacteria populations were around 8.0 log (CFU mL^{-1}) at the end of all fermentations as recommended for probiotic products. The final beverage obtained was considered a non-alcoholic drink since the ethanol degree was lower than 0.5% (*v/v*).

7. Final Remarks

The choice to consume alcohol should be based on individual considerations, taking into account the influence on health and diet, the risk of alcoholism and abuse, the effect on behavior, and other factors that may vary with age and lifestyle.

Fermentation remains the oldest but most prevalent means of food and drinks processing and preservation. Important studies have been carried out aiming to understand certain physiological responses of *Saccharomyces* and non-*Saccharomyces* yeasts during the fermentation process, in order to promote the production and the marketing of new high-performance dehydrated yeasts strains for food and beverages elaboration, that will also be able to produce health-promoting compounds.

Fermented drinks, namely wine and beer, can definitely serve as vehicles for beneficial compounds that play an important role in human health, namely in the prevention of some 21st-century diseases. However, further studies are required to ascertain the combination of grape cultivar and inoculation strain or inoculation strategy that could optimize the concentration of health-promoting compounds so as to realize their potential bioactivity in wine.

In what concerns probiotics, the credibility of specific health claims, and their safety must be established through science-based clinical studies. Maybe, in the future, probiotics will be used as approved drugs that will be prescribed together with, or instead of, antibiotics.

The role of traditional beverages, in the future of the fermented beverage industry, maybe to inspire the expansion of new products and, of course, assess a country's disposition to accept them. Undeniably, with the availability and improvements in technology, and consumers' increasing interest in functional foods, the positioning in the market for fermented beverages is more promising than ever.

Funding: We appreciate the financial support provided to the Research Unit in Vila Real (PEst-OE/QUI/UI0616/2014) by FCT—Portugal and COMPETE.

Conflicts of Interest: The authors declare no conflict of interest.

References

1. Manchester, K.L. Louis Pasteur (1822–1895)—chance and the prepared mind. *Trends Biotechnol.* **1995**, *13*, 511–515. [CrossRef]
2. Truninger, M. The historical development of industrial and domestic food technologies. In *The Handbook of Food Research*; Murcott, A., Belasco, W., Jackson, P., Eds.; Bloomsbury: London, UK, 2013; pp. 82–102.
3. Mishra, S.S.; Ray, R.C.; Panda, S.K.; Montet, D. Technological Innovations in Processing of Fermented Foods. An Overview. In *Fermented Foods, Part II: Technological Intervention*, 1st ed.; Ray, R.C., Montet, D., Eds.; Taylors and Francis, CRC Press: London, UK, 2017; p. 525.
4. Ray, R.C.; Joshi, V.K. Fermented Foods: Past, present, and future scenario. In *Microorganisms and Fermentation of Traditional Foods*; Ray, R.C., Montet, D., Eds.; CRC Press: Boca Raton, FL, USA, 2014; pp. 1–36.
5. Ray, R.C. Fermented foods in health-related issues. *Int. J. Food Ferment. Technol.* **2013**, *3*, 1.
6. Ghosh, J.S. Solid state fermentation and food processing: a short review. *J. Nutr. Food Sci.* **2015**, *6*, 453. [CrossRef]
7. Blaylock, J.; Smallwood, D.; Kassel, K.; Variyam, J.; Aldrich, L. Economics, food choices, and nutrition. *Food Policy* **1999**, *24*, 269–286. [CrossRef]
8. Lähteenmäki, L. Claiming health in food products. *Food Qual. Prefer.* **2013**, *27*, 196–201. [CrossRef]
9. Renaud, S.D.; de Lorgeril, M. Wine, alcohol, platelets, and the French paradox for coronary heart disease. *Lancet* **1992**, *339*, 1523–1526. [CrossRef]
10. Guilford, J.M.; Pezzuto, J.M. Wine and health: A review. *Am. J. Enol. Vitic.* **2011**, *62*, 471–486. [CrossRef]
11. Poli, A.; Marangoni, F.; Avogaro, A.; Barba, G.; Bellentani, S.; Bucci, M.; Cambieri, R.; Catapano, A.L.; Costanzo, S.; Cricelli, C.; et al. Moderate alcohol use and health: A consensus document. *Nutr. Metab. Cardiovasc. Dis.* **2013**, *23*, 487–504. [CrossRef]
12. Chakravarthi, S.; Jessop, C.E.; Bulleid, N.J. The role of glutathione in disulfide bond formation and endoplasmic-reticulum-generated oxidative stress. *EMBO Rep.* **2006**, *7*, 271–275. [CrossRef] [PubMed]
13. Wu, G.; Fang, Y.-Z.; Yang, S.; Lupton, J.R.; Turner, N.D. Glutathione Metabolism and Its Implications for Health. *J. Nutr.* **2004**, *134*, 489–492. [CrossRef] [PubMed]
14. Brosnan, J.T.; Brosnan, M.E. Glutathione and The Sulfur-Containing Amino Acids: An Overview. In *Glutathione and Sulfur Amino Acids in Human Health and Disease*; Masella, R., Mazza, G., Eds.; John Wiley & Sons: Hoboken, NJ, USA, 2009; ISBN 978-0-470-17085-4.
15. Luyckx, J.; Baudouin, C. Trehalose: An intriguing disaccharide with potential for medical application in ophthalmology. *Clin. Ophthalmol.* **2011**, *5*, 577–581. [CrossRef]
16. Eleutherio, E.; Panek, A.; De Mesquita, J.F.; Trevisol, E.; Magalhães, R. Revisiting. yeast trehalose metabolism. *Curr. Genet.* **2015**, *61*, 263–274. [CrossRef]
17. Hornedo-Ortega, R.; Cerezo, A.B.; Troncoso, A.M.; Garcia-Parrilla, M.C.; Mas, A. Melatonin and Other Tryptophan Metabolites Produced by Yeasts: Implications in Cardiovascular and Neurodegenerative Diseases. *Front. Microbiol.* **2016**, *6*, 1565. [CrossRef]
18. Reiter, R.J.; Tan, D.X.; Manchester, L.C.; Pilar-Terron, M.; Flores, L.J.; Koppisepi, S. Medical implications of melatonin: receptor-mediated and receptor-independent actions. *Adv. Med. Sci.* **2007**, *52*, 11–28.
19. Mas, A.; Guillamon, J.M.; Torija, M.J.; Beltran, G.; Cerezo, A.B.; Troncoso, A.M.; Garcia-Parrilla, M.C. Bioactive compounds derived from the yeast metabolism of aromatic amino acids during alcoholic fermentation. *BioMed. Res. Int.* **2014**, 898045. [CrossRef] [PubMed]
20. Galano, A.; Tan, D.X.; Reiter, R.J. Melatonin as a natural ally against oxidative stress: a physicochemical examination. *J. Pineal Res.* **2011**, *51*, 1–16. [CrossRef] [PubMed]
21. Garrido, M.; Paredes, S.D.; Cubero, J.; Lozano, M.; Toribio-Delgado, A.F.; Muñoz, J.L.; Reiter, R.J.; Barriga, C.; Rodríguez, A.B. Jerte valley cherry-enriched diets improve nocturnal rest and increase 6-sulfatoxymelatonin and total antioxidant capacity in the urine of middle-aged and elderly humans. *J. Gerontol. A Biol. Sci. Med. Sci.* **2010**, *65A*, 909–914. [CrossRef] [PubMed]

22. Vitalini, S.; Gardana, C.; Zanzotto, A.; Simonetti, P.; Faoro, F.; Fico, G.; Iriti, M. The presence of melatonin in grapevine (*Vitis vinifera* L.) berry tissues. *J. Pineal Res.* **2011**, *51*, 331–337. [CrossRef] [PubMed]

23. Maldonado, M.D.; Moreno, H.; Calvo, J.R. Melatonin present in beer contributes to increase the levels of melatonin and antioxidant capacity of the human serum. *Clin. Nutr.* **2009**, *28*, 188–191. [CrossRef] [PubMed]

24. Rodriguez-Naranjo, M.I.; Torija, M.J.; Mas, A.; Cantos-Villar, E.; Garcia-Parrilla, M.D. Production of melatonin by *Saccharomyces* strains undergrowth and fermentation conditions. *J. Pineal Res.* **2012**, *53*, 219–224. [CrossRef] [PubMed]

25. Murch, S.J.; Hall, B.A.; Le, C.H.; Saxena, P.K. Changes in the levels of indoleamine phytochemicals during véraison and ripening of wine grapes. *J. Pineal Res.* **2010**, *49*, 95–100. [CrossRef] [PubMed]

26. Rodriguez-Naranjo, M.I.; Gil-Izquierdo, A.; Troncoso, A.M.; Cantos, E.C.; Garcia-Parrilla, M.C. Melatonin: A new bioactive compound present in wine. *J. Food Compost. Anal.* **2011**, *24*, 603–608. [CrossRef]

27. Samuel, S.M.; Thirunavukkarasu, M.; Penumathsa, S.V.; Paul, D.; Maulik, N. Akt/FOXO3a/SIRT1-Mediated Cardioprotection by n-Tyrosolagainst Ischemic Stress in Rat in Vivo Model of Myocardial Infarction: Switching Gears toward Survival and Longevity. *J. Agric. Food Chem.* **2008**, *56*, 9692–9698. [CrossRef]

28. Dudley, J.I.; Lekli, I.; Mukherjee, S.; Das, M.; Bertelli, A.A.; Das, D.K. Does white wine qualify for French paradox? Comparison of the cardioprotective effects of red and white wines and their constituents: resveratrol, tyrosol, and hydroxytyrosol. *J. Agric. Food Chem.* **2008**, *56*, 9362–9373. [CrossRef]

29. Willcox, B.J.; Donlon, T.A.; He, Q.; Chen, R.; Grove, J.S.; Yano, K.; Masaki, K.H.; Willcox, D.C.; Rodriguez, B.; Curb, J.D. FOXO3a Genotype Is Strongly Associated with Human Longevity. *Proc. Natl. Acad. Sci. USA* **2008**, *105*, 13987–13992. [CrossRef]

30. Thirunavukkarasu, M.; Penumathsa, S.V.; Samuel, S.M.; Akita, Y.; Zhan, L.; Bertelli, A.A.; Maulik, G.; Maulik, N. White Wine-induced Cardio Protection against Ischemia-Reperfusion Injury is Mediated by Life-Extending Akt/FOXO3a/NFκB Survival Pathway. *J. Agric. Food Chem.* **2008**, *56*, 6733–6739. [CrossRef]

31. Utsunomiya, H. Flavor terminology and reference standards for sensory analysis of sake. *J. Brew. Soc. Jpn.* **2006**, *101*, 730–739. [CrossRef]

32. Luís, R.S.; Paula, B.A.; Patrícia, V.; Rosa, M.S.; Martha, E.T.; Encarna, V. Analysis of non-colored phenolics in red wine: Effect of *Dekkera bruxellensis*. *Yeast* **2005**, *89*, 185–189. [CrossRef]

33. Gould, B.A.; Mann, S.; Davies, A.B.; Altman, D.G.; Raftery, E.B. α-Adrenoreceptor Blockade with Indoramin in Hypertension. *J. Cardiovasc. Pharmacol.* **1983**, *5*, 343–348. [CrossRef]

34. Kirby, R.S.; Pool, J.L. Alpha adrenoceptor blockade in the treatment of benign prostatic hyperplasia: Past, present, and future. *Br. J. Urol.* **1997**, *80*, 521–532. [CrossRef]

35. Lingappa, B.T.; Prasad, M.; Lingappa, Y.; Hunt, D.F.; Biemann, K. Phenethyl Alcohol, and Tryptophol: Auto-antibiotics Produced by the Fungus *Candida albicans*. *Science* **1969**, *163*, 192–194. [CrossRef] [PubMed]

36. Luna-Solano, G.; Salgado-Cervantes, M.A.; Ramirez-Lepe, M.; Garcia-Alvarado, M.A.; Rodriguez-Jimenes, G.C. Effect of drying type and drying conditions over the fermentative ability of brewer's yeast. *J. Food Process. Eng.* **2003**, *26*, 135–147. [CrossRef]

37. Rapoport, A. Anhydrobiosis and Dehydration of Yeasts. In *Biotechnology of Yeasts and Filamentous Fungi*; Sibirny, A.A., Ed.; Springer International Publishing: Cham, Switzerland, 2017; pp. 87–116.

38. Gervais, P.; Beney, L. Osmotic mass transfer in the yeast *Saccharomyces cerevisiae*. *Cell. Mol. Biol.* **2001**, *47*, 831–840. [PubMed]

39. Dupont, S.; Beney, L.; Ritt, J.-F.; Lherminier, J.; Gervais, P. Lateral reorganization of the plasma membrane is involved in the yeast resistance to severe dehydration. *Biochim. Biophys. Acta Biomembr.* **2010**, *1798*, 975–985. [CrossRef] [PubMed]

40. Garre, E.; Raginel, F.; Palacios, A.; Julien, A.; Matallana, E. Oxidative stress responses, and lipid peroxidation damage are induced during dehydration in the production of dry active wine yeasts. *Int. J. Food Microbiol.* **2010**, *136*, 295–303. [CrossRef] [PubMed]

41. Câmara, A.A., Jr.; Nguyen, T.D.; Jossier, A.; Endrizzi, A.; Saurel, R.; Simonin, H.; Husson, F. Improving total glutathione and trehalose contents in *Saccharomyces cerevisiae* cells to enhance their resistance to fluidized bed drying. *Process. Biochem.* **2018**, *69*, 45–51. [CrossRef]

42. Rahman, I.; Kode, A.; Biswas, S.K. Assay for the quantitative determination of glutathione and glutathione disulfide levels using enzymatic recycling method. *Nat. Protoc.* **2006**, *1*, 3159–3165. [CrossRef]

43. Bourboulou, A.; Shahi, P.; Chakladar, A.; Delrot, S.; Bachhawat, A.K. Hgt1p, a High-Affinity Glutathione Transporter from the Yeast *Saccharomyces cerevisiae*. *J. Biol. Chem.* **2000**, *275*, 13259–13265. [CrossRef]

44. Veal, E.A.; Day, A.M.; Morgan, B.A. Hydrogen peroxide sensing and signaling. *Mol. Cell* **2007**, *26*, 1–14. [CrossRef]
45. Gostimskaya, I.; Grant, C.M. Yeast mitochondrial glutathione is an essential antioxidant with mitochondrial thioredoxin providing a back-up system. *Free Radic. Biol. Med.* **2016**, *94*, 55–65. [CrossRef]
46. Bachhawat, A.K.; Thakur, A.; Kaur, J.; Zulkifli, M. Glutathione transporters. *Biochim. Biophys. Acta Gen. Subj.* **2013**, *1830*, 3154–3164. [CrossRef]
47. Russell, I. Chapter 4—Yeast. In *Brewing Materials and Processes*; Bamforth, C.W., Ed.; Academic Press: London, UK, 2016; pp. 77–96.
48. Herdeiro, R.S.; Pereira, M.D.; Panek, A.D.; Eleutherio, E.C.A. Trehalose protects *Saccharomyces cerevisiae* from lipid peroxidation during oxidative stress. *Biochim. Biophys. Acta Gen. Subj.* **2006**, *1760*, 340–346. [CrossRef]
49. Câmara Jr., A.A.; Maréchal, P.-A.; Tourdot-Maréchal, R.; Husson, F. Dehydration stress responses of yeasts *Torulaspora delbrueckii*, *Metschnikowia pulcherrima* and *Lachancea thermotolerans*: Effects of glutathione and trehalose biosynthesis. *Food Microbiol.* **2019**, *79*, 137–146. [CrossRef]
50. Sprenger, J.; Hardeland, R.; Fuhrberg, B.; Han, S.-Z. Melatonin and other 5-methoxylated indoles in yeast: presence in high concentrations and dependence on tryptophan availability. *Cytologia* **1999**, *64*, 209–213. [CrossRef]
51. Tan, D.-X.; Hardeland, R.; Manchester, L.C.; Korkmaz, A.; Ma, S.; Rosales-Corral, S.; Reiter, R.J. Functional roles of melatonin in plants, and perspectives in nutritional and agricultural science. *J. Exp. Bot.* **2012**, *63*, 577–597. [CrossRef]
52. Germann, S.M.; Jacobsen, S.A.; Schneider, K.; Harrison, S.J.; Jensen, N.B.; Stahlhut, S.G.; Borodina, I.; Luo, H.; Zhu, J.; Maury, J. Glucose-based microbial production of the hormone melatonin in yeast *Saccharomyces cerevisiae*. *Biotech. J.* **2016**, *11*, 717–724. [CrossRef]
53. Manfroi, L.; Silva, P.H.A.; Rizzon, L.A.; Sabaini, P.S.; Glória, M.B.A. Influence of alcoholic and malolactic starter cultures on bioactive amines in Merlot wines. *Food Chem.* **2009**, *116*, 208–213. [CrossRef]
54. Wang, Y.Q.; Ye, D.O.; Zhu, B.Q.; Wu, G.F.; Duan, C.Q. Rapid HPLC analysis of amino acids and biogenic amines in wines during fermentation and evaluation of matrix effect. *Food Chem.* **2014**, *163*, 6–15. [CrossRef]
55. Tan, D.-X.; Hardeland, R.; Back, K.; Manchester, L.C.; Alatorre-Jimenez, M.A.; Reiter, R.J. On the significance of an alternate pathway of melatonin synthesis via 5-methoxytryptamine: comparisons across species. *J. Pineal Res.* **2016**, *61*, 27–40. [CrossRef]
56. Fernández-Cruz, E.; Cerezo, A.; Cantos-Villar, E.; Troncoso, A.; García-Parrilla, M. Time course of l-tryptophan metabolites when fermenting natural grape musts: effect of inoculation treatments and cultivar on the occurrence of melatonin and related indolic compounds. *Aust. J. Grape Wine R.* **2019**, *25*, 92–100. [CrossRef]
57. Rodriguez-Naranjo, M.I.; Gil-Izquierdo, A.; Troncoso, A.M.; Cantos-Villar, E.; Garcia-Parrilla, M.C. Melatonin is synthesised by yeast during alcoholic fermentation in wines. *Food Chem.* **2011**, *126*, 1608–1613. [CrossRef]
58. Fernández-Cruz, E.; Alvarez-Fernández, M.A.; Valero, E.; Troncoso, A.M.; García-Parrilla, M.C. Melatonin and derived tryptophan metabolites produced during alcoholic fermentation by different yeast strains. *Food Chem.* **2017**, *217*, 431–437. [CrossRef] [PubMed]
59. Vilela, A. Use of Nonconventional Yeasts for Modulating Wine Acidity. *Fermentation* **2019**, *5*, 27. [CrossRef]
60. Puertas, B.; Jiménez, M.J.; Cantos-Villar, E.; Cantoral, J.M.; Rodríguez, M.E. Use of *Torulaspora delbrueckii* and *Saccharomyces cerevisiae* in semi-industrial sequential inoculation to improve quality of Palomino and Chardonnay wines in warm climates. *J. Appl. Microbiol.* **2017**, *122*, 733–746. [CrossRef]
61. Kim, D.; Kim, H.; Kim, K.; Roh, S. The Protective Effect of Indole-3-Acetic Acid (IAA) on H_2O_2-Damaged Human Dental Pulp Stem Cells Is Mediated by the AKT Pathway and Involves Increased Expression of the Transcription Factor Nuclear Factor-Erythroid 2-Related Factor 2 (Nrf2) and Its Downstream Target Heme Oxygenase 1 (HO-1). *Oxid. Med. Cell Longev.* **2017**, *2017*, 8639485. [CrossRef]
62. Ehrlich, F. Uber Tryptophol (β-Indolyl-Athylalkohol), Ein Neues Gar Produkt der Hefe Aus Aminosäuren. *Ber. Dtsch. Chem. Ges.* **1912**, *45*, 883–889. [CrossRef]
63. Dickinson, J.R. Nitrogen metabolism. In *The Metabolism and Molecular Physiology of Saccharomyces cerevisiae*, 2nd ed.; Dickinson, J.R., Schweizer, M., Eds.; CRC Press: London, UK, 2004; ISBN 0-415-29900-490000.
64. Hazelwood, L.A.; Jean-Marc, D.; van Maris, A.J.A.; Pronk, J.T.; Dickinson, J.R. The Ehrlich pathway fuel alcohol production: a century of research on *Saccharomyces cerevisiae* metabolism. *Appl. Environ. Microbiol.* **2008**, *74*, 2259–2266. [CrossRef] [PubMed]

65. Banach, A.; Ooi, B. Enhancing the Yields of Phenolic Compounds during Fermentation Using *Saccharomyces cerevisiae* Strain 96581. *Food Nut. Sci.* **2014**, *5*, 2063–2070. [CrossRef]

66. Marsh, A.J.; Hill, C.; Ross, R.P.; Cotter, P.D. Fermented beverages with health-promoting potential: Past and future perspectives. *Trends Food Sci Technol.* **2014**, *38*, 113–124. [CrossRef]

67. Fijan, S. Microorganisms with claimed probiotic properties: an overview of recent literature. *Int J. Environ. Res. Public Health.* **2014**, *11*, 4745–4767. [CrossRef]

68. Kozyrovska, N.O.; Reva, O.N.; Goginyan, V.B.; De Vera, J.-P. Kombucha microbiome as a probiotic: A view from the perspective of post-genomics and synthetic ecology. *Biopolym. Cell.* **2012**, *28*, 103–113. [CrossRef]

69. Teoh, A.L.; Heard, G.; Cox, J. Yeast ecology of Kombucha fermentation. *Int. J. Food Microbiol.* **2004**, *95*, 119–126. [CrossRef] [PubMed]

70. Dutta, D.; Gachhui, R. Nitrogen-fixing, and cellulose-producing *Gluconacetobacter kombuchae* sp. nov., isolated from Kombucha tea. *Int. J. Syst. Evol. Microbiol.* **2007**, *57*, 353–357. [CrossRef] [PubMed]

71. Dutta, D.; Gachhui, R. Novel nitrogen-fixing *Acetobacter nitrogenifigens* sp. nov., isolated from Kombucha tea. *Int. J. Syst. Evol. Microbiol.* **2006**, *56*, 1899–1903. [CrossRef] [PubMed]

72. Freire, A.L.; Ramos, C.L.; da Costa Souza, P.N.; Cardoso, M.G.B.; Schwan, R.F. Nondairy beverage produced by controlled fermentation with potential probiotic starter cultures of lactic acid bacteria and yeast. *Int. J. Food Microbiol.* **2017**, *248*, 39–46. [CrossRef] [PubMed]

 *fermentation*

Review

Climate Changes and Food Quality: The Potential of Microbial Activities as Mitigating Strategies in the Wine Sector

Carmen Berbegal [1,2], **Mariagiovanna Fragasso** [1], **Pasquale Russo** [1], **Francesco Bimbo** [1], **Francesco Grieco** [3], **Giuseppe Spano** [1] **and Vittorio Capozzi** [1,*]

[1] Dipartimento di Scienze Agrarie, degli Alimenti e dell'Ambiente, Università di Foggia, via Napoli 25, 71100 Foggia, Italy; carmen.berbegal@uv.es (C.B.); mariagiovanna.fragasso@gmail.com (M.F.); pasquale.russo@unifg.it (P.R.); francesco.bimbo@unifg.it (F.B.); giuseppe.spano@unifg.it (G.S.)

[2] EnolabERI BioTecMed, Universitat de València, 46100 Valencia, Spain

[3] Istituto di Scienze delle Produzioni Alimentari, Consiglio Nazionale delle Ricerche, Unità Operativa di Supporto di Lecce, 73100 Lecce, Italy; francesco.grieco@ispa.cnr.it

[*] Correspondence: vittorio.capozzi@unifg.it; Tel.: +39-(0)881-589-303

Received: 6 September 2019; Accepted: 19 September 2019; Published: 23 September 2019

Abstract: Climate change threatens food systems, with huge repercussions on food security and on the safety and quality of final products. We reviewed the potential of food microbiology as a source of biotechnological solutions to design climate-smart food systems, using wine as a model productive sector. Climate change entails considerable problems for the sustainability of oenology in several geographical regions, also placing at risk the wine typicity. The main weaknesses identified are: (i) The increased undesired microbial proliferation; (ii) the improved sugars and, consequently, ethanol content; (iii) the reduced acidity and increased pH; (iv) the imbalanced perceived sensory properties (e.g., colour, flavour); and (v) the intensified safety issues (e.g., mycotoxins, biogenic amines). In this paper, we offer an overview of the potential microbial-based strategies suitable to cope with the five challenges listed above. In terms of microbial diversity, our principal focus was on microorganisms isolated from grapes/musts/wines and on microbes belonging to the main categories with a recognized positive role in oenological processes, namely *Saccharomyces* spp. (e.g., *Saccharomyces cerevisiae*), non-*Saccharomyces* yeasts (e.g., *Metschnikowia pulcherrima*, *Torulaspora delbrueckii*, *Lachancea thermotolerans*, and *Starmerella bacillaris*), and malolactic bacteria (e.g., *Oenococcus oeni*, *Lactobacillus plantarum*).

Keywords: climate change; food quality; viticulture; wine; fermentation; yeast; *Saccharomyces*; non-*Saccharomyces*; alcoholic fermentation; lactic acid bacteria; malolactic fermentation

---

1. Introduction

"Climate change threatens our ability to ensure global food security, eradicate poverty and achieve sustainable development. Greenhouse gas (GHG) emissions from human activity and livestock are a significant driver of climate change, trapping heat in the earth's atmosphere and triggering global warming. Climate change has both direct and indirect effects on agricultural productivity including changing rainfall patterns, drought, flooding and the geographical redistribution of pests and diseases. FAO is supporting countries to both mitigate and adapt to the effects of climate change through a wide range of research based and practical programmes and projects, as an integral part of the 2030 agenda and the Sustainable Development Goals." http://www.fao.org/climate-change/en/.

It is clear how widespread and complex the impacts of climate change phenomena associated with global warming on food systems are [1–5]. We can disentangle these extensive and multifaceted influences in different (often interdependent) components, such as agricultural, livestock and fishery yields, food prices, effectiveness of delivery, global food quality, and, a crucial facet of global quality, food safety [6]. Great attention has been placed to many aspects related to food security (e.g., yields reduction, prices rises). Instead, marginal interest has been given to quality issues, including, among others, palatability, hygienic properties, nutritional contributes, and functional contributes. For fermented foods and beverages, microbes' activity associated with the matrices is susceptible to affect all the main aspects contributing to the final product quality [7,8]. Mitigation and adaptation strategies for the effects of climate change belong to different disciplines, such as agricultural sciences, plant and animal biology and breeding, food technology, and food microbiology. In this mini-review article, we use wine as a model matrix to describe the impact of climate changes on the quality of fermented matrices, examining the potential of protechnological microbes as agents capable to 'mitigate' the negative features of this evolving climatic influence.

Within the macro-category of fermented products, wine belongs to the group of fermented alcoholic beverages [7]. Yeasts are responsible for alcoholic fermentation (AF) and more generally, for biochemical changes linked to the chemical transformation of must obtained from grapevine crushing in wine [9–11]. Among oenological yeasts, the following categories can be found: (i) Yeast belonging to the *Saccharomyces* genera, and particularly to the *Saccharomyces cerevisiae* species, which are mainly responsible for alcoholic fermentation in wine [10–12]; and (ii) the heterogeneous category of the so-called non-*Saccharomyces* yeasts [10,11,13]. Within this complex category, we can find both protechnological species/strains [13,14] and spoilage organisms [15,16]. Non-*Saccharomyces* of interest for their oenological aptitude, other than contributing to alcoholic fermentation, can be helpful to solve specific technological/oenological issues (e.g., reduction of volatile acidity) [13,17], to modulate wine aroma [17–19], and/or to exert biocontrol activity against undesired microbes [20–22]. Together with the eukaryotic contribution to wine quality, we have to mention malolactic bacteria to encompass all microbes that positively modulate wine chemistry. Malolactic bacteria and lactic acid bacteria (LAB) are capable of decarboxylating malic in lactic acid, and are responsible for the so-called malolactic fermentation (MLF), a process associated with positive changes in palatability, increased aromatic complexity, and enhanced microbial stability [23].

2. Wine Quality and Climate Change

Climate change affects, to different extents, wine production and quality. About 10 years ago, Mira de Orduña provided an extensive review of the 'climate change-associated effects on grape and wine quality and production' [24]. The review followed a cause-and-effect ratio analysis, and pointed out the effects on viticulture and the corresponding consequences on winemaking. Adopting this point of view, we can examine the main effects of climate change on viticulture and oenology (Table 1).

Table 1. A list of the effects of climate change on viticulture and enology. Often, oenological effects are a consequence of viticultural effects.

Viticultural Effects	Oenological Effects
Harvest dates	Harvest conditions and fruit quality
Grape maturation (effect of temperature, of carbon dioxide and of radiation)	Effects of high sugar and alcohol concentrations
Indirect effects of climate change	Microbial and sensory effects of lower acidities and increased potassium and pH levels
Effects on vine pests	Climate change associated effects on wine chemistry
Effect on root systems	Effect on oak

Modified from Mira de de Orduña [24].

Harvesting is in a double relationship with climate trends; on the one side, harvesting is a function of the seasonal climate, on the other, it provides a criterion to classify different grapevine varieties depending on their relationship to the climate. In general, data from different grapevine production areas offer a picture of prior fruit maturation patterns, with a consequential shift forward of the harvesting time [24]. Considering the different grapevine varieties, recent evidence on early wine grape harvests in France indicates that climate change has profoundly transformed the climatic drivers of the plant, with possible repercussions for viticulture management and wine quality [25,26]. If we consider the general influence of temperature increases, not only on a given phenological phase (i.e., fruit maturity), we have to report an increase in sugar contents, decreased concentration of organic acids/total acidity, and improved potassium content [24,27]. Moving from primary to secondary metabolites, the effort to summarize specific trends becomes more complex, giving that more variables act in the system that are susceptible to influencing the pathways associated with metabolites' biosynthesis: Temperature, carbon dioxide, and radiation [24]. In general, climate change has led to significant modulations in the accumulation of heterogeneous classes of polyphenols and volatile organic compounds [24,28]. In addition to the direct effect, we have to consider the indirect effects, such as enhanced salinity and increased probability of wild bushfires [24]. Present evidence also suggests that climate change can influence the proliferation of certain viticultural pathogens, introducing new insight into pest management in the field [24]. We must also consider the direct effects on the root system imputable to the response of the plant to abiotic heat stress. Finally, the effects on the development and quality of oak, the main wood utilized for wine aging, caused by modifications of carbon dioxide levels and weather patterns have been considered [24].

Shifting from the viticultural to the oenological aspects, we may list the main consequences on the wine quality of the highlighted effects on the raw material. The shift of the harvest date and the impact on grape maturation can intensify oxidative phenomena (e.g., oxidation of specific volatiles) and microbial growth (e.g., increased microbial spoilage proliferation, enhanced risks of starvation during the fermentative process, and increased the content of toxic compounds released by undesired microorganisms, such as mycotoxins) [24,25,27]. The immediate oenological consequence of an increased sugar content is an improved concentration of ethanol in the final product. This phenomenon implies a higher likelihood of stuck/sluggish during the alcoholic fermentation, sharpened microbial stress response, modulation of sensory perception (prominent alcohol sense and a reduced passage of volatiles in the wine headspace, increasing the perception of astringency, masking the perception of esters), and lowered social acceptance of wines, due to the recognized toxic effect of ethylic alcohol (without considering the impact on caloric intake) [28]. Increased pH implies the following: (i) An improved risk of undesired microbial proliferation, from the first fermentative steps (e.g., lactic acid bacteria, spoilage yeasts) up to the aging/finished wines (e.g., *Dekkera/Brettanomyces* yeasts); and (ii) changes in the wine colour, taste, and aroma [28]. Modifications in the wine colour, taste, and aroma can also be addressed by modulation of the compound directly responsible for these perceptions. The phenomena associated with climate change seem to lessen anthocyanins and enhance the proanthocyanidins content, contributing to a reduction of the 'colour potential' and to pronounced astringency [27,28]. In terms of the concern regarding aroma compounds, even if it is difficult to depict clear trends, it is possible to point out some patterns [29,30]. First, it is worth remembering that notes of "green pepper, herbaceous, blackcurrant, blackberry, figs, or prunes are strongly linked with the maturity of the grapes" [31]. The 'cooked' aroma generally increases with temperature. Contrastingly, pyrazine accumulation follows an opposite change (responsible for 'veggie, herbaceous notes') [27,29,32]. The same was found for rotundone contents in grapes (responsible for the peppery aroma) [29]; whereas contrasting results were reported for the terpenol family [29].

It is possible to speculate that the present literature presents findings that are not always harmonic and that it remains difficult to combine direct and indirect effects, both positive and negative. To this purpose, Drappier et al. [28] observed that the remarkably hot 2003 season in Europe offered the opportunity to mimic and test in vivo the climatic condition expected by the conclusion of this century, demonstrating the potential of climate change in clouding wine typicity. With this concern, the authors reported, in light of the recent experimental investigations, the sensory features associated with the different viticultural climates: Enhanced alcohol perception, reduced acidity sense, imbalanced colour development, and perceived aroma [28]. These are sensory defects that are generally coherent with the indications reported in the scientific literature.

3. The Potential of Microbial Activities as Mitigating Technologies

When facing emerging challenges, humans explore different routes in order to find innovative solutions suitable to ensuring the sustainability of resources and productions. This is also true for the problems in food systems triggered by climate change. For example, in the wine sector, the scientific and professional communities have proposed numerous possible approaches susceptible to developing a climate-smart wine system. These potential solutions range from the agronomic and viticultural fields up to applications in the technology and biotechnology branches, with different potentials in terms of performances and temporal horizons. Among other factors, microorganisms can also exert activities to mitigate product depreciation due to climate change. Here, we propose an overview of potential microbial-based strategies able to concretize mitigating biotechnologies, declined in five categories corresponding to the main safety/quality aspects affected by climate changes in oenology.

3.1. Microbial Solution for the Biocontrol of Spoilage Microorganisms in Wine

The main spoilage microbes in enology belong to the yeast genera Brettanomyces (e.g., *B. bruxellensis*), Candida (e.g., *C. stellata*), Hanseniaspora (e.g., *H. vineae*), Pichia (e.g., *P. anomala, P. membranifaciens*), and Zygosaccharomyces (e.g., *Z. bailii, Z. rouxii*); and to the bacterial genera Lactobacillus (e.g., *L. hilgardii*), Leuconostoc (e.g., *L. mesenteroides*), Pediococcus (e.g., *P. damnosus, P. pentosaceus*), Acetobacter (e.g., *A. aceti, A. pasteurianus*), or Gluconobacter (e.g., *G. oxydans*) [33,34]. The increasing incidence of these spoilage microbes could be responsible for considerable economic losses in this sector. In Table 2, we propose an exemplified list of microbial applications potentially suitable to ensuring the control of microbial spoilage.

Table 2. A list of studies that propose microbial-based solutions that can have potential applications in mitigating the development of spoilage microorganisms in wine.

Microorganisms Involved	Microbial-Based Mitigating Strategies	References
Lactococcus lactis (as producer of lacticin 3147)	Use of lacticin 3147 for the biocontrol of lactic acid bacteria in oenology	[35]
Metschnikowia pulcherrima	Biocontrol of spoilage yeasts via iron depletion	[36]
Saccharomyces cerevisiae	Killer activity as biocontrol agents to avoid or reduce wine spoilage	[37]
Enterococcus faecium	Enterocin heat stable, with broad pH range and bactericidal effects	[38]
Pichia membranifaciens	Killer toxin active against spoilage yeast in wine	[39]
Torulaspora delbrueckii	Use as a bio-protective agent alternative to sulphites in winemaking	[40]
Wickerhamomyces anomalus and *Metschnikowia pulcherrima*	Biocontrol activity against spoilage yeasts in winemaking	[22]
Saccharomyces cerevisiae, Candida zemplinina, Hanseniaspora uvarum, Hanseniaspora guilliermondii, Torulaspora delbrueckii, Metschnikowia pulcherrima	Use of co-inoculation of autochthonous yeasts and bacteria in order to control *Brettanomyces bruxellensis* in wine	[21]

Biocontrol provides alternatives to chemical preservatives, such as SO_2, which is associated with adverse reactions in humans [40]. We recognize two different categories of microbial-based solutions: The case when a product of microbial metabolism is added as biopreservatives in the wine chain [34,35,38] or the option to add to the matrix the microorganism itself as a starter/protective culture [20,37,40]. Considering the molecular basis responsible for the antagonistic microbial phenotypes, we highlight two main categories, competition for nutrients and the production of molecules with antimicrobial activities. Concerning the last class, yeasts' killer toxins and bacteriocins are the main reservoirs of this competitive arsenal developed by specific yeasts and bacteria that find potential applications in wine [41].

3.2. Microbial-Based Solutions to Reduce Ethanol Content

High ethanol concentration may reduce the complexity of wine by suppressing the aroma intensity, but also by exalting the perception of 'hotness' and 'bitterness'. Moreover, health considerations combined with market demand make the wine industry actively seek ways to facilitate the production of wines with lower alcohol concentration [42]. Among the possible approaches, microbial strategies present an attractive opportunity to decrease ethanol levels while preserving the quality and aromatic integrity of the wine (Table 3).

Table 3. A list of studies that propose microbial-based solutions that can have potential applications in mitigating an increased ethanol concentration.

Microorganisms Involved	Microbial-Based Mitigating Strategies	References
Saccharomyces cerevisiae	Selection of less ethanol producer yeasts	[43,44]
Saccharomyces cerevisiae	Adaptive evolution to conditions where glycerol synthesis is more favoured than ethanol	[45,46]
Hanseniaspora uvarum, *Schizosaccharomyces pombe,* *Lachancea thermotolerans,* *Saccharomyces kudriavzevii*	Non-*Saccharomyces* sequential inoculation or co-inoculation with *S. cerevisiae*	[14,47–51]
Metschnikowia pulcherrima, *Kluyveromyces* spp., *Candida sake,* *Torulaspora delbrueckii,* *Zygosaccharomyces bailii*	Respiratory consumption of sugars	[52–55]
Saccharomyces cerevisiae	Genetic engineering	[56–58]

S. cerevisiae is efficient at converting sugar to alcohol and has a preeminent tolerance to the stressful conditions encountered during alcoholic fermentation. Thus, one of the methods explored consists in breeding different *S. cerevisiae* strains to select less ethanol producer yeast [43,44]. This strategy could also involve different *Saccharomyces* species, where wine industrial strains can be combined with less known alcoholic species. Indeed, hybrid strains have been described with a reduced efficiency concerning alcohol yields and are able to preserve wine's organoleptic properties after fermentation [43]. Additionally, yeasts could be forced to evolve and adapt to conditions where glycerol synthesis is more favoured than ethanol, for example, conditioning the yeast to higher osmotic pressures [45] or using SO_2 at alkaline pH [46].

Another microbial strategy that has seen growing interest in the last decade involves the use of non-*Saccharomyces* yeasts. These species exhibit physiological properties that are especially relevant during the winemaking process, such as their good fermentative capabilities at low temperatures, resulting in wines with lower alcohol and higher glycerol amounts [10,11]. Several studies have described a reduced ethanol yield (0.2–0.6 % v/v) when using non-*Saccharomyces* and *S. cerevisiae* strains in co-inoculated or sequential cultures [14,47–51,59]. Another alternative to lower the ethanol concentration in wine is to exploit the oxidative metabolism detected in some non-*Saccharomyces* species [52–55]. The supply of oxygen to the fermenters under a controlled flow rate promotes the respiratory consumption of sugars by these non-*Saccharomyces* yeasts.

An additional approach consists in generating low-ethanol yeast strains using metabolic engineering. The principle behind this strategy is the engineering of yeast strains through altered gene expression to modify carbon fluxes in the cell [60]. One of the key target carbon sinks in these approaches has been glycerol, as several research groups have attempted to redirect carbon towards glycerol in order to decrease the flow of carbon to ethanol [56]. Rossouw et al. [43] demonstrated that an alternative metabolite in central carbon metabolism, trehalose, can be targeted as a carbon sink without resulting in the accumulation of undesirable redox-linked metabolites. Besides, the expression in wine yeast of the lactate dehydrogenase gene (*LDH*) from *Lactobacillus casei* has also resulted in reduced ethanol concentration (0.25% v/v less) by diverting carbon to lactic acid production [58].

3.3. Microbial-Based Solutions to Improve Organic Acids Content and to Reduce pH

Among the effects of climate change, the harsh lessening in the acidity of wines has a complex impact on wine quality. Indeed, the low total acidity led to wines with defects in the sensory quality (e.g., less sour/acid taste, changes in the colour) and prone to the implantation of microbial spoilages (reduced wine stability) [24]. These phenomena are likely to be regional-dependent, as recently indicated by Lucio et al. [61], who found an increase of 0.5 units in the pH, also achieving pH values of 3.8–4.0 in the case of wine produced in La Rioja (Spain). Some organic acids are principally associated with fruit composition (tartaric, malic, and citric), while others (succinic, lactic, and acetic acids) are mainly related to the fermentation processes, both to the alcoholic and malolactic [62]. In Table 4, an overview of species/strains selected for their potential of biological acidification of must and wine is given.

Table 4. A list of studies that propose microbial-based solutions that can have potential applications in mitigating the reduced content in organic acids and an increased pH.

Microorganisms Involved	Microbial-Based Mitigating Strategies	References
Candida stellata	Consistent increase in succinic acid content	[63]
Lachancea thermotolerans and *Saccharomyces cerevisiae*	pH reduction and increased total acidity perceived	[59]
Schizosaccharomyces pombe and *Lachancea thermotolerans*	A biotechnological alternative to the traditional malolactic fermentation in red wine production	[64]
Lactobacillus plantarum	Biological acidification using the lactic acid bacterium in pre-alcoholic fermentation	[65]
Candida zemplinina	Moderate production of acetate, succinate, malate, and lactate, with specific nitrogen dependence of acid production	[66]
Lactobacillus plantarum	Selection of MLF starter cultures for high pH must	[67]
Lactobacillus plantarum	Selection of strains to provoke biological acidification in low acidity matrices	[61]
Lactobacillus plantarum	The managing wine acidity depended on the couple LAB/yeast strains co-inoculated	[68]

Non-*Saccharomyces* yeasts and malolactic bacteria are the main reservoirs of microorganisms capable of inducing biological acidification in oenology, due to their physiological features and genetic determinants associated with the production of organic acids [61,69].

The most promising species among non-*Saccharomyces* is *Lachancea thermotolerans* [9,70] due to a considerable aptitude to produce lactic acid [59,64]. Moreover, the use of *L. thermotolerans* has been proposed in combination with *Schizosaccharomyces pombe* [71,72], a yeast capable of converting malic acid in ethanol to mimic classic malolactic fermentation (the decarboxylation of malic acid to lactic acid) [64]. Also, the yeasts *Candida stellata* [73] and *Candida zemplinina* (synonym *Starmerella bacillaris*) [74] have been explored for their possible application in biological acidification in oenological matrices [63,66]. Among malolactic bacteria, *Lactobacillus plantarum*, in reason of the protechnological significance and versatility, extensive applications for their potential to increase the content of lactic acid in the tested matrices have been found [61,65,67,68].

3.4. Microbial-Based Solutions to Modulate/Enhance Sensory Characteristics (Colour, Taste, and Aroma)

The sensory issue represents a more complex matter to provide clear cause–effect solutions. In fact, it is difficult to highlight unambiguous trends associated with climate change (and, consequently, challenging to propose unambiguous microbial-based solutions). However, a plethora of biotechnological solutions that rely on microbial activities are susceptible to applications to cope with the different modifications in sensory attributes addressable to climate change. In Table 5, we provide only a few examples of the microbial-based solutions that are able to modulate/enhance sensory characteristics.

Table 5. A list of studies that propose microbial-based solutions that have potential applications in mitigating modifications of sensory characteristics.

Microorganisms Involved	Microbial-Based Mitigating Strategies	References
Saccharomyces cerevisiae, *Saccharomyces uvarum* and *Saccharomyces montuliensis*	Formation of vinylphenolic pyranoanthocyanins, pigments affecting the colour of the finished wine	[75]
Saccharomyces cerevisiae	Wine yeast are capable to influence volatile sulphur compounds	[76]
Lactobacillus plantarum	Detain enzymes are also involved in improving colour in red wines	[77]
Torulaspora delbrueckii	The yeast in mixed fermentation allows a potential increase of fruity aromas in the wine	[78]
Schizosaccharomyces pombe	The yeast allows increasing the contents of vitisins, especially A type	[78]
Candida zemplinina	The yeast improves vitisin A contents	[79]
Torulaspora delbrueckii and *Saccharomyces cerevisiae*	*T. delbrueckii* in association with *S. cerevisiae* affects the esters content with impact on the aromatic traits of wines.	[80]
Oenococcus oeni and *Saccharomyces cerevisiae*	Co-inoculation of yeasts and lactic acid bacteria as a strategy produces enhancement in wine aroma profile during fermentation	[81]
Saccharomyces cerevisiae	A flor velum *Saccharomyces cerevisiae* strain is able to influence colour and the contents of key aroma compound, susceptible to conceive new red wine types in a climate change scenario.	[82]
Oenococcus oeni	The use of different malolactic starter culture led to modulation in the quality and quantity of volatile organic compounds	[83]
Starmerella bacillaris and *Saccharomyces cerevisiae*	Mixed fermentations could be considered as a tool to enhance the aroma profile	[84]
Hanseniaspora uvarum	Co-inoculation of *Hanseniaspora uvarum* and *Saccharomyces cerevisiae* in order to increase the aromatic profile and lessen the presence of the undesired characters	[85]
Oenococcus oeni	Influence of protechnological and autochthonous strains on compounds relevant for wine aroma, particularly on branched hydroxylated compounds	[86]

3.5. Microbial-Based Solutions to Less Toxic Compounds (Mycotoxins, Biogenic Amines)

During the winemaking process, several microorganisms may cause the depreciation of wine since they can produce undesirable compounds that are toxic to humans, such as biogenic amines (BA) or mycotoxins [7,8,87].

The main microorganisms responsible for BA production in wine are LAB [88] and some non-*Saccharomyces* yeasts [89]. Moreover, several strains of *Enterococcus* spp. and *Staphylococcus* spp. have recently been isolated from must and wine and described as histamine producers [90,91]. Microbial-based solutions that minimize the presence of these toxic compounds in wine are summarized in Table 6.

Table 6. A list of studies that propose microbial-based solutions that can have potential applications in mitigating an increased content in mycotoxin and biogenic amines.

Microorganisms Involved	Microbial-Based Mitigating Strategies	References
Oenococcus oeni	Non-BA producer's selection to carry out the MLF	[92,93]
Schizosaccharomyces pombe	Inhibition of LAB development (and of the consequent BA generation) by removing malic acid and sugars during AF	[94]
Oenococcus oeni, Lactobacillus hilgardii, Lactobacillus brevis	Co-inoculation of *S. cerevisiae* and LAB to control the BA-producing microorganisms	[95,96]
Lactobacillus plantarum, Pediococcus acidilactici	BA degradation	[97–99]
Saccharomyces cerevisiae	OTA reduction by adsorption	[100,101]
Acinetobacter sp., Saccharomyces cerevisiae	OTA degradation by peptidases	[101,102]

One of the main strategies to avoid the presence of BA in wine is the selection of malolactic starter cultures that are unable to produce these toxic compounds [92,93]. Another microbial strategy to reduce the presence of BA in wine is the use of selected yeast strains to induce malic acid consumption, thus avoiding malolactic fermentation and the risks of BA production associated with this phase [94]. Besides, the co-inoculation of yeast and LAB has been proposed as an interesting microbial-based solution to better control BA-producing microorganisms [95,96].

An alternative to the prevention strategies could be the use of BA-degrading microorganisms. Some wine LAB strains belonging to *Lactobacillus* and *Pediococcus* species were demonstrated to be capable of degrading BA, such as histamine, tyramine, and putrescine [97,98]. These strains showed interesting technological properties, suggesting that the ability to degrade BA could also be a criterion to select a new generation of starter cultures [98]. Enzymes isolated and purified from *L. plantarum* and *P. acidilactici* strains, and identified as multicopper oxidases, were able to degrade histamine, tyramine, and putrescine [99]. Such a finding opens a new perspective on the opportunity of employing purified microbial enzymes to deal with the problem of high BA concentrations in wine [103].

Grapes can be infected by mycotoxigenic fungi, of which *Aspergillus* spp. and *Penicillium* spp. producing ochratoxin A (OTA) is of the highest concern [7,8]. Climate is the most important factor in determining contamination once the fungi are established, with high temperatures being a major factor for OTA contamination [104]. Biological decontamination of mycotoxins using microorganisms is one of the well-known strategies to lessen these toxic compounds (Table 6). A promising approach for wine decontamination could be degradation/reduction of OTA by yeasts. Yeasts are efficient bio-sorbents and are used in winemaking to reduce the concentration of harmful substances from the must, which affect alcoholic fermentation [100,101]. Recently, research from Shukla and co-workers [105] suggests that the OTA may also be adsorbed by cells of bacteria, such as *Bacillus subtilis*. Moreover, many different yeast/bacterial strains have been demonstrated to be able to hydrolyze OTA by the action of a putative peptidase [101,102].

4. Conclusions

Climate change threatens food systems, with huge repercussions on food security and on the safety and quality of final products. In this light, it is crucial to develop a "climate-smart food system" [106] tailored to face the complex set of challenges associated with present and future climate trends, in order to ensure food sustainability [107]. We provided here an outline of potential microbial biotechnologies that may be able to tackle the changes in food quality and safety associated with climate change. With this purpose, we used wine production as a model field, considering the socio-economic relevance of this sector and the significant impact not only on the yield and wine quality, but also on the typicity of the wines [108]. Considering on-going research issues and future perspectives, it is always crucial to remember that the food production systems are interdependent structures. In this light, it is mandatory to assess the impact of the proposed biotechnological solution on the technological regimen, on the chemistry of the matrix, and on the protechnological microbiota. In the case of wine, for example, increasing studies are delving into the impact of different non-*Saccharomyces* species/strains on the microbiological [109–111] and chemical [17,112,113] features of wine. One further aspect that deserves attention is the presence of strain-dependent traits that have often been found to be associated with the protechnological and spoilage microbial phenotypes in oenology [16,114,115].

In some cases, biotechnological solutions have been patented, as we recently reviewed in the case of non-*Saccharomyces* yeasts [116]. Microbial-based approaches represent biological methods that can also find application in the production of organic wines. The potential of microbial activities as mitigating strategies in the wine sector renovates interest in the continuous exploration of microbial diversity-associated specific terroirs, autochthonous grapevines, and typical wines [117–119], and on systems that provide rapid, massive, and low-cost screening of the biotechnological potential associated with this microbial diversity [120–123].

Author Contributions: Investigation, C.B., M.F., P.R., F.B., F.G., G.S., and V.C.; conceptualization, C.B., M.F., P.R., F.B., F.G., G.S., and V.C.; literature search, C.B., M.F., P.R., F.B., F.G., G.S., and V.C.; writing—original draft preparation, C.B., M.F. and V.C.; writing—review and editing, P.R., F.B., F.G., and G.S.

Funding: This research was partially funded by the Apulia Region [project DOMINA APULIAE (POR Puglia FESR - FSE 2014-2020-Azione 1.6.–InnoNetwork)] grant number/project code AGBGUK2. Pasquale Russo is the beneficiary of a grant by MIUR in the framework of 'AIM: Attraction and International Mobility' (PON R&I 2014-2020) (practice code D74I18000190001).

Acknowledgments: We thank the Editor and the two anonymous reviewers whose comments have greatly improved this manuscript.

Conflicts of Interest: The authors declare no conflict of interest.

References

1. Rosenzweig, C.; Parry, M.L. Potential impact of climate change on world food supply. *Nature* **1994**, *367*, 133. [CrossRef]
2. Schmidhuber, J.; Tubiello, F.N. Global food security under climate change. *Proc. Natl. Acad. Sci. USA* **2007**, *104*, 19703–19708. [CrossRef] [PubMed]
3. McMichael, A.J.; Powles, J.W.; Butler, C.D.; Uauy, R. Food, livestock production, energy, climate change, and health. *Lancet* **2007**, *370*, 1253–1263. [CrossRef]
4. Lobell, D.B.; Burke, M.B.; Tebaldi, C.; Mastrandrea, M.D.; Falcon, W.P.; Naylor, R.L. Prioritizing Climate Change Adaptation Needs for Food Security in 2030. *Science* **2008**, *319*, 607–610. [CrossRef] [PubMed]
5. DaMatta, F.M.; Grandis, A.; Arenque, B.C.; Buckeridge, M.S. Impacts of climate changes on crop physiology and food quality. *Food Res. Int.* **2010**, *43*, 1814–1823. [CrossRef]
6. Vermeulen, S.J.; Campbell, B.M.; Ingram, J.S.I. Climate Change and Food Systems. *Annu. Rev. Environ. Resour.* **2012**, *37*, 195–222. [CrossRef]
7. Capozzi, V.; Fragasso, M.; Romaniello, R.; Berbegal, C.; Russo, P.; Spano, G. Spontaneous Food Fermentations and Potential Risks for Human Health. *Fermentation* **2017**, *3*, 49. [CrossRef]

8. Russo, P.; Capozzi, V.; Spano, G.; Corbo, M.R.; Sinigaglia, M.; Bevilacqua, A. Metabolites of Microbial Origin with an Impact on Health: Ochratoxin A and Biogenic Amines. *Front. Microbiol.* **2016**, *7*, 482. [CrossRef]

9. Vilela, A. The Importance of Yeasts on Fermentation Quality and Human Health-Promoting Compounds. *Fermentation* **2019**, *5*, 46. [CrossRef]

10. Berbegal, C.; Spano, G.; Tristezza, M.; Grieco, F.; Capozzi, V. Microbial Resources and Innovation in the Wine Production Sector. *S. Afr. J. Enol. Vitic.* **2017**, *38*, 156–166. [CrossRef]

11. Petruzzi, L.; Capozzi, V.; Berbegal, C.; Corbo, M.R.; Bevilacqua, A.; Spano, G.; Sinigaglia, M. Microbial Resources and Enological Significance: Opportunities and Benefits. *Front. Microbiol.* **2017**, *8*, 995. [CrossRef] [PubMed]

12. Du Plessis, H.; Du Toit, M.; Nieuwoudt, H.; Van der Rijst, M.; Hoff, J.; Jolly, N. Modulation of Wine Flavor using *Hanseniaspora uvarum* in Combination with Different *Saccharomyces cerevisiae*, Lactic Acid Bacteria Strains and Malolactic Fermentation Strategies. *Fermentation* **2019**, *5*, 64. [CrossRef]

13. Benito, Á.; Calderón, F.; Benito, S. The Influence of Non-*Saccharomyces* Species on Wine Fermentation Quality Parameters. *Fermentation* **2019**, *5*, 54. [CrossRef]

14. Capozzi, V.; Berbegal, C.; Tufariello, M.; Grieco, F.; Spano, G.; Grieco, F. Impact of co-inoculation of *Saccharomyces cerevisiae, Hanseniaspora uvarum* and *Oenococcus oeni* autochthonous strains in controlled multi starter grape must fermentations. *LWT* **2019**, *109*, 241–249. [CrossRef]

15. Di Toro, M.R.; Capozzi, V.; Beneduce, L.; Alexandre, H.; Tristezza, M.; Durante, M.; Tufariello, M.; Grieco, F.; Spano, G. Intraspecific biodiversity and 'spoilage potential' of *Brettanomyces bruxellensis* in Apulian wines. *LWT Food Sci. Technol.* **2015**, *60*, 102–108. [CrossRef]

16. Avramova, M.; Cibrario, A.; Peltier, E.; Coton, M.; Coton, E.; Schacherer, J.; Spano, G.; Capozzi, V.; Blaiotta, G.; Salin, F.; et al. *Brettanomyces bruxellensis* population survey reveals a diploid-triploid complex structured according to substrate of isolation and geographical distribution. *Sci. Rep.* **2018**, *8*, 4136. [CrossRef]

17. Padilla, B.; Gil, J.V.; Manzanares, P. Past and Future of Non-*Saccharomyces* Yeasts: From Spoilage Microorganisms to Biotechnological Tools for Improving Wine Aroma Complexity. *Front. Microbiol.* **2016**, *7*, 411. [CrossRef]

18. Escribano, R.; González-Arenzana, L.; Portu, J.; Garijo, P.; López-Alfaro, I.; López, R.; Santamaría, P.; Gutiérrez, A.R. Wine aromatic compound production and fermentative behaviour within different non-*Saccharomyces* species and clones. *J. Appl. Microbiol.* **2018**, *124*, 1521–1531. [CrossRef]

19. Escribano-Viana, R.; González-Arenzana, L.; Portu, J.; Garijo, P.; López-Alfaro, I.; López, R.; Santamaría, P.; Gutiérrez, A.R. Wine aroma evolution throughout alcoholic fermentation sequentially inoculated with non-*Saccharomyces/Saccharomyces* yeasts. *Food Res. Int.* **2018**, *112*, 17–24. [CrossRef]

20. Berbegal, C.; Spano, G.; Fragasso, M.; Grieco, F.; Russo, P.; Capozzi, V. Starter cultures as biocontrol strategy to prevent *Brettanomyces bruxellensis* proliferation in wine. *Appl. Microbiol. Biotechnol.* **2018**, *102*, 569–576. [CrossRef]

21. Berbegal, C.; Garofalo, C.; Russo, P.; Pati, S.; Capozzi, V.; Spano, G. Use of Autochthonous Yeasts and Bacteria in Order to Control *Brettanomyces bruxellensis* in Wine. *Fermentation* **2017**, *3*, 65. [CrossRef]

22. Kuchen, B.; Maturano, Y.P.; Mestre, M.V.; Combina, M.; Toro, M.E.; Vazquez, F. Selection of Native Non-*Saccharomyces* Yeasts with Biocontrol Activity against Spoilage Yeasts in Order to Produce Healthy Regional Wines. *Fermentation* **2019**, *5*, 60. [CrossRef]

23. Campbell-Sills, H.; Khoury, M.E.; Gammacurta, M.; Miot-Sertier, C.; Dutilh, L.; Vestner, J.; Capozzi, V.; Sherman, D.; Hubert, C.; Claisse, O.; et al. Two different *Oenococcus oeni* lineages are associated to either red or white wines in Burgundy: Genomics and metabolomics insights. *OENO One* **2017**, *51*, 309. [CrossRef]

24. Mira de Orduña, R. Climate change associated effects on grape and wine quality and production. *Food Res. Int.* **2010**, *43*, 1844–1855. [CrossRef]

25. Campbell, B.M.; Vermeulen, S.J.; Aggarwal, P.K.; Corner-Dolloff, C.; Girvetz, E.; Loboguerrero, A.M.; Ramirez-Villegas, J.; Rosenstock, T.; Sebastian, L.; Thornton, P.K.; et al. Reducing risks to food security from climate change. *Glob. Food Secur.* **2016**, *11*, 34–43. [CrossRef]

26. Cook, B.I.; Wolkovich, E.M. Climate change decouples drought from early wine grape harvests in France. *Nat. Clim. Chang.* **2016**, *6*, 715–719. [CrossRef]

27. Mozell, M.R.; Thach, L. The impact of climate change on the global wine industry: Challenges & solutions. *Wine Econ. Policy* **2014**, *3*, 81–89.

28. Drappier, J.; Thibon, C.; Rabot, A.; Geny-Denis, L. Relationship between wine composition and temperature: Impact on Bordeaux wine typicity in the context of global warming—Review. *Crit. Rev. Food Sci. Nutr.* **2019**, *59*, 14–30. [CrossRef]

29. Van Leeuwen, C.; Darriet, P. The Impact of Climate Change on Viticulture and Wine Quality. *J. Wine Econ.* **2016**, *11*, 150–167. [CrossRef]

30. Scarlett, N.J.; Bramley, R.G.V.; Siebert, T.E. Within-vineyard variation in the 'pepper' compound rotundone is spatially structured and related to variation in the land underlying the vineyard. *Aust. J. Grape Wine Res.* **2014**, *20*, 214–222. [CrossRef]

31. Pons, A.; Allamy, L.; Schüttler, A.; Rauhut, D.; Thibon, C.; Darriet, P. What is the expected impact of climate change on wine aroma compounds and their precursors in grape? *OENO One* **2017**, *51*, 141–146. [CrossRef]

32. Keller, M. Managing grapevines to optimise fruit development in a challenging environment: A climate change primer for viticulturists. *Aust. J. Grape Wine Res.* **2010**, *16*, 56–69. [CrossRef]

33. Cimaglia, F.; Tristezza, M.; Saccomanno, A.; Rampino, P.; Perrotta, C.; Capozzi, V.; Spano, G.; Chiesa, M.; Mita, G.; Grieco, F. An innovative oligonucleotide microarray to detect spoilage microorganisms in wine. *Food Control* **2018**, *87*, 169–179. [CrossRef]

34. Du Toit, M.; Pretorius, I.S. Microbial Spoilage and Preservation of Wine: Using Weapons from Nature's Own Arsenal—A Review. *S. Afr. J. Enol. Vitic.* **2000**, *21*, 74–96. [CrossRef]

35. García-Ruiz, A.; Requena, T.; Peláez, C.; Bartolomé, B.; Moreno-Arribas, M.V.; Martínez-Cuesta, M.C. Antimicrobial activity of lacticin 3147 against oenological lactic acid bacteria. Combined effect with other antimicrobial agents. *Food Control* **2013**, *32*, 477–483. [CrossRef]

36. Oro, L.; Ciani, M.; Comitini, F. Antimicrobial activity of *Metschnikowia pulcherrima* on wine yeasts. *J. Appl. Microbiol.* **2014**, *116*, 1209–1217. [CrossRef]

37. De Ullivarri, M.F.; Mendoza, L.M.; Raya, R.R. Killer activity of *Saccharomyces cerevisiae* strains: Partial characterization and strategies to improve the biocontrol efficacy in winemaking. *Antonie Van Leeuwenhoek* **2014**, *106*, 865–878. [CrossRef]

38. Ndlovu, B.; Schoeman, H.; Franz, C.M.a.P.; Toit, M.D. Screening, identification and characterization of bacteriocins produced by wine-isolated LAB strains. *J. Appl. Microbiol.* **2015**, *118*, 1007–1022. [CrossRef]

39. Alonso, A.; Belda, I.; Santos, A.; Navascués, E.; Marquina, D. Advances in the control of the spoilage caused by *Zygosaccharomyces* species on sweet wines and concentrated grape musts. *Food Control* **2015**, *51*, 129–134. [CrossRef]

40. Simonin, S.; Alexandre, H.; Nikolantonaki, M.; Coelho, C.; Tourdot-Maréchal, R. Inoculation of *Torulaspora delbrueckii* as a bio-protection agent in winemaking. *Food Res. Int.* **2018**, *107*, 451–461. [CrossRef]

41. Mehlomakulu, N.N.; Setati, M.E.; Divol, B. Non-*Saccharomyces* killer toxins: Possible biocontrol agents against *Brettanomyces* in wine? *S. Afr. J. Enol. Vitic.* **2015**, *36*, 94–104. [CrossRef]

42. Varela, C. The impact of non-*Saccharomyces* yeasts in the production of alcoholic beverages. *Appl. Microbiol. Biotechnol.* **2016**, *100*, 9861–9874. [CrossRef]

43. Arroyo-López, F.N.; Pérez-Torrado, R.; Querol, A.; Barrio, E. Modulation of the glycerol and ethanol syntheses in the yeast *Saccharomyces kudriavzevii* differs from that exhibited by Saccharomyces cerevisiae and their hybrid. *Food Microbiol.* **2010**, *27*, 628–637. [CrossRef]

44. Oliveira, B.M.; Barrio, E.; Querol, A.; Pérez-Torrado, R. Enhanced enzymatic activity of glycerol-3-phosphate dehydrogenase from the cryophilic *Saccharomyces kudriavzevii*. *PLoS ONE* **2014**, *9*, e87290. [CrossRef]

45. Tilloy, V.; Ortiz-Julien, A.; Dequin, S. Reduction of Ethanol Yield and Improvement of Glycerol Formation by Adaptive Evolution of the Wine Yeast *Saccharomyces cerevisiae* under Hyperosmotic Conditions. *Appl. Environ. Microbiol.* **2014**, *80*, 2623–2632. [CrossRef]

46. Kutyna, D.R.; Varela, C.; Stanley, G.A.; Borneman, A.R.; Henschke, P.A.; Chambers, P.J. Adaptive evolution of *Saccharomyces cerevisiae* to generate strains with enhanced glycerol production. *Appl. Microbiol. Biotechnol.* **2012**, *93*, 1175–1184. [CrossRef]

47. Benito, Á.; Jeffares, D.; Palomero, F.; Calderón, F.; Bai, F.-Y.; Bähler, J.; Benito, S. Selected *Schizosaccharomyces pombe* Strains Have Characteristics That Are Beneficial for Winemaking. *PLoS ONE* **2016**, *11*, e0151102. [CrossRef]

48. Contreras, A.; Hidalgo, C.; Henschke, P.A.; Chambers, P.J.; Curtin, C.; Varela, C. Evaluation of Non-*Saccharomyces* Yeasts for the Reduction of Alcohol Content in Wine. *Appl. Environ. Microbiol.* **2014**, *80*, 1670–1678. [CrossRef]

49. Rossouw, D.; Bauer, F.F. Exploring the phenotypic space of non-*Saccharomyces* wine yeast biodiversity. *Food Microbiol.* **2016**, *55*, 32–46. [CrossRef]

50. Alonso-del-Real, J.; Contreras-Ruiz, A.; Castiglioni, G.L.; Barrio, E.; Querol, A. The Use of Mixed Populations of *Saccharomyces cerevisiae* and *S. kudriavzevii* to Reduce Ethanol Content in Wine: Limited Aeration, Inoculum Proportions, and Sequential Inoculation. *Front. Microbiol.* **2017**, *8*, 2087. [CrossRef]

51. Ciani, M.; Morales, P.; Comitini, F.; Tronchoni, J.; Canonico, L.; Curiel, J.A.; Oro, L.; Rodrigues, A.J.; Gonzalez, R. Non-conventional Yeast Species for Lowering Ethanol Content of Wines. *Front. Microbiol.* **2016**, *7*, 642. [CrossRef]

52. Contreras, A.; Hidalgo, C.; Schmidt, S.; Henschke, P.A.; Curtin, C.; Varela, C. The application of non-*Saccharomyces* yeast in fermentations with limited aeration as a strategy for the production of wine with reduced alcohol content. *Int. J. Food Microbiol.* **2015**, *205*, 7–15. [CrossRef]

53. Morales, P.; Rojas, V.; Quirós, M.; Gonzalez, R. The impact of oxygen on the final alcohol content of wine fermented by a mixed starter culture. *Appl. Microbiol. Biotechnol.* **2015**, *99*, 3993–4003. [CrossRef]

54. Quirós, M.; Rojas, V.; Gonzalez, R.; Morales, P. Selection of non-*Saccharomyces* yeast strains for reducing alcohol levels in wine by sugar respiration. *Int. J. Food Microbiol.* **2014**, *181*, 85–91. [CrossRef]

55. Rodrigues, A.J.; Raimbourg, T.; Gonzalez, R.; Morales, P. Environmental factors influencing the efficacy of different yeast strains for alcohol level reduction in wine by respiration. *LWT Food Sci. Technol.* **2016**, *65*, 1038–1043. [CrossRef]

56. Remize, F.; Roustan, J.L.; Sablayrolles, J.M.; Barre, P.; Dequin, S. Glycerol Overproduction by Engineered *Saccharomyces cerevisiae* Wine Yeast Strains Leads to Substantial Changes in By-Product Formation and to a Stimulation of Fermentation Rate in Stationary Phase. *Appl. Environ. Microbiol.* **1999**, *65*, 143–149.

57. Rossouw, D.; Heyns, E.H.; Setati, M.E.; Bosch, S.; Bauer, F.F. Adjustment of Trehalose Metabolism in Wine *Saccharomyces cerevisiae* Strains to Modify Ethanol Yields. *Appl. Environ. Microbiol.* **2013**, *79*, 5197–5207. [CrossRef]

58. Dequin, S.; Baptista, E.; Barre, P. Acidification of Grape Musts by *Saccharomyces cerevisiae* Wine Yeast Strains Genetically Engineered to Produce Lactic Acid. *Am. J. Enol. Vitic.* **1999**, *50*, 45–50.

59. Gobbi, M.; Comitini, F.; Domizio, P.; Romani, C.; Lencioni, L.; Mannazzu, I.; Ciani, M. *Lachancea thermotolerans* and *Saccharomyces cerevisiae* in simultaneous and sequential co-fermentation: A strategy to enhance acidity and improve the overall quality of wine. *Food Microbiol.* **2013**, *33*, 271–281. [CrossRef]

60. Varela, C.; Kutyna, D.R.; Solomon, M.R.; Black, C.A.; Borneman, A.; Henschke, P.A.; Pretorius, I.S.; Chambers, P.J. Evaluation of Gene Modification Strategies for the Development of Low-Alcohol-Wine Yeasts. *Appl. Environ. Microbiol.* **2012**, *78*, 6068–6077. [CrossRef]

61. Lucio, O.; Pardo, I.; Krieger-Weber, S.; Heras, J.M.; Ferrer, S. Selection of *Lactobacillus* strains to induce biological acidification in low acidity wines. *LWT* **2016**, *73*, 334–341. [CrossRef]

62. Vilela, A. Use of Nonconventional Yeasts for Modulating Wine Acidity. *Fermentation* **2019**, *5*, 27. [CrossRef]

63. Ciani, M.; Ferraro, L. Enhanced Glycerol Content in Wines Made with Immobilized Candida stellata Cells. *Appl. Environ. Microbiol.* **1996**, *62*, 128–132.

64. Benito, Á.; Calderón, F.; Palomero, F.; Benito, S. Combine Use of Selected *Schizosaccharomyces pombe* and *Lachancea thermotolerans* Yeast Strains as an Alternative to theTraditional Malolactic Fermentation in Red Wine Production. *Molecules* **2015**, *20*, 9510–9523. [CrossRef]

65. Onetto, C.A.; Bordeu, E. Pre-alcoholic fermentation acidification of red grape must using Lactobacillus plantarum. *Antonie Van Leeuwenhoek* **2015**, *108*, 1469–1475. [CrossRef]

66. Magyar, I.; Nyitrai-Sárdy, D.; Leskó, A.; Pomázi, A.; Kállay, M. Anaerobic organic acid metabolism of *Candida zemplinina* in comparison with *Saccharomyces* wine yeasts. *Int. J. Food Microbiol.* **2014**, *178*, 1–6. [CrossRef]

67. Berbegal, C.; Peña, N.; Russo, P.; Grieco, F.; Pardo, I.; Ferrer, S.; Spano, G.; Capozzi, V. Technological properties of *Lactobacillus plantarum* strains isolated from grape must fermentation. *Food Microbiol.* **2016**, *57*, 187–194. [CrossRef]

68. Lucio, O.; Pardo, I.; Heras, J.M.; Krieger, S.; Ferrer, S. Influence of yeast strains on managing wine acidity using *Lactobacillus plantarum*. *Food Control* **2018**, *92*, 471–478. [CrossRef]

69. Su, J.; Wang, T.; Wang, Y.; Li, Y.-Y.; Li, H. The use of lactic acid-producing, malic acid-producing, or malic acid-degrading yeast strains for acidity adjustment in the wine industry. *Appl. Microbiol. Biotechnol.* **2014**, *98*, 2395–2413. [CrossRef]

70. Morata, A.; Loira, I.; Tesfaye, W.; Bañuelos, M.A.; González, C.; Suárez Lepe, J.A. *Lachancea thermotolerans* Applications in Wine Technology. *Fermentation* **2018**, *4*, 53. [CrossRef]
71. Loira, I.; Morata, A.; Palomero, F.; González, C.; Suárez-Lepe, J.A. *Schizosaccharomyces pombe*: A Promising Biotechnology for Modulating Wine Composition. *Fermentation* **2018**, *4*, 70. [CrossRef]
72. Benito, S.; Palomero, F.; Morata, A.; Calderón, F.; Suárez-Lepe, J.A. New applications for *Schizosaccharomyces pombe* in the alcoholic fermentation of red wines. *Int. J. Food Sci. Technol.* **2012**, *47*, 2101–2108. [CrossRef]
73. García, M.; Esteve-Zarzoso, B.; Cabellos, J.M.; Arroyo, T. Advances in the Study of *Candida stellata*. *Fermentation* **2018**, *4*, 74. [CrossRef]
74. Masneuf-Pomarede, I.; Juquin, E.; Miot-Sertier, C.; Renault, P.; Laizet, Y.; Salin, F.; Alexandre, H.; Capozzi, V.; Cocolin, L.; Colonna-Ceccaldi, B.; et al. The yeast *Starmerella bacillaris* (synonym *Candida zemplinina*) shows high genetic diversity in winemaking environments. *FEMS Yeast Res.* **2015**, *15*, fov045. [CrossRef]
75. Morata, A.; González, C.; Suárez-Lepe, J.A. Formation of vinylphenolic pyranoanthocyanins by selected yeasts fermenting red grape musts supplemented with hydroxycinnamic acids. *Int. J. Food Microbiol.* **2007**, *116*, 144–152. [CrossRef]
76. Swiegers, J.H.; Pretorius, I.S. Modulation of volatile sulfur compounds by wine yeast. *Appl. Microbiol. Biotechnol.* **2007**, *74*, 954–960. [CrossRef]
77. Brizuela, N.; Tymczyszyn, E.E.; Semorile, L.C.; Valdes La Hens, D.; Delfederico, L.; Hollmann, A.; Bravo-Ferrada, B. *Lactobacillus plantarum* as a malolactic starter culture in winemaking: A new (old) player? *Electron. J. Biotechnol.* **2019**, *38*, 10–18. [CrossRef]
78. Loira, I.; Morata, A.; Comuzzo, P.; Callejo, M.J.; González, C.; Calderón, F.; Suárez-Lepe, J.A. Use of *Schizosaccharomyces pombe* and *Torulaspora delbrueckii* strains in mixed and sequential fermentations to improve red wine sensory quality. *Food Res. Int.* **2015**, *76*, 325–333. [CrossRef]
79. Romboli, Y.; Mangani, S.; Buscioni, G.; Granchi, L.; Vincenzini, M. Effect of *Saccharomyces cerevisiae* and *Candida zemplinina* on quercetin, vitisin A and hydroxytyrosol contents in Sangiovese wines. *World J. Microbiol. Biotechnol.* **2015**, *31*, 1137–1145. [CrossRef]
80. Renault, P.; Coulon, J.; de Revel, G.; Barbe, J.-C.; Bely, M. Increase of fruity aroma during mixed *T. delbrueckii/S. cerevisiae* wine fermentation is linked to specific esters enhancement. *Int. J. Food Microbiol.* **2015**, *207*, 40–48. [CrossRef]
81. Tristezza, M.; di Feo, L.; Tufariello, M.; Grieco, F.; Capozzi, V.; Spano, G.; Mita, G.; Grieco, F. Simultaneous inoculation of yeasts and lactic acid bacteria: Effects on fermentation dynamics and chemical composition of Negroamaro wine. *LWT Food Sci. Technol.* **2016**, *66*, 406–412. [CrossRef]
82. Moreno, J.; Moreno-García, J.; López-Muñoz, B.; Mauricio, J.C.; García-Martínez, T. Use of a flor velum yeast for modulating colour, ethanol and major aroma compound contents in red wine. *Food Chem.* **2016**, *213*, 90–97. [CrossRef]
83. Campbell-Sills, H.; Capozzi, V.; Romano, A.; Cappellin, L.; Spano, G.; Breniaux, M.; Lucas, P.; Biasioli, F. Advances in wine analysis by PTR-ToF-MS: Optimization of the method and discrimination of wines from different geographical origins and fermented with different malolactic starters. *Int. J. Mass Spectrom.* **2016**, *397–398*, 42–51. [CrossRef]
84. Englezos, V.; Torchio, F.; Cravero, F.; Marengo, F.; Giacosa, S.; Gerbi, V.; Rantsiou, K.; Rolle, L.; Cocolin, L. Aroma profile and composition of Barbera wines obtained by mixed fermentations of *Starmerella bacillaris* (synonym *Candida zemplinina*) and Saccharomyces cerevisiae. *LWT* **2016**, *73*, 567–575. [CrossRef]
85. Tristezza, M.; Tufariello, M.; Capozzi, V.; Spano, G.; Mita, G.; Grieco, F. The Oenological Potential of *Hanseniaspora uvarum* in Simultaneous and Sequential Co-fermentation with *Saccharomyces cerevisiae* for Industrial Wine Production. *Front. Microbiol.* **2016**, *7*, 670. [CrossRef]
86. Gammacurta, M.; Lytra, G.; Marchal, A.; Marchand, S.; Christophe Barbe, J.; Moine, V.; de Revel, G. Influence of lactic acid bacteria strains on ester concentrations in red wines: Specific impact on branched hydroxylated compounds. *Food Chem.* **2018**, *239*, 252–259. [CrossRef]
87. Costantini, A.; Vaudano, E.; Pulcini, L.; Carafa, T.; Garcia-Moruno, E. An Overview on Biogenic Amines in Wine. *Beverages* **2019**, *5*, 19. [CrossRef]
88. Coton, M.; Romano, A.; Spano, G.; Ziegler, K.; Vetrana, C.; Desmarais, C.; Lonvaud-Funel, A.; Lucas, P.; Coton, E. Occurrence of biogenic amine-forming lactic acid bacteria in wine and cider. *Food Microbiol.* **2010**, *27*, 1078–1085. [CrossRef]

89. Tristezza, M.; Vetrano, C.; Bleve, G.; Spano, G.; Capozzi, V.; Logrieco, A.; Mita, G.; Grieco, F. Biodiversity and safety aspects of yeast strains characterized from vineyards and spontaneous fermentations in the Apulia Region, Italy. *Food Microbiol.* **2013**, *36*, 335–342. [CrossRef]

90. Capozzi, V.; Ladero, V.; Beneduce, L.; Fernández, M.; Alvarez, M.A.; Benoit, B.; Laurent, B.; Grieco, F.; Spano, G. Isolation and characterization of tyramine-producing Enterococcus faecium strains from red wine. *Food Microbiol.* **2011**, *28*, 434–439. [CrossRef]

91. Benavent-Gil, Y.; Berbegal, C.; Lucio, O.; Pardo, I.; Ferrer, S. A new fear in wine: Isolation of *Staphylococcus epidermidis* histamine producer. *Food Control* **2016**, *62*, 142–149. [CrossRef]

92. Berbegal, C.; Benavent-Gil, Y.; Navascués, E.; Calvo, A.; Albors, C.; Pardo, I.; Ferrer, S. Lowering histamine formation in a red Ribera del Duero wine (Spain) by using an indigenous O. oeni strain as a malolactic starter. *Int. J. Food Microbiol.* **2017**, *244*, 11–18. [CrossRef] [PubMed]

93. Garofalo, C.; El Khoury, M.; Lucas, P.; Bely, M.; Russo, P.; Spano, G.; Capozzi, V. Autochthonous starter cultures and indigenous grape variety for regional wine production. *J. Appl. Microbiol.* **2015**, *118*, 1395–1408. [CrossRef] [PubMed]

94. Benito, S. The impacts of *Schizosaccharomyces* on winemaking. *Appl. Microbiol. Biotechnol.* **2019**, *103*, 4291–4312. [CrossRef] [PubMed]

95. Izquierdo Cañas, P.M.; Pérez-Martín, F.; García Romero, E.; Seseña Prieto, S.; de los Palop Herreros, M.L. Influence of inoculation time of an autochthonous selected malolactic bacterium on volatile and sensory profile of Tempranillo and Merlot wines. *Int. J. Food Microbiol.* **2012**, *156*, 245–254. [CrossRef]

96. Smit, A.Y.; Engelbrecht, L.; du Toit, M. Managing Your Wine Fermentation to Reduce the Risk of Biogenic Amine Formation. *Front. Microbiol.* **2012**, *3*, 76. [CrossRef] [PubMed]

97. Niu, T.; Li, X.; Guo, Y.; Ma, Y. Identification of a Lactic Acid Bacteria to Degrade Biogenic Amines in Chinese Rice Wine and Its Enzymatic Mechanism. *Foods.* **2019**, *8*, 312. [CrossRef]

98. Capozzi, V.; Russo, P.; Ladero, V.; Fernandez, M.; Fiocco, D.; Alvarez, M.A.; Grieco, F.; Spano, G. Biogenic Amines Degradation by *Lactobacillus plantarum*: Toward a Potential Application in Wine. *Front. Microbiol.* **2012**, *3*, 122. [CrossRef]

99. Callejón, S.; Sendra, R.; Ferrer, S.; Pardo, I. Identification of a novel enzymatic activity from lactic acid bacteria able to degrade biogenic amines in wine. *Appl. Microbiol. Biotechnol.* **2014**, *98*, 185–198. [CrossRef]

100. Caridi, A.; Sidari, R.; Pulvirenti, A.; Meca, G.; Ritieni, A. Ochratoxin A adsorption phenotype: An inheritable yeast trait. *J. Gen. Appl. Microbiol.* **2012**, *58*, 225–233. [CrossRef]

101. Petruzzi, L.; Bevilacqua, A.; Corbo, M.R.; Speranza, B.; Capozzi, V.; Sinigaglia, M. A Focus on Quality and Safety Traits of *Saccharomyces cerevisiae* Isolated from Uva di Troia Grape Variety. *J. Food Sci.* **2017**, *82*, 124–133. [CrossRef]

102. De Bellis, P.; Tristezza, M.; Haidukowski, M.; Fanelli, F.; Sisto, A.; Mulè, G.; Grieco, F. Biodegradation of Ochratoxin A by Bacterial Strains Isolated from Vineyard Soils. *Toxins* **2015**, *7*, 5079–5093. [CrossRef]

103. Callejón, S.; Sendra, R.; Ferrer, S.; Pardo, I. Cloning and characterization of a new laccase from *Lactobacillus plantarum* J16 CECT 8944 catalyzing biogenic amines degradation. *Appl. Microbiol. Biotechnol.* **2016**, *100*, 3113–3124. [CrossRef]

104. Paterson, R.R.M.; Venâncio, A.; Lima, N.; Guilloux-Bénatier, M.; Rousseaux, S. Predominant mycotoxins, mycotoxigenic fungi and climate change related to wine. *Food Res. Int. Ott. Ont.* **2018**, *103*, 478–491. [CrossRef]

105. Shukla, S.; Park, J.H.; Chung, S.H.; Kim, M. Ochratoxin A reduction ability of biocontrol agent Bacillus subtilis isolated from Korean traditional fermented food Kimchi. *Sci. Rep.* **2018**, *8*, 8039. [CrossRef]

106. Wheeler, T.; von Braun, J. Climate Change Impacts on Global Food Security. *Science* **2013**, *341*, 508–513. [CrossRef]

107. Whitfield, S.; Challinor, A.J.; Rees, R.M. Frontiers in Climate Smart Food Systems: Outlining the Research Space. *Front. Sustain. Food Syst.* **2018**, *2*, 2. [CrossRef]

108. Fraga, H.; Malheiro, A.C.; Moutinho-Pereira, J.; Santos, J.A. An overview of climate change impacts on European viticulture. *Food Energy Secur.* **2012**, *1*, 94–110. [CrossRef]

109. Berbegal, C.; Borruso, L.; Fragasso, M.; Tufariello, M.; Russo, P.; Brusetti, L.; Spano, G.; Capozzi, V. A Metagenomic-Based Approach for the Characterization of Bacterial Diversity Associated with Spontaneous Malolactic Fermentations in Wine. *Int. J. Mol. Sci.* **2019**, *20*, 3980. [CrossRef]

110. Balmaseda, A.; Bordons, A.; Reguant, C.; Bautista-Gallego, J. Non-*Saccharomyces* in Wine: Effect Upon *Oenococcus oeni* and Malolactic Fermentation. *Front. Microbiol.* **2018**, *9*, 534. [CrossRef]

111. Nardi, T.; Panero, L.; Petrozziello, M.; Guaita, M.; Tsolakis, C.; Cassino, C.; Vagnoli, P.; Bosso, A. Managing wine quality using *Torulaspora delbrueckii* and *Oenococcus oeni* starters in mixed fermentations of a red Barbera wine. *Eur. Food Res. Technol.* **2019**, *245*, 293–307. [CrossRef]

112. Benito, Á.; Calderón, F.; Benito, S. The Combined Use of *Schizosaccharomyces pombe* and *Lachancea thermotolerans*—Effect on the Anthocyanin Wine Composition. *Molecules* **2017**, *22*, 739. [CrossRef]

113. Escribano, R.; González-Arenzana, L.; Garijo, P.; Berlanas, C.; López-Alfaro, I.; López, R.; Gutiérrez, A.R.; Santamaría, P. Screening of enzymatic activities within different enological non-*Saccharomyces* yeasts. *J. Food Sci. Technol.* **2017**, *54*, 1555–1564. [CrossRef]

114. Capozzi, V.; Di Toro, M.R.; Grieco, F.; Michelotti, V.; Salma, M.; Lamontanara, A.; Russo, P.; Orrù, L.; Alexandre, H.; Spano, G. Viable But Not Culturable (VBNC) state of *Brettanomyces bruxellensis* in wine: New insights on molecular basis of VBNC behaviour using a transcriptomic approach. *Food Microbiol.* **2016**, *59*, 196–204. [CrossRef]

115. Avramova, M.; Vallet-Courbin, A.; Maupeu, J.; Masneuf-Pomarède, I.; Albertin, W. Molecular Diagnosis of *Brettanomyces bruxellensis* Sulfur Dioxide Sensitivity Through Genotype Specific Method. *Front. Microbiol.* **2018**, *9*, 1260. [CrossRef]

116. Roudil, L.; Russo, P.; Berbegal, C.; Albertin, W.; Spano, G.; Capozzi, V. Non-*Saccharomyces* Commercial Starter Cultures: Scientific Trends, Recent Patents and Innovation in the Wine Sector. *Recent Pat. Food Nutr. Agric.* **2019**. [CrossRef]

117. Garofalo, C.; Tristezza, M.; Grieco, F.; Spano, G.; Capozzi, V. From grape berries to wine: Population dynamics of cultivable yeasts associated to "Nero di Troia" autochthonous grape cultivar. *World J. Microbiol. Biotechnol.* **2016**, *32*, 59. [CrossRef]

118. Bokulich, N.A.; Collins, T.S.; Masarweh, C.; Allen, G.; Heymann, H.; Ebeler, S.E.; Mills, D.A. Associations among Wine Grape Microbiome, Metabolome, and Fermentation Behavior Suggest Microbial Contribution to Regional Wine Characteristics. *mBio* **2016**, *7*, e00631-16. [CrossRef]

119. Knight, S.; Klaere, S.; Fedrizzi, B.; Goddard, M.R. Regional microbial signatures positively correlate with differential wine phenotypes: Evidence for a microbial aspect to *terroir. Sci. Rep.* **2015**, *5*, 14233. [CrossRef]

120. Capozzi, V.; Yener, S.; Khomenko, I.; Farneti, B.; Cappellin, L.; Gasperi, F.; Scampicchio, M.; Biasioli, F. PTR-ToF-MS Coupled with an Automated Sampling System and Tailored Data Analysis for Food Studies: Bioprocess Monitoring, Screening and Nose-space Analysis. *J. Vis. Exp. JoVE* **2017**. [CrossRef]

121. Romano, A.; Capozzi, V.; Spano, G.; Biasioli, F. Proton transfer reaction–mass spectrometry: Online and rapid determination of volatile organic compounds of microbial origin. *Appl. Microbiol. Biotechnol.* **2015**, *99*, 3787–3795. [CrossRef]

122. Nieuwoudt, H.H.; Pretorius, I.S.; Bauer, F.F.; Nel, D.G.; Prior, B.A. Rapid screening of the fermentation profiles of wine yeasts by Fourier transform infrared spectroscopy. *J. Microbiol. Methods* **2006**, *67*, 248–256. [CrossRef]

123. Maqueda, M.; Pérez-Nevado, F.; Regodón, J.A.; Zamora, E.; Álvarez, M.L.; Rebollo, J.E.; Ramírez, M. A low-cost procedure for production of fresh autochthonous wine yeast. *J. Ind. Microbiol. Biotechnol.* **2011**, *38*, 459–469. [CrossRef]

 *fermentation*

Review

The Influence of Non-*Saccharomyces* Species on Wine Fermentation Quality Parameters

Ángel Benito, Fernando Calderón and Santiago Benito *

Departamento de Química y Tecnología de Alimentos, Universidad Politécnica de Madrid,
Ciudad Universitaria S/N, 28040 Madrid, Spain
* Correspondence: santiago.benito@upm.es; Tel.: +34-913363710 or +34-913363984

Received: 16 May 2019; Accepted: 28 June 2019; Published: 30 June 2019

Abstract: In the past, some microbiological studies have considered most non-*Saccharomyces* species to be undesirable spoilage microorganisms. For several decades, that belief made the *Saccharomyces* genus the only option considered by winemakers for achieving the best possible wine quality. Nevertheless, in recent decades, some strains of non-*Saccharomyces* species have been proven to improve the quality of wine. Non-*Saccharomyces* species can positively influence quality parameters such as aroma, acidity, color, and food safety. These quality improvements allow winemakers to produce innovative and differentiated wines. For that reason, the yeast strains *Torulaspora delbrueckii*, *Lachancea thermotolerans*, *Metschnikowia pulcherrima*, *Schizosaccharomyces pombe*, and *Pichia kluyveri* are now available on the market. Other interesting species, such as *Starmerella bacillaris*, *Meyerozyma guilliermondii*, *Hanseniospora* spp., and others, will probably be available in the near future.

Keywords: *Torulaspora delbrueckii*; *Lachancea thermotolerans*; *Metschnikowia pulcherrima*; *Schizosaccharomyces pombe*; *Pichia kluyveri*; non-*Saccharomyces*

1. Introduction

Over the last decades, the dry yeast market based on *Saccharomyces cerevisiae* has allowed alcoholic fermentation to start faster than the regular spontaneous methods, reducing the production times. In contrast, non-*Saccharomyces* species have often been inhibited by *S. cerevisiae* inoculations at the industrial level, despite being the predominant species in grapes before fermentation starts to take place [1]. The inoculation of *S. cerevisiae* in large populations exceeding 10^6 cfu/mL and the inhibition of non-*Saccharomyces* species such as *Hanseniorsopara*, *Kloeckera*, or *Candida* by initial sulfur dioxide addition make it difficult for those non-*Saccharomyces* species to influence alcoholic fermentation. However, first-phase non-*Saccharomyces* species play an important role in spontaneous fermentations until the alcohol level reaches 4 % (*v/v*), when most non-*Saccharomyces* species can no longer survive. Temperatures below 30 °C increase the ethanol resistance of species such as *Starmerella stellata* and *Kloeckera apiculate* [2].

The populations of yeast during alcoholic fermentation change over time. During the first stage, yeast from genera with a low resistance to ethanol, such as *Hanseniaspora*, *Candida*, *Rodotorula*, and *Pichia*, predominate [3–5]. Later, some genera with a moderate resistance to ethanol, such as *Lachancea* [6] or *Torulaspora* [1], may persist for longer. In the last stages of fermentation, most authors report that the *Saccharomyces* genus dominates the medium until fermentable sugars are completely metabolized into ethanol [3–5].

Several studies have reported that non-*Saccharomyces* species show advantages that can improve specific parameters of wine quality [7–15], depending on the specific yeast species and strains used (Table 1). Because of these advantages, the most important manufacturers are now commercializing strains of non-*Saccharomyces* [16] from *Torulaspora delbrueckii*, *Lachancea thermotolerans*, *Metschnikowia pulcherrima*, *Schizosaccharomyces pombe*, and *Pichia kluyveri* (Table 2). However, most of these species also show disadvantages, which must be taken into account during their use. The main disadvantages are their low capacity to metabolize sugar into ethanol and their low resistance to additives such as sulfur dioxide in most cases, although some specific genera, such as *Schizosaccharomyces* [17,18], can withstand those disadvantages. The low fermentative activity of some non-*Saccharomyces* species is usually corrected by combining them with a high fermentative commercial *S. cerevisiae* strain able to metabolize all the sugar into ethanol [19]. This combination usually assumes a fermentation delay of a few days compared to pure fermentation inoculation by *S. cerevisiae*. The main positive influences of non-*Saccharomyces* species in modern winemaking are explained in the following paragraphs.

Table 1. Influence of non-*Saccharomyces* species on winemaking quality parameters.

Starmerella bacillaris	Glycerol ↑
Hanseniaspora spp.	Acetate esters ↑, terpenes ↑, Biogenic amines ↓
Hansenula anomala	C6 alcohols ↓
Lachancea thermotolerans	L-lactic acid ↑, Acidification ↑
Metschnikowia pulcherrima	Esters ↑, Terpenes ↑, Thiols ↑, Aroma complexity ↑,
Pichia guillermondii	Color Stability ↑
Pichia kluyveri	Thiols ↑, Esters ↑
Schizosaccharomyces pombe	L-Malic acid ↓, Deacidification ↑
Torulospora delbrueckii	Acetic acid ↓, Esters ↑, Thiols ↑,
Zygosaccharomyces bailii	Polysaccharides ↑
↑, higher activity; ↓, lower activity≈	

Table 2. Main commercial products that contain non-*Saccharomyces* strains.

Product Name	Manufacturer	Species
Biodiva™	Lallemand www.lallemandwine.com (access on 29/06/2019).	*T. delbrueckii*
Concerto™	Chr. Hansen www.chr-hansen.com (access on 29/06/2019).	*L. thermotolerans*
Flavia®	Lallemand www.lallemandwine.com	*M. pulcherrima*
Frootzen®	Chr. Hansen www.chr-hansen.com	*P. kluyveri*
Prelude™	Chr. Hansen www.chr-hansen.com	*T. delbrueckii*
Primaflora® VB BIO	CENOLIA www.sud-et-bio.com	*T. delbrueckii*
ProMalic	Proenol https://www.proenol.com	*S. pombe*
Viniferm NS TD	Agrovin www.agrovin.com	*T. delbrueckii*
Zymaflore® Alpha	Laffort www.laffort.com (access on 29/06/2019).	*T. delbrueckii*

Modern enology looks for strategies to reduce the final content of ethanol in wine. The main causes of this trend are the consumer demand for products with a lower content of ethanol. High polyphenolic maturity usually increases grape sugar due to the delay in harvest. This effect is common in warm viticulture areas where the over-ripening risk is high. There are some alternative methodologies that can be used to reduce the content of ethanol in wine, such as enzyme or osmotic filtration [20]. Non-*Saccharomyces* species allow us to reduce the initial ethanol content by about 1–2% (*v/v*), depending on the yeast species and fermentation conditions [21–23].

Chemical methods based on food quality acid additions, such as tartaric acid, were the classical solution to acidity imbalances in over-ripe grape juices. On the other hand, for excess acidity, which is commonly found in cold areas, the most common solution was the use of calcium carbonate, potassium bicarbonate, or potassium carbonate to deacidify to regular levels. The main inconveniences of these solutions are the costs of these chemical products, which must be certified as being of food-quality. Nevertheless, over the last decade, some microbial alternatives have been proposed. The first alternatives involved some strains of *S cerevisiae* that are able to influence wine acidity [24,25]; however, the influence on the pH was not significant. The use of some non-*Saccharomyces* species has been proven to be able to reduce the pH by 0.5 units [6], while other species are able to increase the pH by up to 0.5 units [26].

Acetic acid is the main acid responsible for the wine fault termed volatile acidity. Although acetic acid influences the total acidity, it is usually considered separately, as it can negatively influence the wine quality. The fault threshold of volatile acidity is considered to be about 0.8 g/L; above this, most consumers can easily identify the negative vinegar characteristic. The main acetic acid ester, ethyl acetate, in concentrations higher than 12 mg/L produces undesired odor faults [27], which are of even more concern than acetic acid. Some non-*Saccharomyces* species, such as *T. delbrueckii* [1] or *L. thermotolerans* [6], can produce wines with lower contents of acetic acid than *S. cerevisiae*, while other species, such as *Schizosaccharomyces* sp. [18], tend to produce wines with concentrations higher than the fault limit. Nevertheless, large strain variability is reported in most cases [7,28–31].

Some studies have shown that specific non-*Saccharomyces* species are able to produce higher concentrations of fruity esters than *S. cerevisiae* (control) [32]. Specific non-*Saccharomyces* strains can increase the release of varietal aromas, such as terpenes or thiols, which are responsible for the quality of some grape varieties, such as Muscat, Gewurztraminer, Sauvignon blanc, and Verdejo [19,33].

The main strategies for increasing the color of red wines are based on obtaining higher final concentrations of total anthocyanins or higher levels of the most stable anthocyanins, such as vitisins or pyranoanthocyanins [34]. pH reduction is another strategy used to increase the color perception [6,35]. The latest studies have developed strategies to avoid malolactic fermentation [6,36], whose effect on wine quality is the reduction of color due to increases in pH and lactic bacteria enzymatic activity [36,37]. However, the necessity of producing stable wines that will not re-ferment in the bottle means that the vast majority of red wines go through malolactic fermentation. The first microbiological approaches used *S. cerevisiae* strains that absorbed reduced amounts of anthocyanins. Later approaches selected *S. cerevisiae* strains in order to obtain higher contents of acetaldehyde and pyruvic acid, which slightly increase the contents of vitisin A and B. Some non-*Saccharomyces* species may produce up to four times higher concentrations of pyruvic acid or acetaldehyde than *S. cerevisiae*. The combination of specific non-*Saccharomyces* species allows the stabilization of wines from a microbiological point of view, avoiding malolactic fermentation and additionally increasing the acidity and color perception [34,35].

Although wine is a safe product from a microbiological health hazard point of view, as no pathological microorganisms such as *Salmonella* or *E. coli* can withstand the wine ethanol concentrations [38], modern enology has discovered toxic compounds that can appear in wine such as biogenic amines, ethyl carbamate, and ochratoxin A. The main strategy employed to avoid biogenic amines in wine is based on the use of selected lactic bacteria from *Oenococcus oeni* species without decarboxylase activity able to convert specific amino acids into biogenic amines. Regarding this fact, some non-*Saccharomyces* species have been reported to produce higher concentrations of amino acids such as histidine that can evolve to histamine if bacteria decarboxylase activity takes place [30]. Other non-*Saccharomyces* species prevent the malolactic fermentation process, where the production of biogenic amines takes place [36]. Ochratoxin A is produced prior to harvest by spoilage fungal attacks. There are several methods that can be used to reduce the ochratoxin A concentration during the winemaking process, such as that which is conducted through the use of amicrobic filtrations of about 0.45 μm that allow the initial concentration to be reduced by up to 80% [39]. A promising biotechnology method is the use of yeast lees to remove ochratoxin A. This method was first tested using *S. cerevisiae*, although newer studies have shown that some non-*Saccharomyces* species are more efficient at removing ochratoxin A, with rates of up to 70% [38,40–42]. Ethyl carbamate is mainly produced by lactic acid bacteria and through the chemical combination of urea with ethanol during wine ageing. The most common type of management in the wine industry is based on the use of a commercial urease enzyme which is able to remove all the urea that can evolve into ethyl carbamate [38]. Nevertheless, the use of non-*Saccharomyces* species with urease activity allows the removal of the main ethyl carbamate precursor from wine, making it virtually impossible for ethyl carbamate to appear during wine ageing.

Polysaccharides have been proven to improve the mouthfeel properties of wine [43–45]. The improvements in wine quality are mainly related to softening the wine astringency [45] or increasing positive aromatic compounds [46]. The most abundant group of polysaccharides is the arabinogalactan proteins, which originate in grapes [43]. Mannoproteins represent the second most abundant group; however, these polysaccharides are formed during alcoholic fermentation or ageing during lees processes [26,47]. Although the first microbiological applications for increasing the content of mannoproteins in wines were based on the use of *S. cerevisiae* strains, later studies showed that some non-*Saccharomyces* species release higher concentrations of mannoproteins than *S. cerevisiae* [47–50]. Other polysaccharides of a different nature than mannoproteins are also reported for some non-*Saccharomyces* species [26,51–53].

All yeast species inevitably produce acetaldehyde during alcoholic fermentation. The highest concentration is reached during the tumultuous phase of alcoholic fermentation. It usually takes place within 48–72 h of alcoholic fermentation, depending on the fermentation power of the yeast species. Concentrations higher than 125 mg/L usually negatively influence the flavor of wine [54,55], and wines are usually described as being oxidized. Some of the descriptors used for wines where acetaldehyde predominates in the aroma are green apples and fresh-cut grass [55]. Such aromas are easy to identify in white wines. Newer studies on red wines have proven that concentrations below the fault threshold of 125 mg/L increase the valuable stable color forms, such as vitisin B [35], which improves wine color, while the aroma of acetaldehyde cannot be identified in a sensory analysis. Some non-*Saccharomyces* species produce lower concentrations of acetaldehyde than *S. cerevisiae* [56], while others produce higher levels [57].

Glycerol can increase the softness and body of wine. *S. cerevisiae* synthesizes glycerol from glucose through glycolysis, where dihydroxyacetone phosphate is reduced to glycerol-3-phosphate and later oxidized to glycerol [58,59]. One of the first reported advantages of using non-*Saccharomyces* species was the increased glycerol concentration in wine and its influence on wine quality [9,48,60]. Depending on the non-*Saccharomyces* species employed, it is possible to achieve increases from a few decimals to 4 g/L compared to *S. cerevisiae* [9,60]. From a biochemical point of view, species other than those of the *Saccharomyces* genus possess less developed alcohol dehydrogenase enzymatic activity, but more developed glycerol-3-phosphate dehydrogenase enzymatic activity. This metabolism deviates to produce higher final concentrations of glycerol during alcoholic fermentation [59].

Several studies attribute the properties of some non-*Saccharomyces* species to improved wine quality. Nevertheless, recent studies have shown large differences, depending on the non-Saccharomyces strain used [31,61,62]. This oenological phenotypical variability is based on the huge number of different populations and the genomic diversity of those species [63–69]. These results suggest the importance of performing selective processes, such as those that were conducted for *S. cerevisiae* strains in the past.

2. *Torulaspora Delbrueckii*

Torulaspora delbrueckii (Figure 1(1)) is the most studied and commercialized of the non-*Saccharomyces* species in winemaking [1]. The management of *T. delbrueckii* is relatively easy compared to other non-*Saccharomyces* species due to its relatively high fermentative power of up to 9–10% (*v/v*) [70], while several non-*Saccharomyces* species, such as *M. pulcherrima*, *P. guillermondii*, *P. kluyveri*, *S. stellata*, and *Hanseniaspora vinae*, do not tolerate ethanol concentrations higher than 4% (*v/v*). Due to this ethanol resistance, this species can notably influence the final wine product during most of the alcoholic fermentation period, although in most wines, a second more fermentative species such as *S. cerevisiae* [48] or *S. pombe* [71] is required to properly end the alcoholic fermentation. Nevertheless, some industries other than conventional winemaking have started to exclusively use *T. delbrueckii* for fermentation; some examples are for the production of beer or sparkling base wine [1].

One of the first advantages attributed to *T. delbrueckii* was the reduction of the volatile acidity concentration in wines. Some authors have reported reductions in the final acetic acid concentration of about 0.14 to 0.28 g/L compared to *S. cerevisiae* [1,70]. The application of *T. delbrueckii* can decrease the final ethanol concentration in wines by up to 1% (*v/v*) [22], while increasing the glycerol concentration from 0.2 to 0.9 g/L [1,72–74]. Several authors report *T. delbrueckii* as being a greater mannoprotein releaser than *S. cerevisiae* and other non-*Saccharomyces* species [47,48]. Moderate malic acid consumption by *T. delbrueckii* has been commonly observed in sequential fermentations in quantities varying from 20% to 25% [43,48].

T. delbrueckii can improve the intensity and quality of wine aroma, increasing the overall impression and the varietal and fruity characters [72]. *T. delbrueckii* is able to diminish the concentrations of higher alcohols when it is used in sequential fermentations with *S. cerevisiae* [48]. This effect contributes to an increase in the varietal character perception. However, an increase in alcohol production has also been reported [75,76]. Several authors have reported the production of higher final concentrations of fruity esters [72,77]. In contrast, other studies have reported the opposite effect [73,76]. These differences in higher alcohols and ester formation have been explained by the high strain variability in these parameters shown by the species [31,62]. *T. delbrueckii* is reported to release conjugated terpenes in some wine varieties characterized by these varietal compounds [78]. In addition, proper *T. delbrueckii* strain selection allows for the release of higher concentrations of thiols, which increase the varietal character of varieties such as Sauvignon blanc or Verdejo [72,79].

A moderate undesirable effect reported by most authors is a delay in sequential fermentation involving *T. delbrueckii* and *S. cerevisiae* compared with the *S. cerevisiae* control.

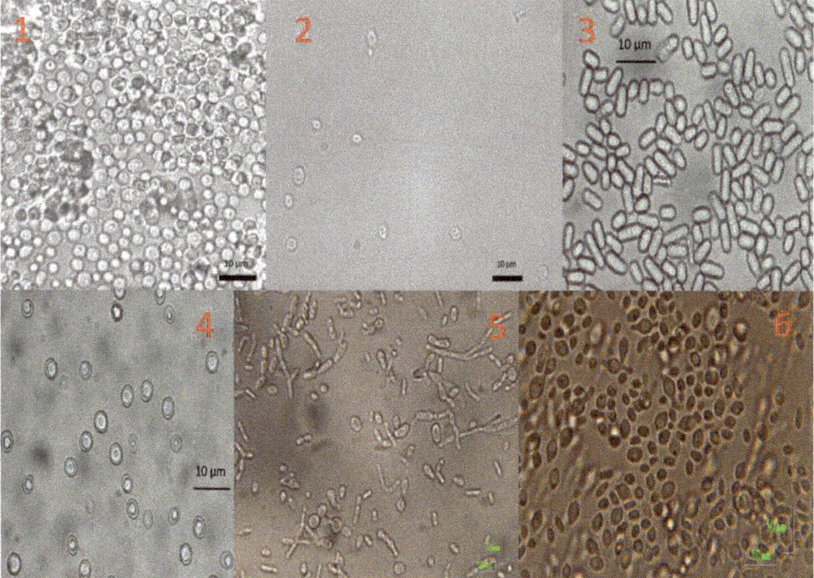

Figure 1. Microscopic observation of Torulaspora delbrueckii (**1**), Lachancea thermotolerans (**2**), Schizosaccharomyces pombe (**3**), Metschnikowia pulcherrima (**4**), Meyerozyma guilliermondii (**5**) and Hanseniospora uvarum (**6**) cells.

3. *Lachancea Thermotolerans*

Lachancea thermotolerans (Figure 1(2)) is the most recommended of the non-*Saccharomyces* species used to acidify grape juices that suffer from a lack of acidity [80]. This ability is very useful in viticultural areas in the south of Europe or any other warm viticulture region [6,81]. *L. thermotolerans* can acidify wines due to its unique ability among yeasts to produce lactic acid during its fermentative metabolism [7,82]. Hranilovic et al. (2018) have shown the pathway of lactate formation from pyruvate through the enzyme lactate dehydrogenase enzyme [83]. The production of lactic acid can vary from a few decimals of g/L to up to almost 10 g/L, depending on factors such as the strain or fermentative temperature used [6,84]. The production of lactic acid without degrading any malic acid directly influences the titratable acidity quality parameter. The production of lactic acid is reported to increase the titratable acidity by up to about 9 g/L [84] compared with the regular *S. cerevisiae* control. Some studies have reported a reduction in pH from about pH 4 to pH 3.5 in low-acid grape juice, which would be considered an acidic wine in most warm viticulture regions [6]. The reduction of pH also positively influences the color of red wine due to the increase in the color intensity of anthocyanins such as the flavylium ion [35,36].

L. thermotolerans also has other interesting properties. Some authors have described *L. thermotolerans* as an interesting resource that can be employed to reduce the final concentration of volatile acidity in wine [85]. Some studies have reported that *L. thermotolerans* fermentations produce lower concentrations of acetic acid than *S. cerevisiae*, by about 0.24 g/L [82,84], while other authors have reported smaller differences in sequential fermentations of about 0.05 g/L [36,86]. Recent studies reported *L. thermotolerans* strains to show a strain variability of up to 49% in acetic acid production [31]. Later studies supported the idea that strain variability shows great biodiversity around the world [69], which translates into large differences in the phenotypic fermentative performance for the different strains [31]. Although the first studies showed lower final concentrations in glycerol than *S. cerevisiae* for single pure fermentations of about 1.5 g/L [84], later studies showed that, in sequential fermentations, those including *L. thermotolerans* often reach higher final levels of glycerol of up to 1 g/L [7,86]. These results, combined with those related to ethanol production, indicate that although *L. thermotolerans* is less fermentative than *S. cerevisiae*, it possesses a more developed glycerol–pyruvic pathway. Nevertheless, the production of glycerol by *L. thermotolerans* also depends on other factors, such as temperature [82], as it produces higher contents at 20 °C than at 30 °C. Later studies reported that the injection of oxygen during *L. thermotolerans* fermentations increases the production of glycerol while reducing the production of ethanol [87].

Some studies have reported that *L. thermotolerans* sequential fermentations produce lower final concentrations of higher alcohols than *S. cerevisiae*—from 13 to 55 mg/L, depending on the study [81,82,88]. Nevertheless, other authors have reported the opposite effect, with fermentations involving *L. thermotolerans* increasing higher alcohol concentrations by up to 100 mg/L [7]. These discrepancies are explained by the great variability in *L. thermotolerans* strains, in terms of higher alcohol production (up to 40%) [31,62] and oxygen availability [87].

Some authors have reported increases in the total ester content of up to 33% [69], while other studies have not observed important differences [7]. Nevertheless, all studies have reported increases in the specific ester ethyl lactate, which is related to lactic acid metabolism [32,34]. *L. thermotolerans* produces lower concentrations of fatty acids than *S. cerevisiae* in pure fermentations, although specific strains of *L. thermotolerans* tend to produce higher concentrations of the specific fatty acid isovaleric acid [62]. Specific *L. thermotolerans* strains are able to release higher amounts of terpenes, depending on their glucosidase activity [7,32].

Some authors have reported higher total final anthocyanin concentrations in sequential fermentations involving *L. thermotolerans* than in the *S. cerevisiae* controls. The differences varied from 8% to 10% [34,89]. Additionally, the higher coloration of anthocyanins produced by lactic acid acidification at a low pH notably influences the final color intensity, which is higher than that of *S. cerevisiae* [6,18,35,36].

Some authors have reported that some specific *L. thermotolerans* strains release up to 100 mg/L higher concentrations of polysaccharides than *S. cerevisiae* [7]. Nevertheless, this ability is strain-dependent, as the observed strain variability between the studied *L. thermotolerans* strains is close to 40% [7].

Studies that perform sensory analyses usually describe the wines fermented by *L. thermotolerans* as being more acidic than the controls fermented by *S. cerevisiae* [36,57]. This perception is not as obvious in highly acidic wines from northern Germany [32], but is very evident in wines fermented from low acidic grape juices from warm areas in southern Europe, where the pH is reduced by about 0.4 units and the total acidity increases by about 3 g/L [81]. The color intensity perception is usually higher due to the increase of visible red and purple colors at lower pH values [6].

Other modern applications of *L. thermotolerans* facilitate the management of ochratoxin A. Some *L. thermotolerans* strains are able to efficiently inhibit the development of ochratoxigenic fungi in the vineyard [90,91]. As the legislation trend is to reduce the applications of pesticides, *L. thermotolerans* appears to be an interesting alternative to the management of ochratoxin A.

The main problems in the industry management of *L. thermotolerans* are its sensitivity to sulfur dioxide and its moderate ethanol tolerance. Among the non-*Saccharomyces* species, *L. thermotolerans* is considered a fermentative species that is able to ferment wines at up to levels slightly higher than 10% (*v/v*) in ethanol [28,82], but it must be combined with a *S. cerevisiae* [86] strain for the production of regular dry wines or with another more fermentative genus, such as *Schizosaccharomyces* [36], to ensure proper alcoholic fermentation cessation. In other fermentative industries such as beer, sweet wine, or sparkling base wines, the fermentative power of *L. thermotolerans* is sufficient to achieve the desired final ethanol concentration [6,51]. Additionally, some studies have observed that sequential fermentations between *L. thermotolerans* and *S. cerevisiae* produce lower final ethanol concentrations varying from 0.2% to 0.4% (*v/v*) [36,69,86].

Another reported problem is the release of higher concentrations of biogenic amine amino acid precursors such as lysine, ornithine, and tyrosine during alcoholic fermentation compared with *S. cerevisiae* [32,57]. Although there is no direct correlation between biogenic amine formation and the presence of amino acid precursors, this fact must be taken into account, especially for wines that will perform malolactic fermentation or barrel ageing [38]. Nevertheless, the acidification performed by *L. thermotolerans* can partially inhibit the capacity of unselected lactic bacterial strains to produce biogenic amines, as the development of lactic bacterial genera is limited at low pH values. This potential food safety problem is also present when *L. thermotolerans* is used in ageing over lees for histidine, tyrosine, ornithine, and lysine amino acids [49].

Although most studies regarding the *Lachancea* genus and oenology have focused on *L. thermotolerans*, other species have started to show promising potential. For example, *L. fermentati* is a higher fermenter than *L. thermotolerans*. Pure inoculations of *L. fermentati* produce wines with lower concentrations of acetaldehyde, SO_2, and H_2S compared to the *S. cerevisiae* controls [80,92].

4. *Schizosaccharomyces* Species

Schizosaccharomyces pombe is the most recommended of the non-*Saccharomyces* species to de-acidify excessively acidic wines from cool areas, such as those from the north of Europe. Indeed, modern studies also employ this species to stabilize wines from a microbiological point of view, for example, in red wines from warm viticulture areas, where the performance of a proper malolactic fermentation process is complicated due to the low levels of malic acid and the high pH [36]. *S. pombe* is able to metabolize malic acid into ethanol and CO_2, consequently reducing the total wine acidity [18]. Benito et al. (2014) have shown the biochemical pathway used to degrade malic acid into ethanol through pyruvate decarboxylase and alcohol dehydrogenase enzymes [93]. In wines with malic acid contents higher than 5 g/L, which are considered very acidic by regular consumers, *S. pombe* can completely remove any malic acid present, decreasing the total acidity by about 4 g/L and the pH by about 0.4 units [94]. Figure 1(3) shows a microscopic observation of *S. pombe* cells during pure alcoholic fermentation.

Recently, the use of *S. pombe* has been suggested in warm viticulture areas where grape juices contain high levels of sugar, pH values are close to 4, and malic acid concentrations are usually less than 1 g/L [26,48]. Under these circumstances, to try to perform malolactic fermentation is dangerous, with a high risk of deviations, such as the production of undesired high levels of volatile acidity or biogenic amines. Nevertheless, if malolactic fermentation is not performed before bottling, it often takes place in the bottle, generating undesired turbidity. In these scenarios, the use of *S. pombe* alone or in small percentages in combined inoculums with *S. cerevisiae* allows the achievement of microbiological stability so that wine can be bottled without the risk of bottle refermentation.

Specific strains of *S. pombe* are the most effective option to remove gluconic acid from wine during alcoholic fermentation, with a removal percentage of up to 91% [95–99]. Gluconic acid can negatively influence the quality of wine, generating microbial instability, as it can be used by lactic acid bacteria to increase volatile acidity, reducing the protective effect of sulfur dioxide.

One of the main problems of using *S. pombe* is that it tends to generate high levels of acetic acid [18,30]. This acid usually produces a quality-detrimental vinegar character, which is not tolerated by consumers of quality wines. This undesirable effect has been solved with different strategies, such as the combined use with *S. cerevisiae* [94], *L. thermotolerans* [6,36], or *T. delbrueckii* [71]; the addition of magnesium [100]; or the use of alginate cells [101] and fed-batch fermentation [102]. These alternatives allow the production of wines with lower acetic acid contents than those produced with *S. cerevisiae*. Another undesirable effect of the use of *S. pombe* is an increase in the ethanol concentration, as the degradation of 2.33 g/L of malic acid produces about 0.1% (*v/v*) of additional ethanol [103]. Although no malolactic fermentation is needed after *S. pombe* alcoholic fermentation, the concentration of amino acids that can evolve to biogenic amines usually increases [30].

The malolactic fermentation process usually reduces the anthocyanin content and color intensity from 10% to 23% [36,37]. This phenomenon takes place due to the cell absorption and glycosidase enzymatic activity of lactic bacteria [35,104]. *S. pombe*-fermented wines show higher contents of total anthocyanins and consequently higher color intensities as malolactic fermentation is not needed. Additionally, *S. pombe* is able to produce up to five times more pyruvic acid than *S. cerevisiae* [105], which translates to the formation of a consequently higher concentration of the stable anthocyanin vitisin A, which contains pyruvic acid [35]. Additionally, the combined use with *L. thermotolerans* increases the color intensity due to the additional reduction of pH that increases the color intensity of flavylium ions [6,18,35] *S. pombe* releases higher amounts of polysaccharides than any other *Saccharomyces* or non-*Saccharomyces* yeast [26,52], consequently improving the wine structure. The nature of these polysaccharides is different than that reported for *S. cerevisie*, including the presence of α-galactomannose and β-glucans in their compositions [53].

S. pombe is characterized by producing significantly lower concentrations of higher alcohols and esters than *S. cerevisiae* and other yeast species [34,57]. This is very interesting when retention of the varietal aroma of grapes is desired more than the fermentative aroma [19,48,72].

Regarding food safety, the use of *S. pombe* allows the control of biogenic amines, as no malolactic fermentation, which is able to produce this toxic compound, is required [38]. Additionally, the urease enzymatic activity developed by *S. pombe* eliminates the main precursor of ethyl carbamate: urea. Indeed, *S. pombe* can remove 70% of the initial concentration of the carcinogen ochratoxin A during alcoholic fermentation [40].

In recent years, other industries have started to use *Schizosaccharomyces* species in products and processes other than the production of grape wine, such as ginger fermentation [106,107], apple wine [108], kei-apple fermentation [109], sparkling wine [110], bioethanol [111], bilberry fermentation [71], plum wine [112], and water purification [18].

Although *S. pombe* is the most studied yeast from the genus *Schizosacchromyces*, *Schizosaccharomyces japonicus* shows similar properties to *S. pombe* and a better performance in specific quality parameters such as glycerol production and polysaccharide release [53].

5. *Metschnikowia Pulcherrima*

Metschnikowia pulcherrima (Figure 1(4)) influences wine quality parameters. It can increase the glycerol concentration by a few decimals in combined fermentations compared to single *S. cerevisiae* controls. It is also able to reduce the malic acid content by about 10% and the acetaldehyde concentration by about 10 mg/L [19]. One modern application is the reduction of the final ethanol concentration. For that purpose, *M. pulcherrima* can be used in order to achieve ethanol reductions down to 1% (*v/v*) [22,113,114].

Following the comparison of sequential fermentations of *M. pulcherrima* and *S. cerevisiae*, some studies have described *M. pulcherrima* as a producer of low higher alcohol concentrations compared to *S. cerevisiae* that vary from 20% to 30% [32]. On some occasions, this effect means that varietal aroma compounds such as terpenes or thiols that are not masked by concentrations of higher alcohols that are higher than the perception threshold have a greater effect on wine aroma [19]. On the other hand, most studies have reported that *M. pulcherrima* is a higher producer of fruity esters [32]. Most studies have reported significant differences, especially for ethyl octanoate, which is produced in higher concentrations varying from 20% to 25% in sequential fermentations involving *M. pulcherrima* than in *S. cerevisiae* [19]. This specific ester increases fruity aromas related to pineapple, which are usually considered pleasant and very positive, in neutral grape varieties that do not possess varietal aroma compounds such as terpenes or thiols.

The most relevant influence on wine quality related to the use of *M. pulcherrima* is the ability of the cystathionine-β-lyase activity of selected strains to cause the release of varietal thiols such as 4-methyl-4-sulfanylpentan-2-one in concentrations six times higher than those in *S. cerevisiae* [19]. This aromatic compound is the most important quality indicator in thiolic wine varieties such as Sauvignon blanc or Verdejo. Figure 1(4) shows a microscopic observation of alcoholic fermentation performed by a pure culture of *M. pulcherrima* and sterilized grape juice.

6. *Meyerozyma Guilliermondii*

The use of *Meyerozyma guilliermondii* (Figure 1(5)) focuses on wine color improvements. *M. guilliermondii* is reported to be the yeast species with the highest hydroxycinnamate decarboxylase enzymatic activity [115]. This enzymatic activity allows the production of pyranoanthocyanin adducts, which condensate with grape anthocyanins to produce highly stable colored compounds that remain for a longer period of time than other anthocyanins. This biological enzymatic activity was first investigated in *S. cerevisiae*; however, although the enzymatic activity improved the color intensity, color stability, and removed ethyl phenol precursors, a maximum activity level of 16% was reached and there was a great dependency on the studied strain [116]. *M. guilliermondii* has been reported to increase hydroxycinnamate decarboxylase enzymatic activity by up to 90%. This type of biotechnology allows us to produce modern wines that contain up to 11-times higher concentrations of vinylphenolic pyranoanthocyanin adducts, which are the most stable color forms reported in winemaking [115].

7. *Pichia Kluyveri*

Some studies have reported the use of *Pichia kluyveri* (Figure 2) in sequential fermentations to produce higher levels of esters than *S. cerevisiae*, such as 2-phenylethyl acetate, by about 20%, or ethyl octanoate, by about 10% [32]. The total terpene concentration was also shown to increase by about 20%; this fact contributed to an increase in the grape variety typicity.

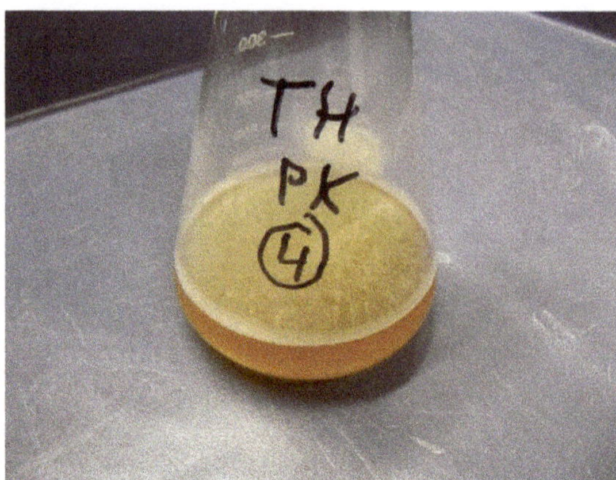

Figure 2. Film produced by *Pichia kluyveri* over grape juice.

8. *Starmerella Bacillaris*

Starmerella bacillaris, formerly *Candida stellata* [59], produces the highest glycerol concentration (up to 14 g/L) of the non-*Saccharomyces* yeasts during alcoholic fermentation [9,117,118], while most *S. cerevisiae* strains have been shown to produce final concentrations that vary from 5 to 8 g/L. These concentrations can improve the mouthfeel sensation and flavor of wine [9]. Another interesting property is its fructophilic character [119,120], in contrast with the glucophlilic character of *S. cerevisiae*.

9. *Hanseniaspora*

Species from the *Hanseniospora* genus possess a characteristic apiculate shape (Figure 1(6)). Most of the yeasts present at the beginning of spontaneous fermentation belong to this genus [121–125]. Although no manufacturer has offered a commercial strain to date, the genus is supposed to make up an important percentage of the yeasts that are in grapes. This indicates that, in traditional fermentations, the *Hanseniospora* genus notably influences alcoholic fermentations during the first phase until alcohol levels of about 4% are reached. At these levels, most *Hanseniospora* strains cannot survive due to their low tolerance to ethanol [126]. In combination with *S. cerevisiae*, which properly ends alcoholic fermentation, strains of the *Hanseniospora* genus can positively influence wine quality [9,125]. The *Hanseniaspora* genus is an interesting source of enzymes for modern winemaking challenges [125]. The most remarkable enzymatic activity is reported for β-glucosidase [127], β-xylosidase [128], glycolytic, and protease [123,129].

From a sensory point of view, the improvements are based on more intense wine flavor and aroma complexity. At an industrial scale, *Hanseniospora guilliermondii*, *Hanseniospora uvarum*, and *Hanseniospora vinae* [130] are the most appropriate species to achieve these purposes [125]. The aroma improvements are explained from a chemical point of view due to the production of higher concentrations of 2-phenylethyl acetate [131,132], acetate esters such as isoamyl acetate [124,127,130,132,133], medium-chain fatty acid ethyl esters [134], benzenoids [135,136], and terpenes [125,127] and reductions in the final concentration of higher alcohols [124,130,133]. Martin et al. 2018 [125] have explained the main metabolic pathways responsible for the ability of some species of *Hanseniaspora/Kloeckera* genera to produce benzenoids, diacetyl-acetoin, lactones, higher alcohols, acetate esters, fatty acids, and isoprenoids.

The most appropriate species to improve the color and polyphenolic composition in red wines from the *Hanseniaspora* genus are *Hanseniaspora clermontiae*, *Hanseniaspora opuntiae*, *H. guilliermondii*, and *H. vinae* [125]. These species can improve quality parameters such as color intensity and total anthocyanins [124]. These color improvements are based on the ability of *Hanseniaspora* species to produce vitisin A [137], vitisin B [138], and malvidin-3-O-glucoside-4-vinylguaiacol [137].

Fermentation **2019**, *5*, 54

10. Conclusions

Non-*Saccharomyces* species can play an important role in winemaking. Depending on the specific type of wine or the enological problem to solve, different non-*Saccharomyces* strains should be selected to attain the desired objective. The combination of non-*Saccharomyces* species with *Saccharomyces* species or even with another high fermentative non-*Saccharomyces* species can also lead to the best solution. At this time, the most commonly used strains in industry are *Torulaspora delbrueckii*, *Lachancea thermotolerans*, *Schizosaccharomyces pombe*, *Metschnikowia pulcherrima*, and *Pichia kluyveri*, which are present in available products. It is likely that over the next few years new species will start to be available on the market, and products that contain combinations of non-*Saccharomyces* species will also be available in order to simulate spontaneous alcoholic fermentations.

Conflicts of Interest: The authors declare no conflict of interest.

References

1. Benito, S. The impact of *Torulaspora delbrueckii* yeast in winemaking. *Appl. Microbiol. Biotechnol.* **2018**, *102*, 3081–3094. [CrossRef] [PubMed]
2. Gao, C.; Fleet, G.H. The effects of temperature and pH on the ethanol tolerance of the wine yeasts, *Saccharomyces cerevisiae, Candida stellata* and *Kloeckera apiculata*. *J. Appl. Bacteriol.* **1988**, *65*, 405–409. [CrossRef]
3. Di Maro, E.; Ercolini, D.; Coppola, S. Yeast dynamics during spontaneous wine fermentation of the Catalanesca grape. *Int. J. Food Microbiol.* **2007**, *117*, 201–210. [CrossRef] [PubMed]
4. Hierro, N.; González, Á.; Mas, A.; Guillamón, J.M. Diversity and evolution of non-*Saccharomyces* yeast populations during wine fermentation: Effect of grape ripeness and cold maceration. *FEMS Yeast Res.* **2006**, *6*, 102–111. [CrossRef] [PubMed]
5. Ortiz, M.J.; Barrajón, N.; Baffi, M.A.; Arévalo-Villena, M.; Briones, A. Spontaneous must fermentation: Identification and biotechnological properties of wine yeasts. *LWT Food Sci. Technol.* **2013**, *50*, 371–377. [CrossRef]
6. Benito, S. The impacts of *Lachancea thermotolerans* yeast strains on winemaking. *Appl. Microbiol. Biotechnol.* **2018**, *102*, 6775–6790. [CrossRef] [PubMed]
7. Comitini, F.; Gobbi, M.; Domizio, P.; Romani, C.; Lencioni, L.; Mannazzu, I.; Ciani, M. Selected non-*Saccharomyces* wine yeasts in controlled multistarter fermentations with *Saccharomyces cerevisiae*. *Food Microbiol.* **2011**, *28*, 873–882. [CrossRef] [PubMed]
8. Ciani, M.; Comitini, F. Non-*Saccharomyces* wine yeasts have a promising role in biotechnological approaches to winemaking. *Ann. Microbiol.* **2011**, *61*, 25–32. [CrossRef]
9. Jolly, N.P.; Varela, C.; Pretorius, I.S. Not your ordinary yeast: Non-*Saccharomyces* yeasts in wine production uncovered. *FEMS Yeast Res.* **2014**, *14*, 215–237. [CrossRef]
10. Varela, C. The impact of non-*Saccharomyces* yeasts in the production of alcoholic beverages. *Appl. Microbiol. Biotechnol.* **2016**, *100*, 9861–9874. [CrossRef]
11. García, M.; Esteve-Zarzoso, B.; Arroyo, T. Non-*Saccharomyces* Yeasts: Biotechnological Role for Wine Production. In *Grape and Wine Biotechnology*; IntechOpen: London, UK, 2016; Volume 11, pp. 1–25.
12. Padilla, B.; Gil, J.V.; Manzanares, P. Past and future of non-*Saccharomyces* yeasts: From spoilage microorganisms to biotechnological tools for improving wine aroma complexity. *Front. Microbiol.* **2016**, *7*, 411–431. [CrossRef]
13. Padilla, B.; Zulian, L.; Ferreres, À.; Pastor, R.; Esteve-Zarzoso, B.; Beltran, G.; Mas, A. Sequential inoculation of native non-*Saccharomyces* and *Saccharomyces cerevisiae* strains for wine making. *Front. Microbiol.* **2017**, *8*, 1293–1305. [CrossRef]
14. Petruzzi, L.; Capozzi, V.; Berbegal, C.; Corbo, M.R.; Bevilacqua, A.; Spano, G.; Sinigaglia, M. Microbial resources and enological significance: Opportunities and benefits. *Front. Microbiol.* **2017**, *8*, 995–1008. [CrossRef]
15. Ciani, M.; Comitini, F. Use of Non-*Saccharomyces* Yeasts in Red Winemaking. In *Red Wine Technology*; Academic Press: Cambridge, MA, USA, 2019; Volume 4, pp. 51–68.

16. Roudil, L.; Russo, P.; Berbegal, C.; Albertin, W.; Spano, G.; Capozzi, V. Non-*Saccharomyces* Commercial Starter Cultures: Scientific Trends, Recent Patents and Innovation in the Wine Sector. *Recent Pat. Food. Nutr. Agric.* **2019**, *10*, 1–13. [CrossRef]

17. Benito, S.; Palomero, F.; Gálvez, L.; Morata, A.; Calderón, F.; Palmero, D.; Suárez-Lepe, J.A. Quality and composition of red wine fermented with *Schizosaccharomyces pombe* as sole fermentative yeast, and in mixed and sequential fermentations with *Saccharomyces cerevisiae*. *Food Technol. Biotechnol.* **2014**, *52*, 376–382.

18. Benito, S. The impacts of *Schizosaccharomyces* on winemaking. *Appl. Microbiol. Biotechnol.* **2019**, *103*, 4291–4312. [CrossRef]

19. Ruiz, J.; Belda, I.; Beisert, B.; Navascués, E.; Marquina, D.; Calderón, F.; Rauhut, D.; Santos, A.; Benito, S. Analytical impact of *Metschnikowia pulcherrima* in the volatile profile of Verdejo white wines. *Appl. Microbiol. Biotechnol.* **2018**, *102*, 8501–8509. [CrossRef]

20. Röcker, J.; Schmitt, M.; Pasch, L.; Ebert, K.; Grossmann, M. The use of glucose oxidase and catalase for the enzymatic reduction of the potential ethanol content in wine. *Food Chem.* **2016**, *210*, 660–670. [CrossRef]

21. Röcker, J.; Strub, S.; Ebert, K.; Grossmann, M. Usage of different aerobic non-*Saccharomyces* yeasts and experimental conditions as a tool for reducing the potential ethanol content in wines. *Eur. Food Res. Technol.* **2016**, *242*, 2051–2070. [CrossRef]

22. Contreras, A.; Hidalgo, C.; Henschke, P.A.; Chambers, P.J.; Curtin, C.; Varela, C. Evaluation of non-*Saccharomyces* yeasts for the reduction of alcohol content in wine. *Appl. Environ. Microbiol.* **2014**, *80*, 1670–1678. [CrossRef]

23. Ciani, M.; Morales, P.; Comitini, F.; Tronchoni, J.; Canonico, L.; Curiel, J.A.; Oro, L.; Rodrigues, A.J.; Gonzalez, R. Non-conventional yeast species for lowering ethanol content of wines. *Front. Microbiol.* **2016**, *7*, 642–655. [CrossRef]

24. Redzepovic, S.; Orlic, S.; Majdak, A.; Kozina, B.; Volschenk, H.; Viljoen-Bloom, M. Differential malic acid degradation by selected strains of *Saccharomyces* during alcoholic fermentation. *Int. J. Food Microbiol.* **2003**, *83*, 49–61. [CrossRef]

25. Zelle, R.M.; De Hulster, E.; Van Winden, W.A.; De Waard, P.; Dijkema, C.; Winkler, A.A.; Geertman, J.M.A.; Van Dijken, J.P.; Pronk, J.T.; Van Maris, A.J.A. Malic acid production by *Saccharomyces cerevisiae*: Engineering of pyruvate carboxylation, oxaloacetate reduction, and malate export. *Appl. Environ. Microbiol.* **2008**, *74*, 2766–2777. [CrossRef]

26. Benito, Á.; Calderón, F.; Benito, S. Mixed alcoholic fermentation of *Schizosaccharomyces pombe* and *Lachancea thermotolerans* and its influence on mannose-containing polysaccharides wine Composition. *AMB Express* **2019**, *9*, 17–25. [CrossRef]

27. Lambrechts, M.G.; Pretorius, I.S. Yeast and its Importance to Wine Aroma - A Review. *S. Afr. J. Enol. Vitic.* **2000**, *21*, 97–129. [CrossRef]

28. Du Plessis, H.W.; du Toit, M.; Hoff, J.W.; Hart, R.S.; Ndimba, B.K.; Jolly, N.P. Characterisation of non-*Saccharomyces* yeasts using different methodologies and evaluation of their compatibility with malolactic fermentation. *S Afr. J. Enol. Vitic.* **2017**, *38*, 46–63. [CrossRef]

29. Benito, S.; Palomero, F.; Calderón, F.; Palmero, D.; Suárez-Lepe, J.A. Selection of appropriate *Schizosaccharomyces* strains for winemaking. *Food Microbiol.* **2014**, *42*, 218–224. [CrossRef]

30. Benito, A.; Jeffares, D.; Palomero, F.; Calderón, F.; Bai, F.-Y.; Bähler, J.; Benito, S. Selected *Schizosaccharomyces pombe* strains have characteristics that are beneficial for winemaking. *PLoS ONE* **2016**, *11*, e0151102. [CrossRef]

31. Escribano, R.; González-Arenzana, L.; Portu, J.; Garijo, P.; López-Alfaro, I.; López, R.; Santamaría, P.; Gutiérrez, A.R. Wine aromatic compound production and fermentative behaviour within different non-*Saccharomyces* species and clones. *J. Appl. Microbiol.* **2018**, *124*, 1521–1531. [CrossRef]

32. Benito, S.; Hofmann, T.; Laier, M.; Lochbühler, B.; Schüttler, A.; Ebert, K.; Fritsch, S.; Röcker, J.; Rauhut, D. Effect on quality and composition of Riesling wines fermented by sequential inoculation with non-*Saccharomyces* and *Saccharomyces cerevisiae*. *Eur. Food Res. Technol.* **2015**, *241*, 707–717. [CrossRef]

33. Belda, I.; Ruiz, J.; Esteban-Fernández, A.; Navascués, E.; Marquina, D.; Santos, A.; Moreno-Arribas, M.V. Microbial contribution to Wine aroma and its intended use for Wine quality improvement. *Molecules* **2017**, *22*, 189. [CrossRef]

34. Chen, K.; Escott, C.; Loira, I.; del Fresno, J.M.; Morata, A.; Tesfaye, W.; Calderon, F.; Suárez-Lepe, J.A.; Han, S.; Benito, S. Use of non-*Saccharomyces* yeasts and oenological tannin in red winemaking: Influence on colour, aroma and sensorial properties of young wines. *Food Microbiol.* **2018**, *69*, 51–63. [CrossRef]

35. Benito, A.; Calderón, F.; Benito, S. The combined use of *Schizosaccharomyces pombe* and *Lachancea thermotolerans* - Effect on the anthocyanin wine composition. *Molecules* **2017**, *22*, 739. [CrossRef]

36. Benito, Á.; Calderón, F.; Palomero, F.; Benito, S. Combine use of selected *Schizosaccharomyces pombe* and *Lachancea thermotolerans* yeast strains as an alternative to the traditional malolactic fermentation in red wine production. *Molecules* **2015**, *20*, 9510–9523. [CrossRef]

37. Mylona, A.E.; Del Fresno, J.M.; Palomero, F.; Loira, I.; Bañuelos, M.A.; Morata, A.; Calderón, F.; Benito, S.; Suárez-Lepe, J.A. Use of *Schizosaccharomyces* strains for wine fermentation-Effect on the wine composition and food safety. *Int. J. Food Microbiol.* **2016**, *232*, 63–72. [CrossRef]

38. Benito, S. The Management of Compounds that Influence Human Health in Modern Winemaking from an HACCP Point of View. *Fermentation* **2019**, *5*, 33. [CrossRef]

39. Gambuti, A.; Strollo, D.; Genovese, A.; Ugliano, M.; Ritieni, A.; Moio, L. Influence of enological practices on ochratoxin A concentration in wine. *Am. J. Enol. Vitic.* **2005**, *56*, 155–162.

40. Cecchini, F.; Morassut, M.; Garcia Moruno, E.; Di Stefano, R. Influence of yeast strain on ochratoxin A content during fermentation of white and red must. *Food Microbiol.* **2006**, *23*, 411–417. [CrossRef]

41. Meca, G.; Blaiotta, G.; Ritieni, A. Reduction of ochratoxin A during the fermentation of Italian red wine Moscato. *Food Control.* **2010**, *21*, 579–583. [CrossRef]

42. Gil-Serna, J.; Vázquez, C.; González-Jaén, M.; Patiño, B. Wine Contamination with Ochratoxins: A Review. *Beverages* **2018**, *4*, 6. [CrossRef]

43. Vidal, S.; Williams, P.; Doco, T.; Moutounet, M.; Pellerin, P. The polysaccharides of red wine: Total fractionation and characterization. *Carbohydr. Polym.* **2003**, *54*, 439–447. [CrossRef]

44. Vidal, S.; Courcoux, P.; Francis, L.; Kwiatkowski, M.; Gawel, R.; Williams, P.; Waters, E.; Cheynier, V. Use of an experimental design approach for evaluation of key wine components on mouth-feel perception. *Food Qual. Prefer.* **2004**, *15*, 209–217. [CrossRef]

45. Vidal, S.; Francis, L.; Williams, P.; Kwiatkowski, M.; Gawel, R.; Cheynier, V.; Waters, E. The mouth-feel properties of polysaccharides and anthocyanins in a wine like medium. *Food Chem.* **2004**, *85*, 519–525. [CrossRef]

46. Juega, M.; Nunez, Y.P.; Carrascosa, A.V.; Martinez-Rodriguez, A.J. Influence of Yeast Mannoproteins in the Aroma Improvement of White Wines. *J. Food Sci.* **2012**, *77*, M499–M504. [CrossRef]

47. Domizio, P.; Liu, Y.; Bisson, L.F.; Barile, D. Use of non-*Saccharomyces* wine yeasts as novel sources of mannoproteins in wine. *Food Microbiol.* **2014**, *43*, 5–15. [CrossRef]

48. Belda, I.; Navascués, E.; Marquina, D.; Santos, A.; Calderon, F.; Benito, S. Dynamic analysis of physiological properties of *Torulaspora delbrueckii* in wine fermentations and its incidence on wine quality. *Appl. Microbiol. Biotechnol.* **2015**, *99*, 1911–1922. [CrossRef]

49. Belda, I.; Navascués, E.; Marquina, D.; Santos, A.; Calderón, F.; Benito, S. Outlining the influence of non-conventional yeasts in wine ageing over lees. *Yeast* **2016**, *33*, 329–338. [CrossRef]

50. García, M.; Apolinar-Valiente, R.; Williams, P.; Esteve-Zarzoso, B.; Arroyo, T.; Crespo, J.; Doco, T. Polysaccharides and Oligosaccharides Produced on Malvar Wines Elaborated with *Torulaspora delbrueckii* CLI 918 and *Saccharomyces cerevisiae* CLI 889 Native Yeasts from D.O. "vinos de Madrid". *J. Agric. Food Chem.* **2017**, *65*, 6656–6664.

51. Domizio, P.; House, J.F.; Joseph, C.M.L.; Bisson, L.F.; Bamforth, C.W. *Lachancea thermotolerans* as an alternative yeast for the production of beer. *J. Inst. Brew.* **2016**, *122*, 599–604. [CrossRef]

52. Domizio, P.; Liu, Y.; Bisson, L.F.; Barile, D. Cell wall polysaccharides released during the alcoholic fermentation by *Schizosaccharomyces pombe* and *S. japonicus*: Quantification and characterization. *Food Microbiol.* **2017**, *61*, 136–149. [CrossRef]

53. Domizio, P.; Lencioni, L.; Calamai, L.; Portaro, L.; Bisson, L.F. Evaluation of the yeast *Schizosaccharomyces japonicus* for use in wine production. *Am. J. Enol. Vitic.* **2018**, *69*, 266–277. [CrossRef]

54. Regodón Mateos, J.A.; Pérez-Nevado, F.; Ramírez Fernández, M. Influence of *Saccharomyces cerevisiae* yeast strain on the major volatile compounds of wine. *Enzyme Microb. Technol.* **2006**, *40*, 151–157. [CrossRef]

55. Aguera, E.; Sire, Y.; Mouret, J.R.; Sablayrolles, J.M.; Farines, V. Comprehensive Study of the Evolution of the Gas-Liquid Partitioning of Acetaldehyde during Wine Alcoholic Fermentation. *J. Agric. Food Chem.* **2018**, *66*, 6170–6178. [CrossRef] [PubMed]

56. Ciani, M.; Beco, L.; Comitini, F. Fermentation behaviour and metabolic interactions of multistarter wine yeast fermentations. *Int. J. Food Microbiol.* **2006**, *10*, 239–245. [CrossRef] [PubMed]

57. Benito, A.; Calderón, F.; Benito, S. Combined use of *S. pombe* and *L. thermotolerans* in winemaking. Beneficial effects determined through the study of wines' analytical characteristics. *Molecules* **2016**, *21*, 1744. [CrossRef] [PubMed]

58. Scanes, K.T.; Hohrnann, S.; Prior, B.A. Glycerol Production by the Yeast *Saccharomyces cerevisiae* and its Relevance to Wine: A Review. *S. Afr. J. Enol. Vitic.* **2017**, *19*, 17–24. [CrossRef]

59. García, M.; Esteve-Zarzoso, B.; Cabellos, J.; Arroyo, T. Advances in the Study of *Candida stellata*. *Fermentation* **2018**, *4*, 74. [CrossRef]

60. Jolly, N.P.; Augustyn, O.P.H.; Pretorius, I.S. The Role and Use of Non-*Saccharomyces* Yeasts in Wine Production. *S. Afr. J. Enol. Vitic.* **2017**, *27*, 15–39. [CrossRef]

61. Escribano, R.; González-Arenzana, L.; Garijo, P.; Berlanas, C.; López-Alfaro, I.; López, R.; Gutiérrez, A.R.; Santamaría, P. Screening of enzymatic activities within different enological non-*Saccharomyces* yeasts. *J. Food Sci. Technol.* **2017**, *54*, 1555–1564. [CrossRef]

62. Escribano-Viana, R.; González-Arenzana, L.; Portu, J.; Garijo, P.; López-Alfaro, I.; López, R.; Santamaría, P.; Gutiérrez, A.R. Wine aroma evolution throughout alcoholic fermentation sequentially inoculated with non-*Saccharomyces/Saccharomyces* yeasts. *Food Res. Int.* **2018**, *112*, 17–24. [CrossRef]

63. Andorra, I.; Monteiro, M.; Esteve-Zarzoso, B.; Albergaria, H.; Mas, A. Analysis and direct quantification of *Saccharomyces cerevisiae* and *Hanseniaspora guilliermondii* populations during alcoholic fermentation by fluorescence in situ hybridization, flow cytometry and quantitative PCR. *Food Microbiol.* **2011**, *28*, 1483–1491. [CrossRef] [PubMed]

64. González, S.S.; Alcoba-Flórez, J.; Laich, F. *Lachancea lanzarotensis* sp. nov., an ascomycetous yeast isolated from grapes and wine fermentation in Lanzarote, Canary Islands. *Int. J. Syst. Evol. Microbiol.* **2013**, *63*, 358–363. [CrossRef] [PubMed]

65. Sipiczki, M.; Pfliegler, W.P.; Holb, I.J. *Metschnikowia* Species Share a Pool of Diverse rRNA Genes Differing in Regions That Determine Hairpin-Loop Structures and Evolve by Reticulation. *PLoS ONE* **2013**, *8*, e67384. [CrossRef] [PubMed]

66. Freel, K.C.; Friedrich, A.; Hou, J.; Schacherer, J. Population genomic analysis reveals highly conserved mitochondrial genomes in the yeast species *Lachancea thermotolerans*. *Genome Biol. Evol.* **2014**, *6*, 2586–2594. [CrossRef] [PubMed]

67. Albertin, W.; Chasseriaud, L.; Comte, G.; Panfili, A.; Delcamp, A.; Salin, F.; Marullo, P.; Bely, M. Winemaking and bioprocesses strongly shaped the genetic diversity of the ubiquitous yeast *Torulaspora delbrueckii*. *PLoS ONE* **2014**, *9*, e94246. [CrossRef] [PubMed]

68. Jeffares, D.C.; Rallis, C.; Rieux, A.; Speed, D.; Převorovský, M.; Mourier, T.; Marsellach, F.X.; Iqbal, Z.; Lau, W.; Cheng, T.M.K.; et al. The genomic and phenotypic diversity of *Schizosaccharomyces pombe*. *Nat. Genet.* **2015**, *47*, 235–241. [CrossRef] [PubMed]

69. Hranilovic, A.; Bely, M.; Masneuf-Pomarede, I.; Jiranek, V.; Albertin, W. The evolution of *Lachancea thermotolerans* is driven by geographical determination, anthropisation and flux between different ecosystems. *PLoS ONE* **2017**, *12*, e0184652. [CrossRef]

70. Bely, M.; Stoeckle, P.; Masneuf-Pomarède, I.; Dubourdieu, D. Impact of mixed *Torulaspora delbrueckii-Saccharomyces cerevisiae* culture on high-sugar fermentation. *Int. J. Food Microbiol.* **2008**, *122*, 312–320. [CrossRef]

71. Liu, S.; Laaksonen, O.; Kortesniemi, M.; Kalpio, M.; Yang, B. Chemical composition of bilberry wine fermented with non-*Saccharomyces* yeasts (*Torulaspora delbrueckii* and *Schizosaccharomyces pombe*) and *Saccharomyces cerevisiae* in pure, sequential and mixed fermentations. *Food Chem.* **2018**, *266*, 262–274. [CrossRef]

72. Belda, I.; Ruiz, J.; Beisert, B.; Navascués, E.; Marquina, D.; Calderón, F.; Rauhut, D.; Benito, S.; Santos, A. Influence of *Torulaspora delbrueckii* in varietal thiol (3-SH and 4-MSP) release in wine sequential fermentations. *Int. J. Food Microbiol.* **2017**, *257*, 183–191. [CrossRef]

73. Puertas, B.; Jiménez, M.J.; Cantos-Villar, E.; Cantoral, J.M.; Rodríguez, M.E. Use of *Torulaspora delbrueckii* and *Saccharomyces cerevisiae* in semi-industrial sequential inoculation to improve quality of Palomino and Chardonnay wines in warm climates. *J. Appl. Microbiol.* **2017**, *122*, 733–746. [CrossRef] [PubMed]

74. González-Royo, E.; Pascual, O.; Kontoudakis, N.; Esteruelas, M.; Esteve-Zarzoso, B.; Mas, A.; Canals, J.M.; Zamora, F. Oenological consequences of sequential inoculation with non-*Saccharomyces* yeasts (*Torulaspora delbrueckii* or *Metschnikowia pulcherrima*) and *Saccharomyces cerevisiae* in base wine for sparkling wine production. *Eur. Food Res. Technol.* **2015**, *240*, 999–1012. [CrossRef]

75. Azzolini, M.; Fedrizzi, B.; Tosi, E.; Finato, F.; Vagnoli, P.; Scrinzi, C.; Zapparoli, G. Effects of *Torulaspora delbrueckii* and *Saccharomyces cerevisiae* mixed cultures on fermentation and aroma of Amarone wine. *Eur. Food Res. Technol.* **2012**, *235*, 303–313. [CrossRef]

76. Azzolini, M.; Tosi, E.; Lorenzini, M.; Finato, F.; Zapparoli, G. Contribution to the aroma of white wines by controlled *Torulaspora delbrueckii* cultures in association with *Saccharomyces cerevisiae*. *World J. Microbiol. Biotechnol.* **2015**, *31*, 277–293. [CrossRef] [PubMed]

77. Renault, P.; Coulon, J.; de Revel, G.; Barbe, J.C.; Bely, M. Increase of fruity aroma during mixed *T. delbrueckii/S. cerevisiae* wine fermentation is linked to specific esters enhancement. *Int. J. Food Microbiol.* **2015**, *207*, 40–48. [CrossRef] [PubMed]

78. Whitener, M.E.B.; Stanstrup, J.; Carlin, S.; Divol, B.; Du Toit, M.; Vrhovsek, U. Effect of non-*Saccharomyces* yeasts on the volatile chemical profile of Shiraz wine. *Aust. J. Grape Wine Res.* **2017**, *23*, 179–192. [CrossRef]

79. Renault, P.; Coulon, J.; Moine, V.; Thibon, C.; Bely, M. Enhanced 3-sulfanylhexan-1-ol production in sequential mixed fermentation with *Torulaspora delbrueckii/Saccharomyces cerevisiae* reveals a situation of synergistic interaction between two industrial strains. *Front. Microbiol.* **2016**, *7*, 293–303. [CrossRef]

80. Porter, T.J.; Divol, B.; Setati, M.E. *Lachancea* yeast species: Origin, biochemical characteristics and oenological significance. *Food Res. Int.* **2019**, *119*, 378–389. [CrossRef]

81. Benito, Á.; Calderón, F.; Palomero, F.; Benito, S. Quality and composition of airén wines fermented by sequential inoculation of *Lachancea thermotolerans* and *Saccharomyces cerevisiae*. *Food Technol. Biotechnol.* **2016**, *54*, 135–144. [CrossRef]

82. Gobbi, M.; Comitini, F.; Domizio, P.; Romani, C.; Lencioni, L.; Mannazzu, I.; Ciani, M. *Lachancea thermotolerans* and *Saccharomyces cerevisiae* in simultaneous and sequential co-fermentation: A strategy to enhance acidity and improve the overall quality of wine. *Food Microbiol.* **2013**, *33*, 271–281. [CrossRef]

83. Hranilovic, A.; Gambetta, J.M.; Schmidtke, L.; Boss, P.K.; Grbin, P.R.; Masneuf-Pomarede, I.; Bely, M.; Albertin, W.; Jiranek, V. Oenological traits of *Lachancea thermotolerans* show signs of domestication and allopatric differentiation. *Sci. Rep.* **2018**, *8*, 14812–14825. [CrossRef] [PubMed]

84. Kapsopoulou, K.; Kapaklis, A.; Spyropoulos, H. Growth and fermentation characteristics of a strain of the wine yeast *Kluyveromyces thermotolerans* isolated in Greece. *World J. Microbiol. Biotechnol.* **2005**, *21*, 1599–1602. [CrossRef]

85. Vilela, A. *Lachancea thermotolerans*, the Non-*Saccharomyces* Yeast that Reduces the Volatile Acidity of Wines. *Fermentation* **2018**, *4*, 56. [CrossRef]

86. Kapsopoulou, K.; Mourtzini, A.; Anthoulas, M.; Nerantzis, E. Biological acidification during grape must fermentation using mixed cultures of *Kluyveromyces thermotolerans* and *Saccharomyces cerevisiae*. *World J. Microbiol. Biotechnol.* **2007**, *23*, 735–739. [CrossRef]

87. Shekhawat, K.; Porter, T.J.; Bauer, F.F.; Setati, M.E. Employing oxygen pulses to modulate *Lachancea thermotolerans*–*Saccharomyces cerevisiae* Chardonnay fermentations. *Ann. Microbiol.* **2018**, *68*, 93–102. [CrossRef]

88. Balikci, E.K.; Tanguler, H.; Jolly, N.P.; Erten, H. Influence of *Lachancea thermotolerans* on cv. Emir wine fermentation. *Yeast* **2016**, *33*, 313–321. [CrossRef] [PubMed]

89. Hranilovic, A.; Li, S.; Boss, P.K.; Bindon, K.; Ristic, R.; Grbin, P.R.; Van der Westhuizen, T.; Jiranek, V. Chemical and sensory profiling of Shiraz wines co-fermented with commercial non-*Saccharomyces* inocula. *Aust. J. Grape Wine Res.* **2018**, *24*, 166–180. [CrossRef]

90. Ponsone, M.L.; Chiotta, M.L.; Combina, M.; Dalcero, A.; Chulze, S. Biocontrol as a strategy to reduce the impact of ochratoxin A and *Aspergillus* section Nigri in grapes. *Int. J. Food Microbiol.* **2011**, *151*, 70–77. [CrossRef] [PubMed]

91. Ponsone, M.L.; Nally, M.C.; Chiotta, M.L.; Combina, M.; Köhl, J.; Chulze, S.N. Evaluation of the effectiveness of potential biocontrol yeasts against black sur rot and ochratoxin A occurring under greenhouse and field grape production conditions. *Biol. Control.* **2016**, *103*, 78–85. [CrossRef]

92. Porter, T.J.; Divol, B.; Setati, M.E. Investigating the biochemical and fermentation attributes of *Lachancea* species and strains: Deciphering the potential contribution to wine chemical composition. *Int. J. Food Microbiol.* **2019**, *290*, 273–287. [CrossRef]

93. Benito, S.; Palomero, F.; Calderón, F.; Palmero, D.; Suárez-Lepe, J.A. *Schizosaccharomyces*; 2014; ISBN 9780123847331.

94. Benito, S.; Palomero, F.; Morata, A.; Calderón, F.; Palmero, D.; Suárez-Lepe, J.A. Physiological features of *Schizosaccharomyces pombe* of interest in making of white wines. *Eur. Food Res. Technol.* **2013**, *236*, 29–36. [CrossRef]

95. Peinado, R.A.; Mauricio, J.C.; Medina, M.; Moreno, J.J. Effect of *Schizosaccharomyces pombe* on aromatic compounds in dry sherry wines containing high levels of gluconic acid. *J. Agric. Food Chem.* **2004**, *52*, 4529–4534. [CrossRef] [PubMed]

96. Peinado, R.A.; Maestre, O.; Mauricio, J.C.; Moreno, J.J. Use of a *Schizosaccharomyces pombe* mutant to reduce the content in gluconic acid of must obtained from rotten grapes. *J. Agric. Food Chem.* **2009**, *57*, 2368–2377. [CrossRef] [PubMed]

97. Peinado, R.A.; Moreno, J.J.; Medina, M.; Mauricio, J.C. Potential application of a glucose-transport-deficient mutant of *Schizosaccharomyces pombe* for removing gluconic acid from grape must. *J. Agric. Food Chem.* **2005**, *53*, 1017–1021. [CrossRef] [PubMed]

98. Peinado, R.A.; Moreno, J.J.; Maestre, O.; Ortega, J.M.; Medina, M.; Mauricio, J.C. Gluconic Acid Consumption in Wines by *Schizosaccharomyces pombe* and Its Effect on the Concentrations of Major Volatile Compounds and Polyols. *J. Agric. Food Chem.* **2007**, *52*, 493–497. [CrossRef] [PubMed]

99. Peinado, R.A.; Moreno, J.J.; Maestre, O.; Mauricio, J.C. Removing gluconic acid by using different treatments with a *Schizosaccharomyces pombe* mutant: Effect on fermentation byproducts. *Food Chem.* **2007**, *104*, 457–465. [CrossRef]

100. Hu, C.K.; Bai, F.W.; An, L.J. Enhancing ethanol tolerance of a self-flocculating fusant of *Schizosaccharomyces pombe* and *Saccharomyces cerevisiae* by Mg2+ via reduction in plasma membrane permeability. *Biotechnol. Lett.* **2003**, *25*, 1191–1194. [CrossRef]

101. Silva, S.; Ramón-Portugal, F.; Andrade, P.; Abreu, S.; Texeira, M.D.F.; Strehaiano, P. Malic acid consumption by dry immobilized cells of *Schizosaccharomyces pombe*. *Am. J. Enol. Vitic.* **2003**, *54*, 50–55.

102. Roca-Domènech, G.; Cordero-Otero, R.; Rozès, N.; Cléroux, M.; Pernet, A.; Mira de Orduña, R. Metabolism of *Schizosaccharomyces pombe* under reduced osmotic stress conditions afforded by fed-batch alcoholic fermentation of white grape must. *Food Res. Int.* **2018**, *113*, 401–406. [CrossRef]

103. Taillandier, P.; Strehaiano, P. The role of malic acid in the metabolism of *Schizosaccharomyces pombe*: Substrate consumption and cell growth. *Appl. Microbiol. Biotechnol.* **1991**, *35*, 541–543. [CrossRef]

104. Burns, T.R.; Osborne, J.P. Loss of pinot noir wine color and polymeric pigment after malolactic fermentation and potential causes. *Am. J. Enol. Vitic.* **2015**, *66*, 130–137. [CrossRef]

105. Benito, S.; Palomero, F.; Morata, A.; Calderón, F.; Suárez-Lepe, J.A. New applications for *Schizosaccharomyces pombe* in the alcoholic fermentation of red wines. *Int. J. Food Sci. Technol.* **2012**, *47*, 2101–2108. [CrossRef]

106. Choi, J.G.; Kim, S.Y.; Jeong, M.; Oh, M.S. Pharmacotherapeutic potential of ginger and its compounds in age-related neurological disorders. *Pharmacol. Ther.* **2018**, *182*, 56–69. [CrossRef] [PubMed]

107. Huh, E.; Lim, S.; Kim, H.G.; Ha, S.K.; Park, H.Y.; Huh, Y.; Oh, M.S. Ginger fermented with: *Schizosaccharomyces pombe* alleviates memory impairment via protecting hippocampal neuronal cells in amyloid beta1-42 plaque injected mice. *Food Funct.* **2018**, *9*, 171–178.

108. Satora, P.; Semik-Szczurak, D.; Tarko, T.; Bułdys, A. Influence of Selected *Saccharomyces* and *Schizosaccharomyces* Strains and Their Mixed Cultures on Chemical Composition of Apple Wines. *J. Food Sci.* **2018**, *83*, 424–431. [CrossRef] [PubMed]

109. Minnaar, P.P.; Jolly, N.P.; Paulsen, V.; Du Plessis, H.W.; Van Der Rijst, M. *Schizosaccharomyces pombe* and *Saccharomyces cerevisiae* yeasts in sequential fermentations: Effect on phenolic acids of fermented Kei-apple (Dovyalis caffra L.) juice. *Int. J. Food Microbiol.* **2017**, *257*, 232–237. [CrossRef] [PubMed]

110. Ivit, N.N.; Loira, I.; Morata, A.; Benito, S.; Palomero, F.; Suárez-Lepe, J.A. Making natural sparkling wines with non-*Saccharomyces* yeasts. *Eur. Food Res. Technol.* **2018**, *244*, 925–935. [CrossRef]

111. Bakhiet, S.; Mahmoud, M. Production of Bio-ethanol from Molasses by *Schizosaccharomyces* Species. *Annu. Res. Rev. Biol.* **2015**, *7*, 45–53. [CrossRef]

112. Miljić, U.; Puškaš, V.; Vučurović, V.; Muzalevski, A. Fermentation Characteristics and Aromatic Profile of Plum Wines Produced with Indigenous Microbiota and Pure Cultures of Selected Yeast. *J. Food Sci.* **2017**, *82*, 1443–1450. [CrossRef]

113. Varela, C.; Barker, A.; Tran, T.; Borneman, A.; Curtin, C. Sensory profile and volatile aroma composition of reduced alcohol Merlot wines fermented with *Metschnikowia pulcherrima* and *Saccharomyces uvarum*. *Int. J. Food Microbiol.* **2017**, *252*, 1–9. [CrossRef]

114. Varela, C.; Sengler, F.; Solomon, M.; Curtin, C. Volatile flavour profile of reduced alcohol wines fermented with the non-conventional yeast species *Metschnikowia pulcherrima* and *Saccharomyces uvarum*. *Food Chem.* **2016**, *209*, 57–64. [CrossRef] [PubMed]

115. Benito, S.; Morata, A.; Palomero, F.; González, M.C.; Suárez-Lepe, J.A. Formation of vinylphenolic pyranoanthocyanins by *Saccharomyces cerevisiae* and *Pichia guillermondii* in red wines produced following different fermentation strategies. *Food Chem.* **2011**, *124*, 15–23. [CrossRef]

116. Benito, S.; Palomero, F.; Morata, A.; Uthurry, C.; Suárez-Lepe, J.A. Minimization of ethylphenol precursors in red wines via the formation of pyranoanthocyanins by selected yeasts. *Int. J. Food Microbiol.* **2009**, *132*, 145–152. [CrossRef] [PubMed]

117. Ciani, M.; Ferraro, L. Combined use of immobilized *Candida stellata* cells and *Saccharomyces cerevisiae* to improve the quality of wines. *J. Appl. Microbiol.* **1998**, *85*, 247–254. [CrossRef] [PubMed]

118. Ciani, M.; Maccarelli, F. Oenological properties of non-*Saccharomyces* yeasts associated with wine-making. *World J. Microbiol. Biotechnol.* **1998**, *14*, 199–203. [CrossRef]

119. Magyar, I.; Tóth, T. Comparative evaluation of some oenological properties in wine strains of *Candida stellata, Candida zemplinina, Saccharomyces uvarum* and *Saccharomyces cerevisiae*. *Food Microbiol.* **2011**, *28*, 94–100. [CrossRef] [PubMed]

120. Di Maio, S.; Genna, G.; Gandolfo, V.; Amore, G.; Ciaccio, M.; Oliva, D. Presence of *Candida zemplinina* in sicilian musts and selection of a strain for wine mixed fermentations. *S. Afr. J. Enol. Vitic.* **2012**, *33*, 80–87. [CrossRef]

121. Pretorius, I.S. Tailoring wine yeast for the new millennium: Novel approaches to the ancient art of winemaking. *Yeast* **2000**, *16*, 675–729. [CrossRef]

122. Zohre, D.E.; Erten, H. The influence of *Kloeckera* apiculata and *Candida pulcherrima* yeasts on wine fermentation. *Process. Biochem.* **2002**, *38*, 319–324. [CrossRef]

123. López, S.; Mateo, J.; Maicas, S. Screening of *Hanseniaspora* Strains for the Production of Enzymes with Potential Interest for Winemaking. *Fermentation* **2015**, *2*, 1. [CrossRef]

124. Lleixà, J.; Martín, V.; Portillo, M.d.C.; Carrau, F.; Beltran, G.; Mas, A. Comparison of fermentation and wines produced by inoculation of *Hanseniaspora vineae* and *Saccharomyces cerevisiae*. *Front. Microbiol.* **2016**, *7*, 338–350.

125. Martin, V.; Valera, M.; Medina, K.; Boido, E.; Carrau, F. Oenological Impact of the *Hanseniaspora/Kloeckera* Yeast Genus on Wines—A Review. *Fermentation* **2018**, *4*, 76. [CrossRef]

126. Pina, C.; Santos, C.; Couto, J.A.; Hogg, T. Ethanol tolerance of five non-*Saccharomyces* wine yeasts in comparison with a strain of *Saccharomyces cerevisiae* - Influence of different culture conditions. *Food Microbiol.* **2004**, *21*, 439–447. [CrossRef]

127. Hu, K.; Jin, G.J.; Xu, Y.H.; Tao, Y.S. Wine aroma response to different participation of selected *Hanseniaspora uvarum* in mixed fermentation with *Saccharomyces cerevisiae*. *Food Res. Int.* **2018**, *108*, 119–127. [CrossRef] [PubMed]

128. Garijo, P.; González-Arenzana, L.; López-Alfaro, I.; Garde-Cerdán, T.; López, R.; Santamaría, P.; Gutiérrez, A.R. Analysis of grapes and the first stages of the vinification process in wine contamination with *Brettanomyces bruxellensis*. *Eur. Food Res. Technol.* **2014**, *240*, 525–532. [CrossRef]

129. Pérez, G.; Fariña, L.; Barquet, M.; Boido, E.; Gaggero, C.; Dellacassa, E.; Carrau, F. A quick screening method to identify β-glucosidase activity in native wine yeast strains: Application of Esculin Glycerol Agar (EGA) medium. *World J. Microbiol. Biotechnol.* **2011**, *27*, 47–55. [CrossRef]

130. Medina, K.; Boido, E.; Fariña, L.; Gioia, O.; Gomez, M.E.; Barquet, M.; Gaggero, C.; Dellacassa, E.; Carrau, F. Increased flavour diversity of Chardonnay wines by spontaneous fermentation and co-fermentation with *Hanseniaspora vineae*. *Food Chem.* **2013**, *141*, 2513–2521. [CrossRef]

131. Viana, F.; Belloch, C.; Vallés, S.; Manzanares, P. Monitoring a mixed starter of *Hanseniaspora vineae-Saccharomyces cerevisiae* in natural must: Impact on 2-phenylethyl acetate production. *Int. J. Food Microbiol.* **2011**, *151*, 235–240. [CrossRef]

132. Tristezza, M.; Tufariello, M.; Capozzi, V.; Spano, G.; Mita, G.; Grieco, F. The oenological potential of hanseniaspora uvarum in simultaneous and sequential co-fermentation with *Saccharomyces cerevisiae* for industrial wine production. *Front. Microbiol.* **2016**, *7*, 670–684. [CrossRef]

133. Hall, H.; Zhou, Q.; Qian, M.C.; Osborne, J.P. Impact of yeasts present during prefermentation cold maceration of pinot noir grapes on wine volatile aromas. *Am. J. Enol. Vitic.* **2017**, *68*, 81–90. [CrossRef]

134. Hu, K.; Jin, G.J.; Mei, W.C.; Li, T.; Tao, Y.S. Increase of medium-chain fatty acid ethyl ester content in mixed *H. uvarum/S. cerevisiae* fermentation leads to wine fruity aroma enhancement. *Food Chem.* **2018**, *239*, 495–501. [CrossRef] [PubMed]

135. Martin, V.; Giorello, F.; Fariña, L.; Minteguiaga, M.; Salzman, V.; Boido, E.; Aguilar, P.S.; Gaggero, C.; Dellacassa, E.; Mas, A.; et al. De novo synthesis of benzenoid compounds by the yeast *Hanseniaspora vineae* increases the flavor diversity of wines. *J. Agric. Food Chem.* **2016**, *64*, 4574–4583. [CrossRef] [PubMed]

136. Martin, V.; Boido, E.; Giorello, F.; Mas, A.; Dellacassa, E.; Carrau, F. Effect of yeast assimilable nitrogen on the synthesis of phenolic aroma compounds by *Hanseniaspora vineae* strains. *Yeast* **2016**, *33*, 323–328. [CrossRef] [PubMed]

137. Medina, K.; Boido, E.; Dellacassa, E.; Carrau, F. Effects of non-*Saccharomyces* yeasts on color, anthocyanin, and anthocyanin-derived pigments of Tannat grapes during fermentation. *Am. J. Enol. Vitic.* **2018**, *69*, 148–156. [CrossRef]

138. Medina, K.; Boido, E.; Fariña, L.; Dellacassa, E.; Carrau, F. Non-*Saccharomyces* and *Saccharomyces* strains co-fermentation increases acetaldehyde accumulation: Effect on anthocyanin-derived pigments in Tannat red wines. *Yeast* **2016**, *33*, 339–343. [CrossRef] [PubMed]

 *fermentation*

Article

Selection of Native Non-*Saccharomyces* Yeasts with Biocontrol Activity against Spoilage Yeasts in Order to Produce Healthy Regional Wines

Benjamín Kuchen [1,2], Yolanda Paola Maturano [1,2,*], María Victoria Mestre [1,2], Mariana Combina [3], María Eugenia Toro [2] and Fabio Vazquez [2]

[1] Consejo Nacional de Investigaciones Científicas y Técnicas (CONICET), Av. Rivadavia 1917, Ciudad Autónoma de Buenos Aires C1033AAJ, Argentina
[2] Instituto de Biotecnología (IBT), Universidad Nacional de San Juan, Av. San Martín 1109 (O), San Juan 5400, Argentina
[3] Estación Experimental Agropecuaria Mendoza, Instituto Nacional de Tecnología Agropecuaria (INTA), San Martin 3853, 5507 Luján de Cuyo Mendoza, Argentina
* Correspondence: paolamaturano@gmail.com; Tel.: +54-264-421-1700 (ext. 410)

Received: 6 June 2019; Accepted: 4 July 2019; Published: 8 July 2019

Abstract: Two major spoilage yeasts in the wine industry, *Brettanomyces bruxellensis* and *Zygosaccharomyces rouxii*, produce off-flavors and gas, causing considerable economic losses. Traditionally, SO_2 has been used in winemaking to prevent spoilage, but strict regulations are in place regarding its use due to its toxic and allergenic effects. To reduce its usage researchers have been searching for alternative techniques. One alternative is biocontrol, which can be used either independently or in a complementary way to chemical control (SO_2). The present study analyzed 122 native non-*Saccharomyces* yeasts for their biocontrol activity and their ability to be employed under fermentation conditions, as well as certain enological traits. After the native non-*Saccharomyces* yeasts were assayed for their biocontrol activity, 10 biocontroller yeasts were selected and assayed for their ability to prevail in the fermentation medium, as well as with respect to their corresponding positive/negative contribution to the wine. Two yeasts that satisfy these characteristics were *Wickerhamomyces anomalus* BWa156 and *Metschnikowia pulcherrima* BMp29, which were selected for further research in application to mixed fermentations.

Keywords: biocontrol application; non-*Saccharomyces* screening; SO_2 reduction

---

1. Introduction

Wine is the product of complex microbial interactions that start on the grape surface and continue throughout the fermentation [1]. Some yeasts generate metabolites that lead to wine faults that affect flavor, haze or CO_2 production in the final product. One of the major spoilage yeasts is *Brettanomyces bruxellensis* [2]. Wines contaminated with this yeast are characterized by the presence of off-flavors [3]. Other spoilage yeasts frequently described in the food industry belong to the *Zygosaccharomyces* genus. They produce gas in food and beverages [4], and they are difficult to control chemically [5]. Spoilage resulting from this yeast is widespread and causes considerable economic losses in the food industry [6,7].

Traditionally, SO_2 has been used in winemaking during non-fermentation stages to control microbial proliferation such as bacteria, yeasts and fungi. Nevertheless, there are strict regulations regarding its use due to its toxic and allergenic effects on human health [8]. International organizations such as the *Organisation Internationale de la vigne et du vin* encourage SO_2 reduction [9]. Moreover, modern consumers prefer more natural and healthy foods and beverages that are minimally processed and free of preservatives [4,10].

Biocontrol is an alternative proposal that can be used either independently or in a complementary way to chemical control (SO_2). Some *Saccharomyces* and non-*Saccharomyces* yeasts have the ability to biosuppress other yeasts through different mechanisms such as the production of toxic compounds [2], competition for limiting substrates [11] and/or cell to cell contact [1].

At present, a re-evaluation of the role of non-*Saccharomyces* yeasts in winemaking and their use as selected starters in mixed fermentations with *Saccharomyces cerevisiae* is being carried out [12,13]. Non-*Saccharomyces* yeasts are supposed to enhance the wine quality [14]. Nowadays there is a special interest in yeast strains associated with specific geographical locations as they may introduce a regional character or 'terroir' to the winemaking process [12,15].

Yeast growth parameters such as specific growth rate, lag phase duration, product yield and metabolic rates of substrates and products may provide useful information to understand their biocontrol mechanisms and how to use them during the fermentation process. Taking into account that yeast bio-suppression can be associated with substrate competition and secretion of toxic substances, it is important to understand the growth parameters of non-*Saccharomyces* yeasts during fermentation, in order to plan co-inoculation or sequential mixed inoculation with *Saccharomyces* [16–19].

Several authors have analyzed indirect values like "fermentation rate" (CO_2 release) [20,21]. However, there are no reports related to selection of non-*Saccharomyces* yeast for vinification that have studied the prevalence of yeasts with clearly defined kinetic parameters. The aim of the present study was to analyze the biocontrol ability of 122 native non-*Saccharomyces* yeasts against two of the most relevant wine spoilage yeast species, *Z. rouxii* and *B. bruxellensis*. Subsequently, biocontrolling yeasts were characterized for their ability to be employed under fermentation conditions and their capacity to generate positive or negative enological traits, in order to reduce SO_2 and improve the quality of regional wines.

2. Materials and Methods

2.1. Microorganisms

One hundred and twenty-two non-*Saccharomyces* yeasts (Table 1), previously isolated from enological environments from San Juan and Mendoza, Argentina (Cuyo region), were obtained from the Culture Collection of Autochthonous Microorganisms of the Institute of Biotechnology, School of Engineering, UNSJ, San Juan, Argentina, and used in the present study. The yeasts had been used in previous studies by our research group [22,23].

Table 1. Non-*Saccharomyces* yeast isolates assayed.

Species	N° of Isolates	Strain Nomenclature
Candida apis	1	BCa80
Candida cantarelli	1	BCca78
Candida catenulata	1	BCct79
Candida diversa	1	BCd75
Candida famata	4	BCf84
Candida intermedia	1	BCi85
Candida membranifaciens	3	BCm69, BCm70, BCm71
Candida pararugosa	1	BCp73
Candida rugosa	1	BCr81
Candida sake	6	BCsa74, BCsa82, BCsa83, BCsa86, BCsa88, BCsa95
Candida steatolytica	1	BCse76

Table 1. *Cont.*

Species	N° of Isolates	Strain Nomenclature
Candida stellata	1	BCst68
Clavispora lusitaniae	1	BCl157
Cryptococcus albidus	1	BCra158
Debaryomyces hansenii	5	BDb150, BDb152, BDb153, BDb154, BDb155
Debaryomycesvanrijiae	1	BDv151
Hanseniaspora guilliermondii	6	BHg42, BHg44, BHg45, BHg46, BHg47, BHg48
Hanseniaspora osmophila	1	BHo51
Hanseniaspora uvarum	27	BHu1, BHu2, BHu3, BHu5, BHu8, BHu9, BHu10, BHu11, BHu12, BHu13, BHu17, BHu18, BHu19, BHu20, BHu21, BHu23, BHu24, BHu26, BHu27, BHu28, BHu30, BHu31, BHu32, BHu38, BHu40, BHu41
Hanseniaspora vineae	2	BHv43, BHv50
Issatchenkia orientalis	1	BIo160
Metschnikowia pulcherrima	6	BMp4, BMp29, BMp49, BMp151, BMp144, BMp145
Pichia fabianii	1	BPf127
Pichia guilliermondii	1	BPg138
Pichia kluyveri	3	BPkl130, BPkl131, BPkl133
Pichia kudriavzevii	5	BPku128, BPku129, BPku132, BPku134, BPku135
Pichia manshurica	1	BPm125
Pichia membranifaciens	1	BPm136
Pichia occidentalis	21	BPo96, BPo97, BPo98, BPo100, BPo101, BPo102, BPo104, BPo108, BPo110, BPo111, BPo112, BPo113, BPo114, BPo115, BPo116, BPo117, BPo120, BPo121, BPo122, BPo123, BPo124
Starmerella bacillaris	12	BSb52, BSb53, BSb54, BSb55, BSb56, BSb57, BSb58, BSb59, BSb62, BSb63, BSb66, BSb67
Torulaspora delbrueckii	3	BTd147, BTd148, BTd149
Wickerhamomyces anomalus	1	BWa156
TOTAL	122	

Eight spoilage yeasts (4 *Brettanomyces bruxellensis* isolates and 4 *Zygosaccharomyces rouxii* isolates) were obtained from the EEA INTA culture collection, Lujan, Mendoza, Argentina, and used in the study [4,24]. *Saccharomyces cerevisiae* BSc114 [23] was used as positive control with regard to fermentative performance. Isolates were identified through biochemical, physiological and morphological methods [6] as well as molecular methods [25].

2.2. Culture Media

Propagation was carried out in YEPD broth (g/L): Yeast extract 10, peptone, glucose 20, pH: 4.5 adjusted with HCl 1N.

Viable yeast counting was carried out on YEPD-agar (g/L): Yeast extract 10, peptone 20, glucose 20, agar-agar 20, pH: 4.5.

Biocontrol was carried out on YMB-agar supplemented with 0.2 M citrate-phosphate buffer, pH 4.5 (g/L): Glucose 10, yeast extract 3, malt extract 3, peptone 5, NaCl 30, methylene blue 0.030, glycerol 10% *v/v* [26] (modified).

Kinetics and tolerance assays were carried out with concentrated grape must (65 °Brix), diluted at 21 °Brix with 1 g/L of yeast extract added, pH: 4.

Inocula for biocontrol and plate assays were obtained with YEPD broth pH: 4.5, 24 h incubation period.

Inocula for kinetic and tolerances assays were obtained with concentrated grape must (65 °Brix), diluted at 21 °Brix with 1 g/L of yeast extract added, pH: 4, 24 h incubation period.

Complex media was sterilized at 121 °C for 20 min and grape must media at 111 °C for 20 min.

2.3. Screening for Biocontrol Ability of Yeasts

Each spoilage yeast was incorporated at a concentration of 10^6 cells/mL in liquid YMB-agar biocontrol medium at 45 °C, mixed to uniform, and immediately poured into sterile petri dishes. Potential biocontrolling non-*Saccharomyces* yeasts were inoculated as a drop (20 μL) on the agar surface, and plates were incubated at 25 °C until a well-developed lawn. Killer activity was visualized as zone of growth inhibition (more than 1 mm) around the spotted killer yeast colony on plates [2].

Biocontrolling activity against the spoilage species was calculated in 2 ways: (a) Intraspecific inhibition: the proportion at which one biocontrolling strain inhibited the spoilage isolates belonging to one species. In addition, (b) Total inhibition: the proportion at which one biocontroller yeast strain inhibited the spoilage isolates belonging to both species.

2.4. Fermentative Performance

2.4.1. Growth Kinetics during Fermentation

Yeasts were separately cultured in Erlenmeyer flasks (250 mL) containing 200 mL of growth medium. Each isolate was seeded at a concentration of 10^6 cells/mL and incubated at 25 °C for 21 days under static conditions, according to [27] for growth determination by viable cell count. Samples were taken on days 1, 2, 3, 6, 15 and 21. Plates were used for viable cell counts and the experimental data was used to construct a growth curve, which was used to determine kinetic parameters. μmax was calculated as described Monod [28] and the lag phase as described Lodge and Hinshelwood [29], which are the most widespread methods according to [16,17].

2.4.2. Tolerance to Different Stress

Low temperature (15 °C), High concentrations of reducing sugar (30 °Brix), and different Ethanol concentrations (8, 10, 12 and 14% *v/v*) and different molecular SO_2 concentrations (0.1, 0.15, 0.2, 0.3 and 0.4 ppm) were carried out according to Vazquez et al. [27] in tolerance medium for each strain. Growth was monitored with Durham tubes (CO_2 production). Gas production in the Durham tubes was monitored one day after the positive control of each strain. Molecular SO_2 was calculated according to [30,31]. Control was performed at 25 °C, 21 °Brix, 0% *v/v* ethanol and 0 ppm of SO_2.

2.4.3. Plate Assays

SH_2 production: Yeasts were spot-inoculated and evaluated as semi-quantitative over BigGy-agar (BBL[TM], Becton, Dickinson and Company, Sparks, USA, Le Pont de Claix, France) following elaboration instructions. Incubation: 3 days at 25 °C. An arbitrary scale was used for the color of the colony from 1, white color (no production); 2, light brown; 3, brown; 4, dark brown; 5, dark brown/black (high production) [20].

β– glucosidase activity: was performed according to Strauss et al. [32]. Medium containing (g/L): yeast nitrogen base 6.7 (YNB, Difco[TM], Becton, Dickinson and Company, Sparks, USA), Arbutin 5 (Sigma[TM], Sigma-Aldrich, Saint Louis, USA) and agar-agar 2, pH: 5, then autoclaved (121 °C, 20 min). 2 mL of filtered 1% ammonium ferric citrate solution was added to 100 mL media before pouring into plates. Yeast was spot-inoculated. Incubation: 5 days at 30 °C. Activity was positive when a discolored halo of hydrolysis was observed.

Protease activity: was performed according to Comitini et al. [20]. The medium contained (g/L): yeast extract 3, malt extract 3, peptone 5, glucose 10, NaCl 5 and agar-agar 15. In a separate vessel, an equal volume of skimmed milk was prepared with sterile water at 10% p/v. After sterilization both solutions were mixed and poured into sterile petri dishes. Yeast was spot-inoculated. Incubation: 3 days at 25 °C. Activity was observed as a clear halo of hydrolysis.

Pectinase activity: was performed according to Fernandez-Salomäo et al. [33]. The medium contained (g/L): citrus pectin 2, yeast extract 1, KH_2PO_4 0.2, $CaCl_2$ 0.05, $(NH_4)_2SO_4$ 1, $MgSO_4.7H_2O$ 0.8, $MnSO_4$ 0.05, agar-agar 20, pH: 4.5. After sterilization (121 °C, 20 min), it was poured into sterile petri dishes and yeast was spot-inoculated. Incubation: 3 days at 30 °C. After incubation, Lugol solution was added and pectin degradation was observed as a clear halo of hydrolysis.

Pathogenicity: hemolysin production of yeasts was performed according to Manns et al. [34] and Menezes et al. [35], which used Blood agar medium in petri dishes (Britania™, CABA, Argentina) for this purpose. Yeast was spot-inoculated. Incubation: 2 days at 37 °C. Positive activity was observed as a clear zone of hydrolysis.

All assays were carried out using *Saccharomyces cerevisiae* BSc114 as a positive control for biocontrol, sensitivity to inhibition of selected isolates, tolerance to low temperature, and high concentrations of reducing sugars, ethanol and SO_2, and as a negative control for H_2S, β–glucosidase, protease and pectinase activity [23]. The prokaryote *Pseudomonas aeruginosa* BPa987 was used as positive control for hemolysin production of the yeasts [36,37].

2.5. Data Analysis

Each assay was performed independently in triplicate and results are represented as the average of three determinations with the corresponding standard deviation (±SD). Data were tested for normality, homoscedasticity and independence. Parametrical data and significant differences were analyzed by Fisher test. Principal components analysis (PCA) was used to simplify interpretation of the yeast behavior data and is presented in a biplot graph. InfoStat™ -Professional software version 1.5 was used for data analysis.

3. Results and Discussion

To be used as co-inocula together with *Saccharomyces cerevisiae* in wine fermentations, biocontroller yeasts must possess a good specific growth rate and a short lag phase during anaerobiosis to predominate in the medium [18]. In addition, they should not produce any negative attributes to wine, but instead, they should contribute with positive attributes.

3.1. Biocontrol Screening

Non-*Saccharomyces* yeasts are considered to improve the wine complexity and enhance positive traits of regional wines. Several authors have reported that a rational selection of non-*Saccharomyces* yeasts as *S. cerevisiae* co-inoculum improves the quality of wines [21,23,38]. In the present study, 122 non-*Saccharomyces* yeasts belonging to 10 genera and isolated from different enological environments were screened to assess their ability to biocontrol wine spoilage yeasts belonging to *Zygosaccharomyces rouxii* (4 isolates) and *Brettanomyces bruxellensis* (4 isolates) species.

Bioassaying showed that 23 non-*Saccharomyces* yeasts belonging to 6 genera inhibited growth of at least one isolate of either *Brettanomyces bruxellensis* or *Zygosaccharomyces rouxii* (Table 2).

None of the selected biocontrollers inhibited the control (BSc114) lawn development. This fact would allow the application of these yeast isolations in co-inocula with this strain of *S. cerevisiae*. Some of the species used in this work have already been used as biocontrollers of non-*Saccharomyces* yeasts and did not inhibit the development of *S.cerevisiae* [14].

Table 2. Biocontrol of spoilage yeasts by non-*Saccharomyces* yeast isolates.

Yeast specie	Isolate	BBb1	BBb11	BBb20	BBb29	BZr4	BZr6	BZr9	BZr10	BSc114	Intraspecific *B. bruxellensis* inhibition	Intraspecific *Z. rouxii* inhibition	Total Inhibition
Candida intermedia	BCi85	−	−	+	−	+	−	−	−	−	0.25	0.25	0.25
Candida membranifaciens	BCm70	−	−	−	+	−	−	−	−	−	0.25	0	0.13
Candida sake	BCs88	−	−	−	−	+	+	−	−	−	0	0.5	0.25
Candida sake	BCs95	−	−	+	−	+	+	−	−	−	0.25	0.5	0.38
Hanseniaspora uvarum	BHu5	+	−	+	+	+	+	−	−	−	0.75	0.5	0.63
Hanseniaspora uvarum	BHu18	−	−	−	−	+	+	+	−	−	0	0.75	0.38
Hanseniaspora uvarum	BHu23	−	−	−	−	+	+	+	+	−	0	1	0.5
Hanseniaspora uvarum	BHu27	+	+	−	−	−	−	+	−	−	0.5	0.25	0.38
Hanseniaspora uvarum	BHu31	−	−	−	+	−	+	−	+	−	0.25	0.5	0.38
Hanseniaspora uvarum	BHu32	−	−	+	−	−	+	−	+	−	0.25	0.5	0.38
Issatchenkia orientalis	BIo160	−	−	−	−	−	+	−	−	−	0	0.25	0.13
Metschnikowia pulcherrima	BMp4	−	+	+	+	+	+	+	−	−	0.75	0.75	0.75
Metschnikowia pulcherrima	BMp29	+	−	+	−	−	+	+	−	−	0.5	0.5	0.5
Metschnikowia pulcherrima	BMp49	+	+	+	+	+	+	+	−	−	1	0.75	0.88
Metschnikowia pulcherrima	BMp145	−	−	+	−	+	+	+	+	−	0.25	1	0.63
Metschnikowia pulcherrima	BMp151	−	−	−	−	+	−	−	−	−	0	0.25	0.13
Pichia occidentalis	BPo104	−	−	+	−	+	+	+	−	−	0.25	0.75	0.5
Pichia occidentalis	BPo108	−	−	+	−	+	+	+	+	−	0.25	1	0.63
Pichia occidentalis	BPo120	−	−	−	−	−	−	−	+	−	0	0.25	0.13
Pichia guilliermondii	BPg138	+	+	+	+	+	−	−	−	−	1	0.25	0.63
Starmerella bacillaris	BSb57	−	−	+	−	+	+	−	−	−	0.25	0.5	0.38
Starmerella bacillaris	BSb58	−	−	+	−	−	−	−	−	−	0.25	0	0.13
Wickerhamomyces anomalus	BWa156	−	−	+	+	+	+	−	−	−	0.5	0.5	0.5

(+): inhibition of the spoilage yeast, (−): no inhibition of the spoilage yeast ($n = 3$).

Yeast species that showed biocontrol activity in our laboratories have also been cited in other studies as antagonists of different spoilage yeasts, with different mechanisms being involved. *Pichia guilliermondii*, associated with killer toxin production, has been proven to interact with *Penicillium expansum* [39]. *Wickerhamomyces anomalus* has been cited as a *B. bruxellensis* biocontroller, confirming the observations in the present study [40]. Different *W. anomalus* strains have been associated with three killer toxins [39]. This species has also been found to kill a broad range of organisms, including bacteria, hyphomycetes and yeasts [41].

Metschnikowia pulcherrima has been commented on by Oro et al. [14] because of its biocontrol capacity to a wide spectrum of genera like *Hanseniaspora*, *Pichia*, *Torulaspora*, *Zygosaccharomyces*, *Saccharomycodes*, *Candida*, *Issatchenkia*, *Brettanomyces* and *Schizosaccharomyces*, which also confirms our results. The biocontrol mechanism for *M. pulcherrima* would be iron depletion from the medium through binding to pulcherrimic acid [14].

These results can be analyzed from two perspectives: from the spoilage yeast or the antagonistic yeast point of view. Considering spoilage yeasts, the *B. bruxellensis* and *Z. rouxii* isolates analyzed in our study showed different sensitivity to *Candida sake*, *Hanseniaspora uvarum*, *Metschnikowia pulcherrima*, *Pichia occidentalis* and *Starmerella bacillaris* species. Oro et al. [14] reported a similar behavior of spoilage yeasts with different sensitivities to *M. pulcherrima* strains.

Regarding biocontrol isolates, intraspecific biocontrol was observed for BMp49 and BPg138 against all *B. bruxellensis* strains assayed. In the case of *Z. rouxii*, all 4 strains assayed were inhibited by BHu23, BMp145 and BPo108. BHu5, BHu27, BMp4, BMp29, BMp49, BPg138 and BWa156 showed an intraspecific inhibition higher than 0.5 against *B. bruxellensis*, whereas Cs88, Cs95, BHu5, BHu18, BHu23, BHu31, BHu32, BMp4, BMp29, BMp49, BMp145, BPo104, BPo108 and BWa156 demonstrated the same inhibition against *Z. rouxii*. The relevance of wide intraspecific inhibition is the possibility of avoiding adaptation of the spoilage yeast to a particular action mechanism by the antagonistic yeast [42]. In addition, wide interspecific/intergeneric inhibition is also considered a positive factor, because it may control other potential spoilage yeasts not detected in the spoilage analysis [43]. Interspecific/intergeneric biocontrol behavior against *B. bruxellensis* and *Z. rouxii* species was observed for BCi85, BCs95, BHu5, BHu27, BHu31, BHu32, BMp4, BMp29, BMp49, BMp145, BPo104, BPo108, BPg138, BSb57 and BWa156 yeasts; they biocontrolled at least one isolate from each species. Most of the yeasts with interspecific/intergeneric biocontrol showed an intraspecific inhibition of 0.5 or higher. This could be related to a common site of action of the killer toxin [42] or a common biocontrol mode of action affecting yeasts in general, like substrate competition [1].

Hanseniaspora uvarum isolates BHu5 and BHu23, *Metschnikowia pulcherrima* isolates BMp4, BMp29, BMp49 and BMp145, *Pichia guilliermondii* BPg138, *Pichia occidentalis* isolates BPo104 and BPo108 and *Wickerhamomyces anomalus* BWa156 were selected because they showed a total inhibition of 0.50 or more. Except for BHu23, all biocontroller yeasts inhibited at least one strain of both spoilage yeasts. In addition, the 10 isolates were evaluated for their enological characteristics.

3.2. Behavior of the Antagonistic Yeasts

3.2.1. Kinetic Parameters

When selecting non-*Saccharomyces* yeasts for oenological fermentations as a co-inoculum with *S. cerevisiae*, special attention should be paid to their beneficial characteristics to enhance wine quality besides their biocontrolling properties.

To achieve these goals, predominance of the selected non-*Saccharomyces* yeasts in the medium during the first stage of the fermentation is very important. Anaerobic growth kinetics of non-*Saccharomyces* yeasts possess important parameters to elucidate such predominance. Duration of the lag phase (or adaptation) and maximal growth rate are two relevant anaerobiosis parameters, which are described below [16–19].

The present study examined the kinetic parameters of each individual yeast. Nevertheless, in mixed fermentations with grape must, when limiting substrate availability is more prominent compared with the saturation constant, each species will growth at its maximum rate. This is the main parameter to ensure predominance [16], but only when the previous state of the culture (growth stage, age and size of the inoculum) is homogeneous for all experiments [44]. Moreover, this will be governed by the chemical and physical characteristics of the environment unless one of the interacting species produces inhibitory agents against the other [16]. It is also known that the yeast complexes behave differently because of competition, antagonism or cooperation and this could result in the predominance of different yeasts [45,46].

A fermentation growth curve of each antagonist yeast was performed. Viable cell data were recorded to calculate the specific maximal growth rate (μmax) and lag phase. Most of the non-*Saccharomyces* yeasts assayed reached a specific maximal growth rate near 0.04 h^{-1} (Figure 1). BMp29, BMp49 and BMp145 showed a higher μmax which was significantly different. BPg138 and BPo104 displayed a lower μmax which was also significantly different. High specific growth rates are desired because they are a relevant factor in the prevalence of an organism during fermentation [18]. *M. pulcherrima* isolates presented the highest specific growth rates. This behavior could be related to the fact that the mode of action of this species is through the competition of limited substrates and not through a killer factor [14]. The killer factor has been found to consume metabolic energy, reducing the fitness of the yeast that possesses the factor [47], and hence it could decrease the fitness of the other non-*Saccharomyces* yeasts. Growth rates of 0.29 h^{-1} [48], 0.31 h^{-1} [49] and even 0.5 h^{-}1 have been found for *Saccharomyces cerevisiae* during anaerobiosis [50]. Therefore, prevalence of the non-*Saccharomyces* in the medium at the start of the fermentation should be considered for sequential co-inoculation with *S. cerevisiae*.

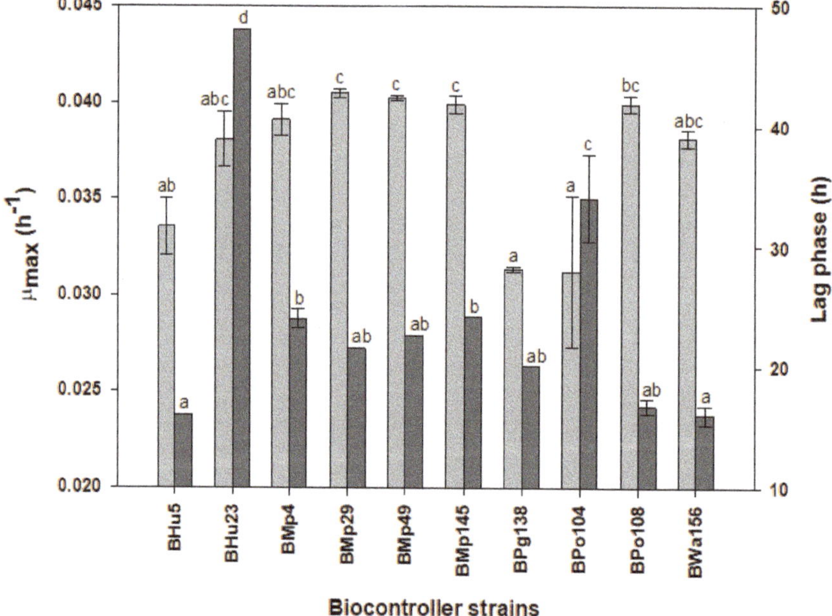

Figure 1. Specific growth rate (light grey) and lag phase (dark grey) of the non-*Saccharomyces* yeasts assayed. Rates are means with standard deviation (*n* = 3). Values with different letters are significantly different.

A successful predominance of the biocontroller during the fermentation start should demonstrate a short lag phase [10]. Most strains showed a lag time of about 20 h (Figure 1). BHu5 and BWa156 showed the shortest lag phases, about 15 h, and they were significantly different. BPo104 and BHu23 showed a significant longer lag phase of about 35 h and 50 h, respectively. A reduced lag phase increments the possibility of non-*Saccharomyces* to predominate the medium, since the lag phase is defined as the period prior of reaching the specific growth rate [16]. The lag phase is also relevant for the non-*Saccharomyces* strains to achieve a constant number for a determined period of time prior to inoculation of *S. cerevisiae* in a sequential mixed fermentation.

3.2.2. Enological Characterization of Yeasts: Tolerance to Molecular SO_2, Ethanol, High Reducing Sugar Concentrations and Low Temperature

The control yeast, *S. cerevisiae* BSc114, was able to ferment grape must at 30 °Brix and 15 °C and tolerated 14% *v/v* of ethanol and 0.4 ppm of molecular SO_2 (Table 3). With respect to SO_2, the non-*Saccharomyces* yeasts BMp29, BMp49, BMp145, BPg138, BPo104, BPo108 and BWa156 showed higher tolerance to SO_2 (0.4 ppm) (Table 3). Although a reduction in SO_2 is a goal of this study, it is relevant to evaluate resistance of the selected isolates to typical SO_2 concentrations used in wineries at the beginning of the process. The chemical could be present after yeast production, but never more than 100 ppm of total SO_2 [51]. Additionally, when non-*Saccharomyces* yeasts are used in integrated management (biocontrol yeasts—SO_2 application) they should be able to tolerate certain SO_2 concentrations. Typical winemaking generally uses at least 0.5 ppm of molecular SO_2 and in order to avoid any microbial contamination, this can increase to a final molecular SO_2 concentration of 0.8 ppm [52]. This means that BMp4, which showed the lowest tolerance and did not show any growth at molecular SO_2 concentrations above 0.1 ppm, would not be suitable for integrated management. Typically, non-*Saccharomyces* yeasts have been cited to be low SO_2 tolerant, but this sensitivity could also be linked to the combination of several factors such as ethanol, SO_2 and temperature [53].

Table 3. Tolerance of individual strains to high sugar concentration and low temperature, and different concentrations of molecular SO_2 and ethanol.

Molecular SO_2 (ppm)	BHu5	BHu23	BMp4	BMp29	BMp49	BMp145	BPg138	BPo104	BPo108	BWa156	BSc114
0 (Control)	+	+	+	+	+	+	+	+	+	+	+
0.1	+	+	−	+	+	+	+	+	+	+	+
0.15	+	+	−	+	+	+	+	+	+	+	+
0.2	+	+	−	+	+	+	+	+	+	+	+
0.3	+	+	−	+	+	+	+	+	+	+	+
0.4	−	−	−	+	+	+	+	+	+	+	+
Ethanol (% *v/v*)											
0 (Control)	+	+	+	+	+	+	+	+	+	+	+
8	+	+	+	+	+	+	+	+	+	+	+
10	+	+	+	+	+	+	+	−	+	+	+
12	−	−	−	+	+	−	−	−	−	−	+
14	−	−	−	−	−	−	−	−	−	−	+
High Sugar concentration (°Brix)											
21 (Control)	+	+	+	+	+	+	+	+	+	+	+
30	−	−	−	+	−	+	−	−	−	+	+
Low Temperature (°C)											
25 (Control)	+	+	+	+	+	+	+	+	+	+	+
15	−	+	+	+	+	+	+	+	+	+	+

Tolerance of yeast strains to different SO_2 and Ethanol concentrations, and to High Reducing Sugar Concentration and Low Temperature. CO_2 production: (+) gas production in Durham tubes and (–) no gas production in Durham tubes. Fermentation results for each strain and treatment were taken one day after the start of the control fermentation. Tubes with SO_2 = 0 ppm, Ethanol = 0% *v/v*, Sugars = 21 °Brix and Temperature = 25 °C were used as controls.

Regarding ethanol tolerance, BPo104 was the least tolerant strain and did not present growth above 10% *v/v* ethanol (Table 3). The most tolerant strains were BMp29 and BMp49 (12% *v/v* of ethanol), and the remaining isolates tolerated 10% *v/v*. None of the isolates were able to grow at 14% *v/v*. Tolerance of the non-*Saccharomyces* yeasts to ethanol is especially important with increasing permanence in the fermentation medium, as the growing *Saccharomyces sp.* population produces high amounts of ethanol [1]. All isolates seemed to tolerate 8% *v/v* during the first fermentation stages [45]. However, high ethanol tolerance could be a problem, because *S. cerevisiae* uses this method to biocontrol other native microbiota [1,54]. The presence of some non-*Saccharomyces* species like killer yeasts for prolonged periods of time could negatively modify the sensory quality of wine and cause stuck fermentation [41]. Even, the effect of metabolic interactions between non-*Saccharomyces* and *S. cerevisiae* wine yeasts could affect the growth and fermentation behavior of *S. cerevisiae* during fermentation [22]. Despite the fact that the high ethanol tolerance and wide biocontrol spectrum described for *M. pulcherrima* could be a potential risk for the normal fermentation process of *S. cerevisiae*, Oro et al. [14] mentioned that *Metschnikowia pulcherrima* does not biocontrol *S. cerevisiae*.

BMp29, BMp145 and BWa156 were able to carry out fermentation at high sugar concentrations (Table 3) whereas the remaining isolates were not. Tolerance to high sugar concentrations is relevant, because must from the Cuyo region (San Juan and Mendoza provinces) usually possesses a high sugar concentration [13]. As *Z. rouxii* yeasts are highly osmotolerant, it is extra important that *Z. rouxii* antagonists develop well under similar conditions [4,55].

With regard to tolerance to low temperature, BHu5 was the only isolate that did not grow (-). The remaining isolates were considered tolerant to low temperatures at the start of the fermentation (+). This is also a relevant factor when the biocontroller yeast is used during white wine fermentations or fermentations carried out at low temperature to preserve aromas [56].

3.2.3. Enzyme and H_2S Production

Control strain BSc114 reported low H_2S production and did not present any of the desired enzymatic activities assayed (Table 4). All non-*Saccharomyces* isolates evaluated except for BMp4 demonstrated desired protease activity (Table 4). This activity contributes to the degradation of proteins that could cause haze in the wine, thus facilitating the process of clarification and filtration [2]. Only BWa156 showed pectinase activity. This activity is another positive attribute that enables degradation of structural grape polysaccharides, increasing juice extraction and improving wine clarification and filtration [57]. It facilitates the release of aromatic precursors from the cells of the skin, seeds and flesh of the grape to the must [22,58]. Pectinase activity could be linked to a substrate colonization role or a trophic role [59]. Regarding yeast development and sugar consumption, firstly, BWa156 could be able to obtain sugars from the intracellular matrix of plant cells. In red wine fermentations with BWa156, this could generate a competitive advantage of the strain in the grape skin layer. Secondly, the yeast could consume galacturonic acid [59] as an alternative to glucose, which is quickly consumed by *S. cerevisiae* [60]. This would extend the time of this energy source for BWa156 and therefore result in a long-term competitive advantage. Although the activity is strain-dependent [22], pectinase production has already been associated with *W. anomalus* [61,62]. Nevertheless, more research is needed. None of the assayed yeasts showed β- glucosidase activity [22].

Table 4. Non-*Saccharomyces* attributes.

Isolate	Positive Traits			Negative Trait
	Protease	*Pectinase*	*β-Glucosidase*	*H_2S production*
BHu5	+	−	−	3
BHu23	+	−	−	3
BMp4	−	−	−	3
BMp29	+	−	−	3
BMp49	+	−	−	3
BMp145	+	−	−	3
BPg138	+	−	−	4
BPo104	+	−	−	3
BPo108	+	−	−	4
BWa156	+	+	−	2
BSc114	−	−	−	2

Means ($n = 3$). Arbitrary H_2S production scale [20]: 1: no production, 5: high production. Enzymes: (+): activity; (−): no activity.

Regarding the possible contribution to negative wine characteristics, most of the assayed yeasts showed medium H_2S production (3 or less on the scale in Table 3). Only BPg138 and BPo108 showed a higher production, 4, which is not desirable. Lowest production was produced by BWa156 (2 on the scale). H_2S production is highly relevant in winemaking and thus very important for the yeast selection because it is associated with the negative persistent odor described as "rotten egg" [27,38].

None of the isolates displayed hemolysin production. This is an important phenotypic characteristic of pathogenicity because it is related to lysis of erythrocytes [34].

Principal components analysis explained 62% of the variation among components (Figure 2). Desirable and undesirable characteristics can be clearly differentiated on the main axis (explaining 36.4%). Desirable characteristics observed were: high growth rate, tolerance to low temperature and high concentrations of ethanol, SO_2 and reducing sugars, and production of positive enzymes such as protease and pectinase. Prolonged adaptation time (Lag phase) and high H_2S production were undesirable characteristics.

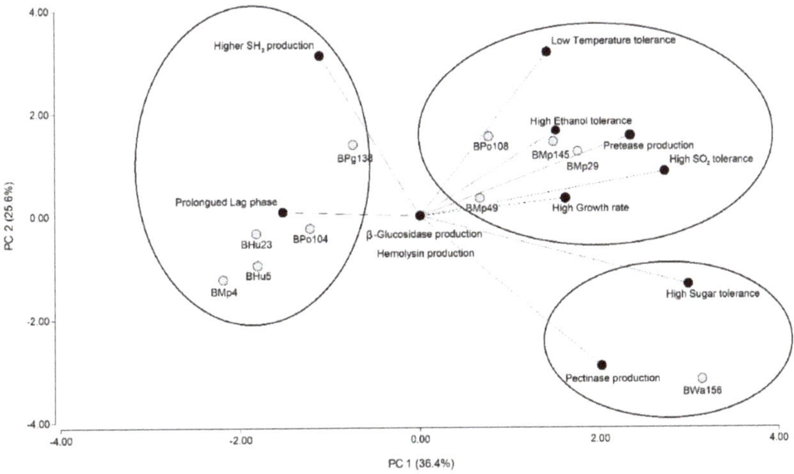

Figure 2. Principal Components Analysis (PCA) of yeast characteristics. References: Ellipses represent clusters obtained from hierarchical cluster analysis (HCA).

In the biplot it can be observed that BWa156 has the ability to prevail in the medium during the early stage of the fermentation. Compared with the other non-*Saccharomyces* assayed, this yeast possesses a short lag phase and high growth rate. The latter characteristic is related to the cellular multiplication and enables the release of killer toxins that may be constitutive [63], incrementing the possibilities of the biocontrol yeast. BWa156 also releases enzymes that could allow utilization of alternative energy sources. Our study also showed its capacity to grow in adverse must conditions such as high sugar and high SO_2 concentrations and the ability to grow at low temperature. In addition, it should be highlighted that the strain may positively contribute to the wine quality through the release of grape compounds because of its protease and pectinase production; these enzymes are not produced by BSc114. As a consequence, it could help intensify the color and enhance aromatic characteristics of the wine. Another advantage of the strain is the low H_2S production.

The biplot demonstrates that BMp29 presents more possibilities to prevail in the medium compared with the other non-*Saccharomyces* isolates assayed, because of its high growth rate and short lag phase. The strain is also able to grow under adverse conditions of grape must such as high sugar and high SO_2 concentrations and low temperature. Its high ethanol tolerance facilitates its growth and possible biocontrol during the fermentation. In the cluster, BMp49 presented a similar behavior to that of BMp29, but it did not develop at high reducing sugar concentrations. BMp145 and BPo108 also showed similar characteristics, but the first had a prolonged lag phase and the second strain the disadvantage of a higher potential to produce H_2S.

BWa156 and BMp29 demonstrated a wide biocontrol spectrum. *Wickerhamomyces anomalus* and *Metchnikowia pulcherrima* have already been used in co-inocula with *Saccharomyces cerevisiae* by Comitini et al. [11] and Oro et al. [14]. Albertin et al. [19] described positive flavor attributes related to *M. pulcherrima*, which supports the possibility of using such species as co-inocula. However, further research is necessary to determine the biocontrol application of the two selected strains [41].

4. Conclusions

The selected non-*Saccharomyces* yeasts BWa156 and BMp29 are highly applicable antagonistic yeasts that positively contribute to the wine process. They are active against relevant spoilage yeasts in the wine industry and can be used to produce wines with reduced SO_2 concentration. The present study is part of a comprehensive research project focusing on the application of non-*Saccharomyces* biocontroller yeasts. The biocontrolling sources and the conditions of implantation, prevalence and biocontrol kinetics is the projection of future research.

Author Contributions: Conceptualization, B.K., F.V., M.E.T., Y.P.M. and M.C.; methodology and experimental design, F.V., M.C. and Y.P.M.; software, B.K. and M.V.M.; validation, F.V. and M.E.T.; formal analysis, B.K., M.V.M. and F.V.; investigation, B.K., Y.P.M., M.V.V., M.C., M.E.T. and F.V.; data curation, B.K., writing—original draft preparation, B.K.; writing—review and editing, B.K., F.V., M.E.T. and Y.P.M.; visualization, B.K.; supervision, F.V., M.E.T. and M.C.; project administration, Y.P.M.; funding acquisition, Y.P.M. All authors discussed the results of the experiments and contributed to the final manuscript.

Funding: This work was supported by the Research Project IDeA 1400- SECITI 0041-2014.

Acknowledgments: The authors want to thank the Chemical Engineer Martha Dina Vallejo for the suggested corrections, the chemical donations and the accompaniment.

Conflicts of Interest: The authors declare no conflict of interest. The funders had no role in the design of the study; in the collection, analyses, or interpretation of data; in the writing of the manuscript, or in the decision to publish the results.

References

1. Ciani, M.; Capece, A.; Comitini, F.; Canonico, L.; Siesto, G.; Romano, P. Yeast interactions in inoculated wine fermentation. *Front. Microbiol.* **2016**, *7*, 555. [CrossRef] [PubMed]

2. Mehlomakulu, N.N.; Setati, M.E.; Divol, B. Characterization of novel killer toxins secreted by wine-related non-*Saccharomyces* yeasts and their action on *Brettanomyces* spp. *Int. J. Food Microbiol.* **2014**, *188*, 83–91. [CrossRef] [PubMed]

3. Oelofse, A.; Lonvaud-Funel, A.; Du Toit, M. Molecular identification of *Brettanomyces bruxellensis* strains isolated from red wines and volatile phenol production. *Food Microbiol.* **2009**, *26*, 377–385. [CrossRef] [PubMed]

4. Rojo, M.C.; López, F.A.; Lerena, M.C.; Mercado, L.; Torres, A.; Combina, M. Evaluation of different chemical preservatives to control *Zygosaccharomyces rouxii* growth in high sugar culture media. *Food Control.* **2015**, *50*, 349–355. [CrossRef]

5. Stratford, M.; Steels, H.; Nebe-von-Caron, G.; Avery, S.V.; Novodvorska, M.; Archer, D.B. Population heterogeneity and dynamics in starter culture and lag phase adaptation of the spoilage yeast *Zygosaccharomyces bailii* to weak acid preservatives. *Int. J. Food Microbiol.* **2014**, *181*, 40–47. [CrossRef] [PubMed]

6. Fleet, G.H. Yeast Spoilage of Food and Beverages. In *The Yeasts: A Taxonomic Study*; Kurtzman, C.P., Fell, J.W., Boekhout, T., Eds.; Elsevier Science: London, UK; Burlington, MA, USA, 2011; pp. 53–63.

7. Escott, C.; del Fresno, J.M.; Loira, I.; Morata, A.; Suárez-Lepe, J.A. *Zygosaccharomyces rouxii*: Control Strategies and Applications in Food and Winemaking. *Fermentation* **2018**, *4*, 69. [CrossRef]

8. Cejudo-Bastante, M.J.; Sonni, F.; Chinnici, F.; Versari, A.; Perez-Coello, M.S.; Riponi, C. Fermentation of sulphite-free white musts with added lysozyme and oenological tannins: Nitrogen consumption and biogenic amines composition of final wines. *LWT-Food Sci. Technol.* **2010**, *43*, 1501–1507. [CrossRef]

9. Ferrer-Gallego, R.; Puxeu, M.; Martín, L.; Nart, E.; Hidalgo, C.; Andorrà, I. Microbiological, Physical, and Chemical Procedures to Elaborate High-Quality SO$_2$-Free Wines. In *Grapes and Wines-Advances in Production, Processing, Analysis and Valorization*; IntechOpen: Madrid, Spain, 2017; pp. 171–193.

10. Swinnen, I.A.M.; Bernaerts, K.; Dens, E.J.; Geeraerd, A.H.; Van Impe, J.F. Predictive modelling of the microbial lag phase: A review. *Int. J. Food Microbiol.* **2004**, *94*, 137–159. [CrossRef]

11. Comitini, F.; De Ingeniis, J.; Pepe, L.; Mannazzu, I.; Ciani, M. *Pichia anomala* and *Kluyveromyces wickerhamii* killer toxins as new tools against *Dekkera/Brettanomyces* spoilage yeasts. *FEMS Microbiol. Lett.* **2004**, *238*, 235–240. [CrossRef]

12. Fleet, G.H. Wine yeasts for the future. *FEMS Yeast Res.* **2008**, *8*, 979–995. [CrossRef]

13. Maturano, Y.P.; Mestre, M.V.; Kuchen, B.; Toro, M.E.; Mercado, L.A.; Vazquez, F.; Combina, M. Optimization of fermentation-relevant factors: A strategy to reduce ethanol in red wine by sequential culture of native yeasts. *Int. J. Food Microbiol.* **2019**, *289*, 40–48. [CrossRef] [PubMed]

14. Oro, L.; Ciani, M.; Comitini, F. Antimicrobial activity of *Metschnikowia pulcherrima* on wine yeasts. *J. Appl. Microbiol.* **2014**, *116*, 1209–1217. [CrossRef] [PubMed]

15. Drumonde-Neves, J.; Franco-Duarte, R.; Lima, T.; Schuller, D.; Pais, C. Yeast biodiversity in vineyard environments is increased by human intervention. *PLoS ONE* **2016**, *11*. [CrossRef] [PubMed]

16. Pirt, S.J. Parameters of growth and analysis of growth data. In *Principles of Microbe and Cell Cultivation*; Blackwell Scientific Publications: Hoboken, NJ, USA, 1975; pp. 4–14.

17. Duarte, A. *Introducción a la Ingeniería Bioquímica*; Universidad Nacional de Colombia: Bogotá, Colombia, 1998; p. 198.

18. Stanbury, P.F.; Whitaker, A.; Hall, S.J. Microbial growth kinetics. In *Principles of Fermentation Technology*; Elsevier Science: Kent, UK, 2014; pp. 21–74.

19. Albertin, W.; Zimmer, A.; Miot-Sertier, C.; Bernard, M.; Coulon, J.; Moine, V.; Colonna-Ceccaldi, B.; Bely, M.; Marullo, P.; Masneuf-Pomarede, I. Combined effect of the *Saccharomyces cerevisiae* lag phase and the non-*Saccharomyces* consortium to enhance wine fruitiness and complexity. *Appl. Microbiol. Biotechnol.* **2017**, *101*, 7603–7620. [CrossRef] [PubMed]

20. Comitini, F.; Gobbi, M.; Domizio, P.; Romani, C.; Lencioni, L.; Mannazzu, I.; Ciani, M. Selected non-*Saccharomyces* wine yeasts in controlled multistarter fermentations with *Saccharomyces cerevisiae*. *Food Microbiol.* **2011**, *28*, 873–882. [PubMed]

21. Domizio, P.; Romani, C.; Lencioni, L.; Comitini, F.; Gobbi, M.; Mannazzu, I.; Ciani, M. Outlining a future for non-*Saccharomyces* yeasts: Selection of putative spoilage wine strains to be used in association with *Saccharomyces cerevisiae* for grape juice fermentation. *Int. J. Food Microbiol.* **2011**, *147*, 170–180. [CrossRef] [PubMed]

22. Maturano, Y.P.; Rodriguez, L.A.; Toro, M.E.; Nally, M.C.; Vallejo, M.; Castellanos de Figueroa, L.I.; Combina, M.; Vazquez, F. Multi-enzyme production by pure and mixed cultures of *Saccharomyces* and non-*Saccharomyces* yeasts during wine fermentation. *Int. J. Food Microbiol.* **2012**, *155*, 43–50. [CrossRef]

23. Mestre, V.; Maturano, Y.P.; Combina, M.; Mercado, L.A.; Toro, M.E.; Vazquez, F. Selection of non-*Saccharomyces* yeasts to be used in grape musts with high alcoholic potential: A strategy to obtain wines with reduced ethanol content. *FEMS Yeast Res.* **2017**, *17*. [CrossRef]

24. Sturm, M.E.; Assof, M.; Fanzone, M.; Martinez, C.; Ganga, M.A.; Jofré, V.; Ramirez, M.L.; Combina, M. Relation between coumarate decarboxylase and vinylphenol reductase activity with regard to the production of volatile phenols by native *Dekkera bruxellensis* strains under 'wine-like' conditions. *Int. J. Food Microbiol.* **2015**, *206*, 51–55. [CrossRef]

25. Esteve-Zarzoso, B.; Belloch, C.; Uruburu, F.; Querol, A. Identification of yeasts by RFLP analysis of the 5.8 S rRNA gene and the two ribosomal internal transcribed spacers. *Int. J. Syst. Evol. Microbiol.* **1999**, *49*, 329–337. [CrossRef]

26. Santos, A.; San Mauro, M.; Bravo, E.; Marquina, D. PMKT2, a new killer toxin from *Pichia membranifaciens*, and its promising biotechnological properties for control of the spoilage yeast *Brettanomyces bruxellensis*. *Microbiology* **2009**, *155*, 624–634. [CrossRef] [PubMed]

27. Vazquez, F.; Figueroa, L.; Toro, M. Enological characteristics of yeasts. In *Methods in Biotechnology: Food Microbiology Protocols*; Spencer, J.F.T., Spencer, A.L., Eds.; Humana Press: Totowa, NJ, USA, 2001; Volume 14, pp. 297–306.

28. Monod, J. *Recherches ser la Croissance des Cultures Bactériennes*, 2nd ed.; Hermann: Paris, France, 1942.

29. Lodge, R.M.; Hinshelwood, C.N. Physicochemical aspects of bacterial growth. Part IX. The lag phase of Bac. lactis aerogenes. *J. Chem. Soc.* **1943**, *51*, 213–219. [CrossRef]

30. Sudraud, P.; Chauvet, S. Activite antilevure de l'anhydride sulfureux moleculaire. *Conn. Vigne Vin.* **1985**, *19*, 31–40. [CrossRef]

31. Ribéreau-Gayon, J.; Peynaud, E.; Ribéreau-Gayon, P.; Sudraud, P. Sciences et Techniques du Vin. In *Clarification et Stabilisation. Matériels et Installation*; Dunod: Paris, France, 1977; Volume 4.

32. Strauss, M.L.A.; Jolly, N.P.; Lambrechts, M.G.; Van Rensburg, P. Screening for the production of extracellular hydrolytic enzymes by non-*Saccharomyces* wine yeasts. *J. Appl. Microbiol.* **2001**, *91*, 182–190. [CrossRef]

33. Fernandez-Salomäo, T.; Amorim, A.C.; Chaves-Alves, V.; Coelho, J.; Silva, D.; Araujo, E. Isolation of pectinase hyperproducing mutants of *Penicillium expansum*. *Rev. Microbiol.* **1996**, *27*, 15–18.

34. Manns, J.M.; Mosser, D.M.; Buckley, H.R. Production of a hemolytic factor by *Candida albicans*. *Infect. Immun.* **1994**, *62*, 5154–5156.

35. Menezes, A.G.T.; Ramos, C.L.; Cenzi, G.; Melo, D.S.; Dias, D.R.; Schwan, R.F. Probiotic Potential, Antioxidant Activity, and Phytase Production of Indigenous Yeasts Isolated from Indigenous Fermented Foods. *Probiotics Antimicrob. Prot.* **2019**. [CrossRef]

36. Iglewsky, B.H. *Medical Microbiology*, 4th ed.; Chapter 27: Pseudomonas; Baron, S., Ed.; University of Texas Medical Branch at Galveston: Galveston, TX, USA, 1996.

37. Luo, G.; Samaranayake, L.P.; Yau, J.Y. *Candida* species exhibit differential in vitro hemolytic activities. *J. Clin. Microbiol.* **2001**, *39*, 2971–2974. [CrossRef]

38. Massera, A.; Assof, M.; Sturm, M.E.; Sari, S.; Jofré, V.; Cordero-Otero, R.; Combina, M. Selection of indigenous *Saccharomyces cerevisiae* strains to ferment red musts at low temperature. *Ann. Microbiol.* **2012**, *62*, 367–380. [CrossRef]

39. Liu, G.L.; Chi, Z.; Wang, G.Y.; Wang, Z.P.; Li, Y.; Chi, Z.M. Yeast killer toxins, molecular mechanisms of their action and their applications. *Crit. Rev. Biotechnol.* **2013**, *35*, 222–234. [CrossRef]

40. De Ingeniis, J.; Raffaelli, N.; Ciani, M.; Mannazzu, I. *Pichia anomala* DBVPG 3003 secretes a ubiquitin-like protein that has antimicrobial activity. *Appl. Environ. Microbiol.* **2009**, *75*, 1129–1134. [CrossRef] [PubMed]

41. Starmer, W.T.; Lachance, M.A. Yeast Ecology. In *The Yeasts: A Taxonomic Study*; Kurtzman, C.P., Fell, J.W., Boekhout, T., Eds.; Elsevier Science: London, UK; Burlington, MA, USA, 2011; pp. 65–83.

42. De Vries, S.; von Dahlen, J.K.; Schnake, A.; Ginschel, S.; Schulz, B.; Rose, L.E. Broad-spectrum inhibition of *Phytophthora infestans* by fungal endophytes. *FEMS Microbiol. Ecol.* **2018**, *94*, fiy037. [CrossRef] [PubMed]

43. Odds, F.C.; Brown, A.J.P.; Gow, N.A.R. Antifungal agents: Mechanisms of action. *Trends Microbiol.* **2003**, *11*, 272–279. [CrossRef]

44. Ginovart, M.; Prats, C.; Portell, X.; Silbert, M. Exploring the lag phase and growth initiation of a yeast culture by means of an individual-based model. *Food Microbial.* **2011**, *28*, 810–817. [CrossRef] [PubMed]

45. Combina, M.; Elía, A.; Mercado, L.; Catania, C.; Ganga, A.; Martinez, C. Dynamics of indigenous yeast populations during spontaneous fermentation of wines from Mendoza, Argentina. *Int. J. Food Microbiol.* **2005**, *99*, 237–243. [CrossRef] [PubMed]

46. Rivero, D.; Berná, L.; Stefanini, I.; Baruffini, E.; Bergerat, A.; Csikász-Nagy, A.; De Filippo, C.; Cavalieri, D. Hsp12p and *PAU* genes are involved in ecological interactions between natural yeast strains. *Environ. Microbiol.* **2015**, *17*, 3069–3081. [CrossRef] [PubMed]

47. Nally, M.C. Función del carácter *killer* en la competencia por el dominio del substrato: Caracterización de mutantes de cepas de levadura *Saccharomyces cerevisiae*. In *Tesis de Licenciatura en Biología*; Universidad Nacional de San Juan: San Juan, Argentina, 2005.

48. Auling, G.; Bellgardt, K.H.; Diekmann, M.; Thoma, M. Dynamics of the growth in batch and continuous culture of *Saccharomyces cerevisiae* during a shift from aerobiosis to anaerobiosis and reverse. *Appl. Microbiol. Biotechnol.* **1984**, *19*, 353–357. [CrossRef]

49. Verduyn, C.; Postma, E.; Scheffers, W.A.; van Dijken, J.P. Physiology of *Saccharomyces cerevisiae* in anaerobic glucoselimited chemostat cultures. *J. Gen. Microbiol.* **1990**, *136*, 395–403. [CrossRef] [PubMed]

50. Jain, V.K. Relationship between energy metabolism and growth. I. Glucose dependence of the exponential growth rate of *Saccharomyces cerevisiae*. *Arch. Microbiol.* **1970**, *72*, 252–259.

51. Andorrà, I.; Martín, L.; Nart, E.; Puxeu, M.; Hidalgo, C.; Ferrer-Gallego, R. Effect of grape juice composition and nutrient supplementation on the production of sulfur dioxide and carboxylic compounds by *Saccharomyces cerevisiae*. *Aust. J. Grape Wine Res.* **2018**, *24*, 260–266. [CrossRef]

52. Milheiro, J.; Filipe-Ribeiro, L.; Vilela, A.; Cosme, F.; Nunes, F.M. 4-Ethylphenol, 4-ethylguaiacol and 4-ethylcatechol in red wines: Microbial formation, prevention, remediation and overview of analytical approaches. *Crit. Rev. Food Sci. Nutr.* **2017**, *59*, 1367–1391. [CrossRef] [PubMed]

53. Jolly, N.P.; Augustyn, O.P.H.; Pretorius, I.S. The role and use of non-*Saccharomyces* yeasts in wine production. *S. Afr. J. Enol. Vitic.* **2006**, *27*, 15–29. [CrossRef]

54. Pretorius, I.S. Tailoring wine yeast for the new millennium: Novel approaches to the ancient art of winemaking. *Yeast* **2000**, *16*, 675–729. [CrossRef]

55. Wang, H.; Hu, Z.; Long, F.; Guo, C.; Niu, C.; Yuan, Y.; Yue, T. Combined effect of sugar content and pH on the growth of a wild strain of *Zygosaccharomyces rouxii* and time for spoilage in concentrated apple juice. *Food Control.* **2016**, *59*, 298–305. [CrossRef]

56. Molina, A.M.; Swiegers, J.H.; Varela, C.; Pretorius, I.S.; Agosin, E. Influence of wine fermentation temperature on the synthesis of yeast-derived volatile aroma compounds. *Appl. Microbiol. Biotechnol.* **2007**, *77*, 675–687. [CrossRef] [PubMed]

57. Charoenchai, C.; Fleet, G.H.; Henschke, P.A.; Todd, B.E.N. Screening of non-*Saccharomyces* wine yeasts for the presence of extracellular hydrolytic enzymes. *Aust. J. Grape Wine Res.* **1997**, *3*, 2–8. [CrossRef]

58. Bautista-Ortín, A.B.; Fernández-Fernández, J.I.; López-Roca, J.M.; Gómez-Plaza, E. The effects of enological practices in anthocyanins, phenolic compounds and wine colour and their dependence on grape characteristics. *J. Food Compos. Anal.* **2007**, *20*, 546–552. [CrossRef]

59. Blanco, P.; Sieiro, C.; Villa, T.G. Production of pectic enzymes in yeasts. *FEMS Microbiol. Lett.* **1999**, *175*, 1–9. [CrossRef] [PubMed]

60. Apel, A.R.; Ouellet, M.; Szmidt-Middleton, H.; Keasling, J.D.; Mukhopadhyay, A. Evolved hexose transporter enhances xylose uptake and glucose/xylose co-utilization in *Saccharomyces cerevisiae*. *Sci. Rep.* **2016**, *6*, 19512. [CrossRef]

61. Masoud, W.; Jespersen, L. Pectin degrading enzymes in yeasts involved in fermentation of *Coffea arabica* in East Africa. *Int. J. Food Microbiol.* **2006**, *110*, 291–296. [CrossRef] [PubMed]

62. Martos, M.A.; Butiuk, A.P.; Rojas, N.L.; Hours, R.A. Cultivo por lote de *Wickerhamomyces anomalus* en un biorreactor a escala laboratorio para la producción de una poligalacturonasa. *Rev. Colomb. Biotecnol.* **2014**, *16*, 68–73. [CrossRef]

63. Schmidt, S.; Zimmermann, S.Y.; Tramsen, L.; Koehl, U.; Lehrnbecher, T. Natural killer cells and antifungal host response. *Clin. Vaccine Immunol.* **2013**, *20*, 452–458. [CrossRef] [PubMed]

 *fermentation*

Article

Effect of Sequential Inoculation with Non-*Saccharomyces* and *Saccharomyces* Yeasts on Riesling Wine Chemical Composition

Ophélie Dutraive [1], Santiago Benito [2,*], Stefanie Fritsch [1], Beata Beisert [1], Claus-Dieter Patz [3] and Doris Rauhut [1]

1   Bereich Getränkewissenschaften & Lebensmittelsicherheit, Institut für Mikrobiologie und Biochemie, Hochschule Geisenheim, Von-Lade-Straße 1, 65366 Geisenheim, Germany; ophelie.dutraive@hotmail.fr (O.D.); Stefanie.Fritsch@hs-gm.de (S.F.); Beata.Beisert@hs-gm.de (B.B.); Doris.Rauhut@hs-gm.de (D.R.)
2   Departamento de Química y Tecnología de Alimentos, Universidad Politécnica de Madrid, Ciudad Universitaria S/N, 28040 Madrid, Spain
3   Bereich Getränkewissenschaften & Lebensmittelsicherheit, Institut für Getränkeforschung, Hochschule Geisenheim, Von-Lade-Straße 1, 65366 Geisenheim, Germany; Claus.Patz@hs-gm.de
*   Correspondence: santiago.benito@upm.es; Tel.: +34-913363710 or +34-913363984

Received: 28 June 2019; Accepted: 29 August 2019; Published: 1 September 2019

Abstract: In recent years, studies have reported the positive influence of non-*Saccharomyces* yeast on wine quality. Many grape varieties under mixed or sequential inoculation show an overall positive effect on aroma enhancement. A potential impact by non-*Saccharomyces* yeast on volatile and non-volatile compounds should benefit the flavor of Riesling wines. Following this trend, four separate sequential fermentations (using the non-*Saccharomyces* yeasts *Torulaspora delbrueckii*, *Metschnikowia pulcherrima*, *Pichia kluyveri*, and *Lachancea thermotolerans* with *Saccharomyces cerevisiae*) were carried out on Riesling must and compared to a pure culture of *S. cerevisiae*. Sequential fermentations influenced the final wine aroma. Significant differences were found in esters, acetates, higher alcohols, fatty acids, and low volatile sulfur compounds between the different trials. Other parameters, including the production of non-volatile compounds, showed significant differences. This fermentation process not only allows the modulation of wine aroma but also chemical parameters such as glycerol, ethanol, alcohol, acidity, or fermentation by-products. These potential benefits of wine diversity should be beneficial to the wine industry.

Keywords: sequential inoculation; *Saccharomyces*; non-*Saccharomyces*; Riesling; aroma compound; *Torulaspora delbrueckii*; *Pichia kluyveri*; *Lachancea thermotolerans*

---

1. Introduction

Wine is considered to be one of the most complex aromatic products. It is composed of a large amount of aroma compounds and yet only a fraction of them are responsible for the bouquet. These compounds can be volatile, with these being the fragrant compounds, or non-volatile, with these being the compounds which are responsible for the taste sensations [1]. The final aromatic bouquet of the wine is the result of various factors: the raw material (variety, climate, vine management, and ripeness, etc.), the vinification choices, and the effect of microorganisms (yeast and lactic bacteria) [2]. In addition to the aromatic composition of the wine, it is important to note that each human has their own flavor perception. This perception is an interaction between three factors: the food properties, the in-mouth environment, and the psycho-social effects. As the aroma flavors lead to preferences for particular choices, the final bouquet of the wine plays a non-negligible role for the consumer taste [3].

Riesling is an aromatic grape variety which has been linked to various aroma compounds. Specific terpenes have been associated to this cultivar [1], such as monoterpenes, which are considered to be principally responsible for the floral and citrus character in wine [4]. The monoterpenes geraniol (rose and geranium notes) [5] and linalool (flower and lavender notes) [5] have been identified as typical for Riesling wines [6]. The thiols can also contribute to the bouquet of Riesling [7]. These compounds have a very low threshold, in the order of 1 ng/L, and express grapefruit, passion fruit, box tree, or black current notes [8].

Two different kinds of aroma can be distinguished: the varietal aromas that are grape-variety-specific and the by-products of the alcoholic fermentation that are aromas produced by yeast metabolism.

Varietal aroma compounds are intrinsic to the grape variety. Currently, only a few varieties have been linked to certain aroma compounds such as the Muscat cultivars and its relatives, like Riesling, with specific terpenes [1]; Sauvignon Blanc [9], with thiols, or Cabernet Sauvignon, with methoxypyrazines [10]. However, only some free forms of aroma compounds, such as monoterpenes or methoxypyrazines, are present in the grape juice. Most of the aroma compounds occur in their linked form, which makes them non-volatile and hence odorless [2]. However, the linked aromas can be liberated through several processes by specific enzymes [11], yeast [12–14], or lactic bacteria [15]. Even if the differences of the aroma compound amount between varieties can seem small, it can have an enormous impact on the final product.

Three families of aromas are known as by-products of the alcoholic fermentations. The esters are a result of various pathways that appear during the alcoholic fermentation as a by-product of the fermentation by yeast [16]. Their proportion in wines is yeast-strain-specific and depends on the fermentation conditions (temperature and pH), and evolves, positively or negatively, during wine aging [17]. The fatty acids can be separated into two groups: the straight chain fatty acids and the branched-chain fatty acids. They are both synthesized during the alcoholic fermentation through the yeast metabolism, but through different pathways [18]. The higher alcohols are synthesized by the yeast as an intermediate in amino acids metabolism. Thus, they are formed from various amino acids through two different pathways: the catabolic and the anabolic [18].

Since yeasts have an impact on the formation and the liberation of aroma compounds during alcoholic fermentation, yeast strains may play a role on the quantitative and qualitative production of these aromas. Nowadays, various strains are commercially available and promise aroma enhancement.

Only a few strains of non-*Saccharomyces* can complete alcoholic fermentation due to their sensitivity to high levels of ethanol. Moreover, they often produce undesirable secondary metabolites such as acetic acid, ethyl acetate, ethyl phenols, aldehydes, and acetoin [19,20]. Their oenological interest to the wine industry has been even less common given that they are SO_2 sensitive and their fermentation rate is low [21,22]. However, since climate change is impacting the level of sugar in grapes, non-*Saccharomyces* have become quite popular for their ability to reduce the level of alcohol [23,24]. Renewed interest is being shown for their ability to enhance certain aromas and produce more complex wines [12,25,26]. As their pure fermentation is not oenologically interesting, they must be used along with a *Saccharomyces* spp. The sequential inoculation target is then to imitate the spontaneous fermentation by providing at the early stage non-*Saccharomyces* and at the middle stage *Saccharomyces*, while avoiding the disadvantages of indigenous yeast, namely the production of off-flavors as well as sluggish or stuck fermentation [19,27,28].

Although some yeast strains can influence some specific volatile compounds to be over their sensory thresholds, wine is a complex matrix where individual compounds are occasionally difficult to perceive depending on the diversity of the matrix.

The aim of this work is to show the impact of non-*Saccharomyces* yeast on the wine's aroma modulation and to find the best strains to improve Riesling wine's flavor. To that purpose, four non-*Saccharomyces* yeasts were selected to be vinified under a sequential inoculation with a *Saccharomyces cerevisiae*. Their aroma profile was determined and compared to a wine fermented with only one *S. cerevisiae* strain.

2. Materials and Method

2.1. Yeast Strains

Commercial strains from different companies were used for this trial: *S. cerevisiae* Level 2® (Lallemand, Montreal, QC, Canada), *Torulaspora delbrueckii* Level 2® (Lallemand, Montreal, Canada), *Metschnikowia pulcherrima* Flavia® (Lallemand, Montreal, Canada), *Pichia kluyveri* FrootZen™ (Hansen, Hørsholm, Denmark), and *Lachancea thermotolerans* Concerto™ (Hansen, Hørsholm, Denmark).

2.2. Vinification

Riesling grape juice from Hochschule Geisenheim University (Germany) was used to lead the microvinification after autoclaving (115 °C, 15 min, 20 psi). Component concentration in the must were: sugar, 225.2 g/L; total dry extract, 237.8 g/L; pH, 2.9; total acidity, 8.7 g/L; tartaric acid, 4.7 g/L; acetic acid <0.1 g/L; malic acid, 3.0 g/L; ethanol <0.1 g/L; gluconic acid <0.1 g/L; glycerol <0.1 g/L; available primary amino acids 58.8 mg/L; ammonium 48 mg/L. In order to provide good conditions for the yeast to grow, the total nitrogen was adjusted to 250 mg/L with Vitamon® A (Erbslöh, Geisenheim, Germany) and nutrients were provided by Optimum White™ (Lallemand, Montreal, Canada), 30 g/hL. Bactiless™ (Lallemand, Montreal, Canada), 30 g/hL, was added to avoid any bacterial contamination.

Five assays, in quadruplicate, were carried out in 2-liter bottles with 1.8 L of juice following a similar methodology to that described before but adapted to the scale [29]. A single inoculation with *S. cerevisiae* Level 2® (10^6 CFU/mL) was performed for the control trial (1) and sequential inoculation was performed for the other fermentations by inoculating first the juice with the non-*Saccharomyces* yeast strain (10^6 CFU/mL) and at 48 h (*T. delbrueckii* Level 2® (2), *M. pulcherrima* Flavia® (3), and *L. thermotolerans* Concerto™ (5) trials) or 96 h (*P. kluyveri* FrootZen™ (4) trial) after with the *S. cerevisiae* Level 2® (10^6 CFU/mL).

The inocula of the first inoculation were carried out in 100 mL of must and 1 mL of YEPD liquid media in a 250 mL Erlenmeyer flask. The YEPD liquid media was prepared with yeast extract (10 g/L), peptone from casein (20 g/L), and glucose (20 g/L), and its pH was adjusted to 6.5. Yeast extract and glucose were provided by Merck (Darmstadt, Germany) and peptone by Roth® (Karlsruhe, Germany). The solution YEPD-must was autoclaved at 121 °C for 30 min. It was inoculated with the dry yeast strains when the media was at ambient temperature under sterile conditions. Only the *P. kluyveri* FrootZen™ strain was previously rehydrated. This strain was a frozen product. The frozen solution was rehydrated in the must in a 250-mL Erlenmeyer flask and incubated at 26 °C for 24 h before the inoculation of the YEPD liquid media. The inocula were cultivated in a chamber at 26 °C for 7 days. Their final concentration was 10^8 CFU/mL. Forty-five milliliters of inocula per bottle were used to inoculate the juice. The bottles were sealed by fermenting bung in order to allow gas to be released and then stored at 20 °C. The inoculum for the second inoculation (*S. cerevisiae* Level 2®) was performed according to recommendations of the yeast producer. Five grams of dry yeast were rehydrated in 50 mL of 35 °C water for 20 min. The inoculation (3.6 mL/bottle) was carried out when the inoculum (10^9 CFU/mL) was about 20 °C. At the end of the fermentation monitored by weight loss, the bottles were stored in a 4 °C room to let the yeasts settle down. Then, the young wines were racked after 48 h and filled in brown glass bottles, according to the same procedure. Furthermore, potassium disulfite (Merck, Darmstadt, Germany) was added to each sample in a concentration of 80 mg/L sulfur dioxide before the bottles were locked with screw caps. Afterwards, the filled bottles were again stored at 4 °C.

2.3. Fermentation Kinetics

Yeast strain kinetic growth was followed on two different agar media: a YEPD medium and a lysine medium [29]. The two media were chosen in order to differentiate *Saccharomyces* and non-*Saccharomyces* growth. The *Saccharomyces* yeast grows only on YEPD media whereas the non-*Saccharomyces* yeast grows on both. Aliquots were taken at regular intervals from the bottles of the quadruplicates of each assay. They were plated on both media in three different dilutions and incubated in a chamber at 26 °C for 48 h before counting.

In order to estimate aliquot dilutions, their concentration in yeast (CFU/mL) was estimated by using a Thoma cell counting chamber (Marienfeld, Germany) under the microscope (objective x40). The Thoma cell counting chamber was also used to estimate the inocula concentration. The weight of each bottle was taken regularly to follow the fermentation progress. The fermentation progress was calculated by the difference of the initial weight at day 0 and the weight from the day.

2.4. Chemical Compounds Analysis

2.4.1. HPLC

Organic acids measurements were performed by HPLC following the method described by Schneider et al. (1987) [30], which was adjusted and improved by Semmler et al. (2017) [31]. Measurements were carried out on the initial juice and obtained wines by the Department of Microbiology and Biochemistry at Hochschule Geisenheim University (HGU).

2.4.2. FTIR

Fourier transform middle infrared spectroscopy (FT-MIR) was used to assess density, alcohol, extract, sugars, pH, glycerol, and SO_2 measurements on initial juice and obtained wines. The method applied followed Baumgartner et al. (2001) [32], Patz et al. (1999) [33], and the Standard Operating Procedure SOP-WG1-84 of the HGU Department of Beverage Research.

2.5. Volatile Compounds Analysis

Volatile compound analysis was conducted by the HGU Department of Microbiology and Biochemistry. Fatty acids, higher alcohols, and esters were measured according to the method described by Rapp et al. (1994) [34] and modified by Fritsch et al. (2017a) [35]. A gas chromatography and mass spectrometer were performed to assess these volatile compounds. Terpenes were determined by the application of solid phase extraction (SPE), gas chromatography, and mass spectrometry following the method of Schüttler et al. (2015) [6], modified by Fritsch et al. (2017b) [36]. Headspace injection and the use of gas chromatography and pulsed flame photometric detection were achieved to measure low volatile compounds following the publication Rauhut et al. (2005) [37], adapted and modified by Rauhut, Beisert (2017) [38].

2.6. Statistical Analysis

R software (version 3.4.1, R Foundation for Statistical Computing, Vienna, Austria, 2017) and its package R-commander were used to perform all the statistical analyses. The significance level $p < 0.05$ was chosen for all the applied tests.

3. Results and Discussion

3.1. Fermentation Kinetics

3.1.1. Yeast Population Kinetics

The yeast populations of the various fermentations processes are shown in Figure 1. Non-*Saccharomyces* yeasts started to decline between 2 to 4 days after the second inoculation with *S. cerevisiae*. Previous studies show similar trends on the decline of the non-*Saccharomyces* population shortly after inoculation with *Saccharomyces* spp. strain [39]. This phenomenon could have numerous explanations. It has previously been shown that yeast growth can be inhibited by yeast metabolites such as ethanol [40,41], medium chain fatty acids [42], and acetaldehyde [43]. Killer toxins produced during the exponential phase by specific strain can also have an inhibitory impact on some yeast growth [44–46]. More recently, it was found that *S. cerevisiae* could secrete peptides who inhibit non-*Saccharomyces* yeast growth [47]. Other parameters such as nutrient limitation and possible physical changes such as temperature and redox potential could also inhibit yeast developing.

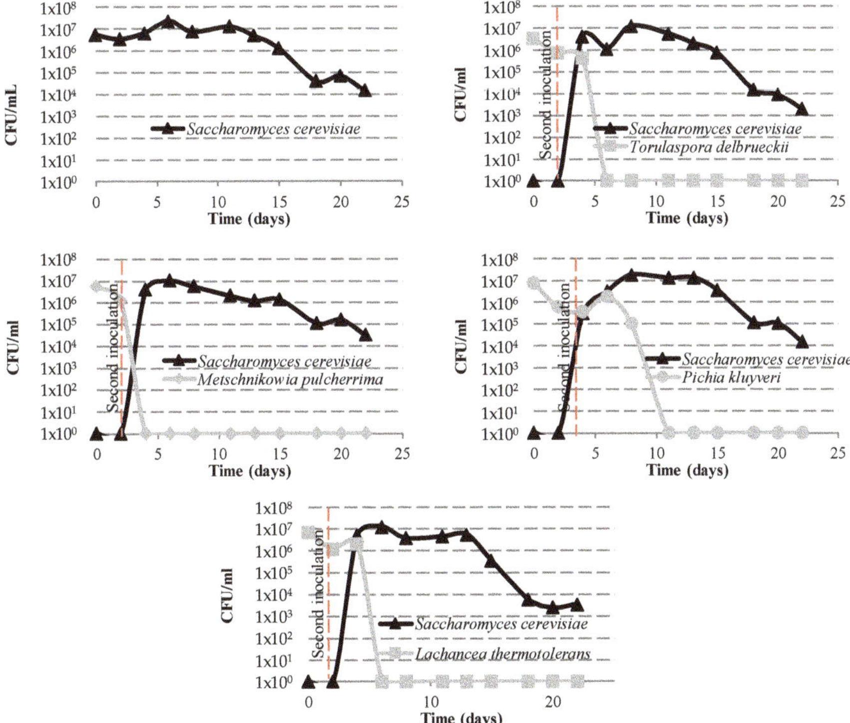

Figure 1. Population development of *Saccharomyces cerevisiae*, *Torulaspora delbrueckii*, *Metschnikowia pulcherrima*, *Pichia kluyveri*, and *Lachancea thermotolerans* during the fermentation process of the different trials.

3.1.2. Alcoholic Fermentation Kinetics

Figure 2 shows the alcoholic fermentation curves based on the weight losses of the replicates and trials during the fermentation process. For each assay, quadruplicates followed the same trend. The curves follow typical fermentation curves which have already been reported in the literature [48]. Fermentation dynamics show differences in the speed of sugar consumption. *S. cerevisiae* fermentation and *M. pulcherrima* sequential fermentation consumed sugars the fastest at the beginning of the alcoholic fermentation. However, other trails consumed the sugars faster during the following steps because all the fermentations reached the stationary phase at the same time. This difference of speed at the beginning could be attributed to the population density of *M. pulcherrima*, which declined immediately after the second inoculation, and the low fermentation rate and power of non-*Saccharomyces* species. As the other non-*Saccharomyces* populations declined later, their low fermentation power and rate, as has been reported by previous studies [49,50], could be associated with a slower fermentation process [51].

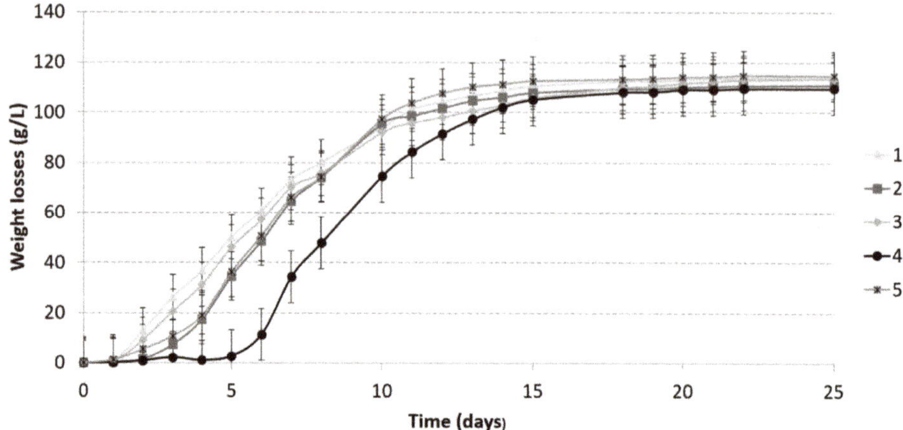

Figure 2. Alcoholic fermentation curve of *S. cerevisiae* Level 2® (1), and sequential fermentations with *S. cerevisiae* Level 2® and *T. delbrueckii* Level 2® (2), *M. pulcherrima* Flavia® (3), *P. kluyveri* FrootZen™ (4), and *L. thermotolerans* Concerto™ (5).

3.2. Chemical Monitoring

Tartaric acid was initially 5.2 g/L. Thus, all the fermentations show a decrease in tartaric acid in the wine between 0.48 and 0.77 g/L (Table 1). This phenomenon can be explained by the precipitation of tartaric acid as calcium or potassium tartrate [52]. Tartaric acid can also be degraded by yeasts by up to 23% [53]. In this trial, the *S. cerevisiae* degraded the tartaric acid by the smallest amount and the sequential inoculation with *T. delbrueckii*, *M. pulcherrima*, and *P. kluyveri* by the greatest amount. It seems that the latter used more tartaric acid for their metabolisms.

Table 1. Final analysis, before adding SO_2, of *S. cerevisiae* Level 2® (1), and sequential fermentations with *S. cerevisiae* Level 2® and *T. delbrueckii* Level 2® (2), *M. pulcherrima* Flavia® (3), *P. kluyveri* FrootZen™ (4), and *L. thermotolerans* Concerto™ (5).

Compounds	1	2	3	4	5
Tartaric acid (g/L)	4.72 ± 0.18a	4.38 ± 0.02b	4.43 ± 0.01b	4.43 ± 0.01b	4.63 ± 0.28ab
Malic acid (g/L)	2.28 ± 0.00d	2.21 ± 0.00c	2.10 ± 0.01a	2.21 ± 0.01c	2.17 ± 0.01b
Shikimic acid (mg/L)	50.19 ± 0.06b	49.94 ± 0.36ab	49.36 ± 0.25a	49.86 ± 0.34ab	49.64 ± 0.37ab
Lactic acid (g/L)	0.21 ± 0.01a	0.17 ± 0.01a	0.21 ± 0.01a	0.19 ± 0.00a	1.51 ± 0.04b
Acetic acid (g/L)	0.25 ± 0.03ab	0.21 ± 0.03a	0.30 ± 0.03b	0.23 ± 0.02a	0.31 ± 0.02b
Citric acid (g/L)	0.14 ± 0.02a	0.14 ± 0.01a	0.14 ± 0.00a	0.15 ± 0.02a	0.13 ± 0.01a
Residual sugars (g/L)	4.4 ± 0.25b	2.8 ± 0.23a	4.5 ± 0.30b	2.9 ± 0.16a	3.0 ± 0.11a
pH	3.1 ± 0.00a	3.2 ± 0.00b	3.2 ± 0.00b	3.2 ± 0.00b	3.1 ± 0.00a
Ethanol (% *v/v*)	13.20 ± 0.19b	13.17 ± 0.44b	12.98 ± 0.35a	13.04 ± 0.28a	12.96 ± 0.31a
Glycerol (g/L)	5.8 ± 0.04a	6.6 ± 0.04b	7.0 ± 0.05c	7.1 ± 0.05c	7.4 ± 0.07d

Average values of the quadruplicates ± standard deviation. The letters represent significantly different statistical groups ($p < 0.05$).

Previous studies have shown that some yeasts are able to consume malic acid such as *Issatchenkia orientalis*, *Saccharomyces* spp., or *Schizosaccharomyces* spp. By up to 45% of the initial level [19,54–57]. The malic acid levels from 2.10 to 2.28 g/L (Table 1) according to the assay are indeed lower than the initial malic acid level of the juice, which was 3.00 g/L. All the trials presented significant differences in their malic acid amount except for the sequential fermentation with *T. delbrueckii* and *P. kluyveri*, which can be seen to have had exactly the same production. The sequential fermentation with *M. pulcherrima* and the single fermentation with *S. cerevisiae* had the lowest and highest levels, respectively. Thus, *M. pulcherrima* shows a higher ability to consume malic acid than the other strains and *S. cerevisiae* shows a lower ability to consume malic acid.

Shikimic acid can be a precursor of aroma compounds such as ethyl cinnamate or benzaldehyde. Differences of the shikimic acid amount are shown. The level in the wines was found to be between 49.4 and 50.2 mg/L (Table 1). The fermentation with *S. cerevisiae* and the sequential fermentation with *M. pulcherrima* have the highest and the lowest amounts of shikimic acid, respectively. Thus, *M. pulcherrima* could have a positive impact on the formation of aroma compounds from this precursor.

Lactic acid concentrations were between 0.17 and 1.51 g/L, with the sequential fermentation with *T. delbrueckii* and *L. thermotolerans* having respectively the lowest and the highest levels (Table 1). Previous studies have reported high production of lactic acid by *L. thermotolerans* during its growth phase and its ability to acidify the must [28,58–62]. Assay 5, fermented with *L. thermotolerans* and *S. cerevisiae*, shows significantly higher production of lactic acid than the other trials, whereas the other trials do not show significant differences in their lactic acid production.

The acetic acid production (Table 1) varied from 0.21 to 0.31 g/L, according to the assay. No statistical differences are shown between the single fermentation with *S. cerevisiae* and the other trials, as has already been reported [49]. The sequential fermentations fermented with *T. delbrueckii* and *P. kluyveri* produced significantly lower acetic acid compared to the other non-*Saccharomyces* yeasts, which differs from previous studies. Nevertheless, the total amount of produced acetic acid does not negatively impact the wine quality [63].

The ethanol level ranges from 12.96 and 13.20% vol. (Table 1). Some strains are known for their ability to produce less alcohol [64,65]. Similar to what has been reported in the literature, the *S. cerevisiae* batch produced the highest ethanol level, albeit one that is not significantly different from that resulting from the sequential fermentation with *T. delbrueckii* [49]. The lowest level of ethanol was produced by the sequential inoculation with *L. thermotolerans*, though this level is not significantly different from the one produced by the sequential inoculations with *M. pulcherrima* and *P. kluyveri*. Similar results have already been found in the literature, showing that these yeasts can be used to produce low-alcohol wines [59,65]. In this trial, *S. cerevisiae* fermentation produced the highest level of ethanol, which confirms its high fermentative purity.

The level of pH was found to be not drastically different from one wine to another and ranged from 3.1 to 3.2 (Table 1). However, the statistical analysis shows significant differences. The fermentation with *S. cerevisiae* and the sequential fermentation with *L. thermotolerans* have lower pH levels, whereas the sequential fermentations with *T. delbrueckii*, *M. pulcherrima* and *P. kluyveri* have higher levels. Higher pH differences of *L. thermotolerans* with the other fermentations may be expected because of its high production of lactic acid [28,66,67]. Some studies confirm that even if lactic acid production is significantly higher for mixed fermentation with *L. thermotolerans* and *S. cerevisiae* than a single inoculation with the latter, the pH level can be significantly the same [58,59].

Glycerol production varies from 5.8 to 7.4 g/L for fermentation with *S. cerevisiae* and sequential fermentation with *L. thermotolerans*, having, respectively, the lowest and the highest production (Table 1). All of the other trials produced significantly higher levels of glycerol, which means that non-*Saccharomyces* have a positive impact on glycerol production. Higher production of this compound from sequential inoculation with non-*Saccharomyces* and *Saccharomyces* spp. has already been shown in the literature [28,68–71].

3.3. Volatile Compounds

3.3.1. Esters

The total esters produced were between 124,842 and 194,053 µg/L (Table 2). *P. kluyveri* sequential fermentation produced the highest level of esters and *L. thermotolerans* sequential fermentation produced an equally significant amount of esters. On the contrary, *T. delbrueckii* and *M. pulcherrima* sequential fermentations produced the lowest amount of esters.

Table 2. Volatile compounds measured after alcoholic fermentations of *S. cerevisiae* Level 2® (1), and sequential fermentations with *S. cerevisiae* Level 2® and *T. delbrueckii* Level 2® (2), *M. pulcherrima* Flavia® (3), *P. kluyveri* FrootZen™ (4), and *L. thermotolerans* Concerto™ (5).

Compounds	1	2	3	4	5
Esters					
Ethyl esters					
Ethyl lactate (mg/L)	nq	nq	nq	nq	52.26 ± 3.27
i-Ethyl butanoate (µg/L)	nd	nd	nd	nd	nd
Ethyl butanoate (µg/L)	432.19 ± 31.18b	308.35 ± 16.36a	332.37 ± 20.24a	278.46 ± 26.92a	304.86 ± 15.42a
Ethyl hexanoate (µg/L)	1876.06 ± 69.54bc	1798.42 ± 45.10b	1994.32 ± 45.97c	1551.1898.47a	1518.12 ± 28.25a
Ethyl octanoate (µg/L)	1472.32 ± 89.10ab	1367.63 ± 33.80a	1672.57 ± 186.55b	1440.66 ± 122.99ab	1364.97 ± 96.58a
Ethyl decanoate (µg/L)	442.63 ± 57.49a	437.23 ± 37.76a	553.12 ± 87.48a	546.90 ± 60.22a	581.35 ± 47.92a
Diethyl succinate (µg/L)	nq	nq	nq	nq	nq
Ethyl 2-hydroxy-4-methyl valerate (µg/L)	nq	nq	nq	nq	nq
Total ethyl esters (µg/L)	4223.20 ± 152.74a	3911.63 ± 18.66a	4552.38 ± 305.74a	3817.20 ± 282.13a	56031.95 ± 3282.57b
Acetates					
Ethyl acetate (mg/L)	159.18 ± 7.20b	135.02 ± 9.10a	115.53 ± 6.80a	184.01 ± 5.02c	127.97 ± 10.25a
Isoamyl acetate and 2-methyl butyl acetate (µg/L)	3927.41 ± 186.64a	4446.21 ± 205.56bc	3779.80 ± 90.54a	4751.71 ± 257.79c	4154.59 ± 163.59ab
Hexyl acetate (µg/L)	694.15 ± 29.53b	666.82 ± 11.07b	626.61 ± 38.94ab	590.44 ± 34.53a	592.86 ± 23.73a
Ethyl phenylacetate (µg/L)	nq	nq	nq	nq	nq
2-Phenyl-ethyl acetate (µg/L)	429.43 ± 9.15b	510.21 ± 16.05c	352.56 ± 14.12a	885.06 ± 28.39d	375.91 ± 8.12a
Total acetates (µg/L)	164229.49 ± 7358.44c	140647.20 ± 9289.39b	120290.10 ± 6811.04a	190236.09 ± 5279.61d	133095.47 ± 10391.01ab
Total esters (µg/L)	168452.69 ± 7369.92b	144558.83 ± 9282.59a	124842.48 ± 6791.85a	194053.29 ± 5518.21c	189127.42 ± 9985.27c
Higher alcohols					
3-Methyl-butanol and 2-methyl-butanol (mg/L)	188.62 ± 9.08a	230.96 ± 8.93b	212.96 ± 11.85ab	224.83 ± 15.01b	212.78 ± 10.67ab
2-Phenyl-ethanol (mg/L)	13.74 ± 1.20a	23.42 ± 0.95c	18.00 ± 0.60b	28.62 ± 1.39d	23.38 ± 1.00c
Hexanol (µg/L)	1131.24 ± 37.40b	1208.40 ± 53.76bc	1277.45 ± 71.62c	677.55 ± 15.71a	1415.87 ± 43.94d
Total higher alcohols (mg/L)	203.48 ± 9.95a	255.59 ± 9.61b	232.23 ± 11.84ab	254.14 ± 15.91b	237.58 ± 11.63b
Fatty acids					
Hexanoic acid (mg/L)	11.13 ± 0.30cd	10.78 ± 0.17bc	11.38 ± 0.14d	10.41 ± 0.15ab	9.90 ± 0.21ab
Octanoic acid (mg/L)	11.99 ± 0.20c	10.96 ± 0.25bc	11.56 ± 0.75c	10.31 ± 0.35b	9.05 ± 0.41a
Decanoic acid (µg/L)	4209.93 ± 141.19c	3975.70 ± 119.78bc	4233.22 ± 178.75c	3434.45 ± 229.12a	3718.80 ± 67.31ab
i-Valeric acid (µg/L)	1446.64 ± 12.53a	1493.16 ± 21.54a	1458.47 ± 20.22a	2093.35 ± 19.51b	1491.55 ± 19.36a
Total fatty acids (mg/L)	28.79 ± 0.38c	27.21 ± 0.39bc	28.64 ± 1.03c	26.26 ± 0.67b	24.15 ± 0.51a

Table 2. *Cont.*

Compounds	1	2	3	4	5
Terpenes					
Linalool oxide 1 (µg/L)	nq	nq	nq	nq	nq
Linalool oxide 2 (µg/L)	nq	nq	nq	nq	nq
Linalool (µg/L)	54.52 ± 1.96a	56.06 ± 1.16a	56.97 ± 1.49a	56.61 ± 2.17a	57.85 ± 2.54a
α-Terpineol (µg/L)	37.27 ± 0.64a	37.46 ± 0.89a	36.92 ± 0.61a	38.51 ± 0.90a	37.81 ± 0.60a
Total terpenes (µg/L)	91.80 ± 2.15a	93.52 ± 1.05a	93.89 ± 2.01a	95.12 ± 2.65a	95.65 ± 2.91a
Low volatile sulfur compounds					
H_2S	3.48 ± 028a	5.85 ± 0.15b	6.28 ± 0.29bc	7.65 ± 0.87cd	7.78 ± 0.82d
MeSH	nq	nq	nq	nq	nq

Average values of the quadruplicates ± standard deviation. The letters represent significantly different statistical groups ($p < 0.05$). Legend: nd, not detected; nq, not quantifiable (traces).

While previous studies are in accordance with this *M. pulcherrima* sequential fermentations result regarding the concentration of esters, others have found higher production of ester compounds than *Saccharomyces* spp. single fermentation [29,72]. Nevertheless, *M. pulcherrima* sequential fermentation produced the highest level of ethyl hexanoate and ethyl octanoate. These compounds confer apple peel and fruit flavors to the wine. Similar results have already been reported [14,73]. In previous studies, ethyl acetate production was reported higher for *M. pulcherrima* than the *S. cerevisiae* control [72,74]. In the present study, *Metschnikowia pulcherrima* produced the lowest amount of ethyl acetate (Table 2), as has been demonstrated by other authors [29].

T. delbrueckii produced higher levels of acetate compounds than the *S. cerevisiae* fermentation, producing higher concentrations of compounds such as isoamyl acetate, 2-methyl butyl acetate, and 2-phenyl-ethyl acetate. Other authors have reported a higher production of 2-phenyl-ethyl acetate [13,27,75].

Previous studies have highlighted the ability of *P. kluyveri* to produce ester compounds [76,77]. Total esters were produced the most by this strain. While its production of acetate esters was significantly higher than the *S. cerevisiae* control, its ethyl esters production did not differ. This result is not in agreement with the literature [29]. This difference could be explained by the length of time between the first and the second inoculation. If this is the case, a longer fermentation with *P. kluyveri* should improve the production of acetate ester. The different results can also be explained by the interaction between yeasts that can influence wine aroma [64,78]. The *P. kluyveri* trial also produced higher levels of isoamyl acetate, 2-methyl butyl acetate, and 2-phenyl-ethyl acetate than the other variants.

L. thermotolerans produced the highest level of ethyl esters. This high level is mainly related to the production of ethyl lactate. The trial produced the lowest amount of all the ethyl esters measured but it was the only producer of ethyl lactate (Table 2). However, the final concentration of ethyl lactate was lower than the sensory threshold of 60 mg/L [28,79]. Some studies have previously reported higher production of ethyl lactate by *L. thermotolerans* sequential fermentation than by *Saccharomyces* spp. single fermentation [59]. Ethyl lactate production might be favored by the formation of lactic acid which can produce ethyl lactate by an esterification reaction with ethanol [80]. Total esters were higher than the *S. cerevisiae* control. The total amount of esters is also influenced by the production of ethyl lactate. All the measured acetates were lower or equal to the *S. cerevisiae* control (Table 2). Thus, *L. thermotolerans* enhanced the milky and fruity flavors associated with ethyl lactate [28,79].

3.3.2. Higher Alcohols

Four higher alcohols were analyzed for the wines and shown in Table 2. Three of them were detected at levels above their odor threshold [5], with the exception of hexanol. *T. delbrueckii* sequential fermentation produced the highest level of higher alcohols, 256 mg/L (Table 2), but this trial was not significantly different to the other sequentially fermented assays.

P. kluyveri produced significantly more alcohols than the *S. cerevisiae* control. It produced significantly the highest level of 2-phenylethanol than the other trials, giving pleasant flavors such as honey, spice, rose, and lilac [5]. It also produced a significantly higher level of 3-methyl-butanol and 2-methylbutanol, which confer whiskey, malt, and burnt notes.

L. thermotolerans produced higher levels of higher alcohols in total than the *S. cerevisiae* control. In particular, it produced more of the 2-phenyl-ethanol, honey, spice, rose, and lilac fragrances than the *S. cerevisiae* control, and the greatest amount of the hexanol compound, resin, flower, and green fragrances [5]. While levels of both compounds are under their perception threshold, the level of higher alcohol contributes positively to the wine aroma [5].

3.3.3. Fatty Acids

Fatty acids were detectable via the analysis (Table 2). Four of them were analyzed and found in higher quantities relative to their odor threshold [5]. The total amount of fatty acid production ranged from 24 to 29 mg/L (Table 2). *S. cerevisiae* fermentation produced the highest levels of fatty acids, followed by the sequential fermentation with *M. pulcherrima*. Both trials' levels were significantly equal according to the statistic. On the contrary, *L. thermotolerans* sequential fermentation produced the lowest levels.

P. kluyveri produced less fatty acids than the control. However, it produced significantly higher levels of i-valeric acid than other fermentations. This compound confers unpleasant notes such as sweat, acid, and a rancid flavor [5].

3.3.4. Terpenes

Four terpene compounds were analyzed but only two of them were quantifiable. Moreover, no significant differences were found between them. Levels of linalool, known for its flower and lavender notes [5], ranged between 55 and 58 μg/L (Table 2). This compound was produced in higher quantities than its perception threshold, 25.2 μg/L [5], and therefore contributed to the bouquet of the wine.

3.3.5. Low Volatile Sulfur Compounds

Among the analyzed low volatile sulfur compounds, only H_2S was detectable and quantifiable. Levels ranged from 3 to 8 μg/L (Table 2). According to the literature, the H_2S threshold varies across a wide range and a negative contribution to the flavor can already be expected at low microgram concentrations [81]. Thus, it could be sensorially detectable. *S. cerevisiae* fermentation produced a significantly lower amount of H_2S than sequential fermentations with non-*Saccharomyces*. On the contrary, *L. thermotolerans* produced the highest amount of H_2S. The high variability between strains and yeasts [49,82–84] and the limited information on H_2S production from sequential inoculation fermentation do not allow the confirmation of those results.

4. Conclusions

In this work, the ability of various yeast strains to enhance the aroma of Riesling wines was investigated. Five commercial wine yeast strains were chosen in order to compare their impact on the wine chemical composition. Four non-*Saccharomyces* species strains were selected due to their ability to produce specific aromas compared to a classic fermentation with a *S. cerevisiae*. Sequential inoculation allowed the modulation of the wines' non-volatile compound production. Significant differences between the trials of chemical parameters such as ethanol, pH, and glycerol were found in this study. Differences in volatile compounds were observed for esters, higher alcohols, fatty acids, and low volatile sulfur compounds. *T. delbrueckii* sequential fermentation produced the lowest acetic acid concentration and the highest concentration of higher alcohols. *M. pulcherrima* sequential fermentation produced the highest levels of ethyl hexanoate, ethyl octanoate, and shikimic acid. The *P. kluyveri* trial produced the highest concentrations of total esters, glycerol, and i-valeric acid. *L. thermotolerans* produced the highest concentrations of lactic acid, ethyl lactate, and H_2S.

Author Contributions: D.R., S.B. and O.D. developed the experimental design; O.D. performed the vinifications; O.D. and S.B. performed the formal data analysis and supervised the project; O.D., S.B. and D.R. wrote the article; D.R., S.F. and B.B. performed gas chromatographic analysis; C.-D.P. introduced to the FTIR analysis and assisted to data acquisition and statistical analysis. All authors discussed the results and contributed to the final manuscript.

Funding: This project has been co-funded by the Erasmus+ program of the European Union, through Vinifera Euromaster (www.vinifera-euromaster.eu). Funding for Santiago Benito was provided by Ossian Vides y Vinos S. L under the framework of the project FPA1720300120 (Centre for the Development of Industrial Technology (CDTI), Spain).

Acknowledgments: Anja Giehl (Department of Beverage Research, HGU) is gratefully acknowledged for her assistance in FTIR-measuring, Carmen Dost for her help to control the fermentation trial, and Heike Semmler for her support in the HPLC analysis (Department of Microbiology and Biochemistry, HGU).

Conflicts of Interest: The authors declare no conflict of interest.

References

1. Rapp, A.; Mandery, H. ChemInform Abstract: Wine Aroma. *Chem. Informationsd.* **1986**, *42*, 873. [CrossRef]
2. Fischer, U. Wine aroma. In *Flavours and Fragrances: Chemistry, Bioprocessing and Sustainability*; Springer: Berlin/Heidelberg, Germany, 2007; ISBN 9783540493389.
3. German, J.B.; Yeritzian, C.; Tolstoguzov, V.B. Olfaction, where nutrition, memory and immunity intersect. In *Flavours and Fragrances: Chemistry, Bioprocessing and Sustainability*; Springer: Berlin/Heidelberg, Germany, 2007; ISBN 9783540493389.
4. Martin, D.M.; Chiang, A.; Lund, S.T.; Bohlmann, J. Biosynthesis of wine aroma: Transcript profiles of hydroxymethylbutenyl diphosphate reductase, geranyl diphosphate synthase, and linalool/nerolidol synthase parallel monoterpenol glycoside accumulation in Gewürztraminer grapes. *Planta* **2012**, *236*, 919–929. [CrossRef] [PubMed]
5. Francis, I.L.; Newton, J.L. Determining wine aroma from compositional data. *Aust. J. Grape Wine Res.* **2005**, *11*, 114–126. [CrossRef]
6. Schüttler, A.; Friedel, M.; Jung, R.; Rauhut, D.; Darriet, P. Characterizing aromatic typicality of riesling wines: Merging volatile compositional and sensory aspects. *Food Res. Int.* **2015**, *69*, 26–37. [CrossRef]
7. Tominaga, T.; Masneuf, I.; Dubourdieu, D. Powerful Aromatic Volatile Thiols in Wines Made from Several Vitis vinifera Grape Varieties and Their Releasing Mechanism. In *Nutraceutical Beverages: Chemistry, Nutrition, and Health Effects*; American Chemical Society: Washington, DC, USA, 2009.
8. Peña-Gallego, A.; Hernández-Orte, P.; Cacho, J.; Ferreira, V. S-Cysteinylated and S-glutathionylated thiol precursors in grapes. A review. *Food Chem.* **2012**, *131*, 1–13. [CrossRef]
9. Tominaga, T.; Murat, M.-L.; Dubourdieu, D. Development of a Method for Analyzing the Volatile Thiols Involved in the Characteristic Aroma of Wines Made from *Vitis vinifera* L. Cv. Sauvignon Blanc. *J. Agric. Food Chem.* **1998**, *46*, 1044–1048. [CrossRef]
10. Roujou de Boubee, D.; Van Leeuwen, C.; Dubourdieu, D. Organoleptic impact of 2-methoxy-3-isobutylpyrazine on red Bordeaux and Loire wines. Effect of environmental conditions on concentrations in grapes during ripening. *J. Agric. Food Chem.* **2000**, *48*, 4830–4834. [CrossRef] [PubMed]
11. Rusjan, D.; Strlič, M.; Košmerl, T.; Prosen, H. The response of monoterpenes to different enzyme preparations in Gewürztraminer (*Vitis vinifera* L.) wines. *S. Afr. J. Enol. Vitic.* **2009**, *30*, 56–64. [CrossRef]
12. Gamero, A.; Quintilla, R.; Groenewald, M.; Alkema, W.; Boekhout, T.; Hazelwood, L. High-throughput screening of a large collection of non-conventional yeasts reveals their potential for aroma formation in food fermentation. *Food Microbiol.* **2016**, *60*, 147–159. [CrossRef] [PubMed]
13. Belda, I.; Ruiz, J.; Beisert, B.; Navascués, E.; Marquina, D.; Calderón, F.; Rauhut, D.; Benito, S.; Santos, A. Influence of *Torulaspora delbrueckii* in varietal thiol (3-SH and 4-MSP) release in wine sequential fermentations. *Int. J. Food Microbiol.* **2017**, *257*, 183–191. [CrossRef]
14. Ruiz, J.; Belda, I.; Beisert, B.; Navascués, E.; Marquina, D.; Calderón, F.; Rauhut, D.; Santos, A.; Benito, S. Analytical impact of *Metschnikowia pulcherrima* in the volatile profile of Verdejo white wines. *Appl. Microbiol. Biotechnol.* **2018**, *102*, 8501–8509. [CrossRef] [PubMed]
15. Grimaldi, A.; Bartowsky, E.; Jiranek, V. Screening of *Lactobacillus* spp. and *Pediococcus* spp. for glycosidase activities that are important in oenology. *J. Appl. Microbiol.* **2005**, *95*, 1061–1069. [CrossRef] [PubMed]
16. Saerens, S.M.G.; Verstrepen, K.J.; Van Laere, S.D.M.; Voet, A.R.D.; Van Dijck, P.; Delvaux, F.R.; Thevelein, J.M. The *Saccharomyces cerevisiae* EHT1 and EEB1 genes encode novel enzymes with medium-chain fatty acid ethyl ester synthesis and hydrolysis capacity. *J. Biol. Chem.* **2006**, *281*, 4446–4456. [CrossRef] [PubMed]
17. Molina, A.M.; Swiegers, J.H.; Varela, C.; Pretorius, I.S.; Agosin, E. Influence of wine fermentation temperature on the synthesis of yeast-derived volatile aroma compounds. *Appl. Microbiol. Biotechnol.* **2007**, *77*, 675–687. [CrossRef] [PubMed]
18. Ugliano, M.; Henschke, P.A. Yeasts and wine flavour. In *Wine Chemistry and Biochemistry*; Springer: New York, NY, USA, 2009; ISBN 9780387741161.
19. Benito, S. The impacts of *Schizosaccharomyces* on winemaking. *Appl. Microbiol. Biotechnol.* **2019**, *103*, 4291–4312. [CrossRef] [PubMed]
20. Benito, S.; Palomero, F.; Morata, A.; Calderón, F.; Suárez-Lepe, J.A. A method for estimating *Dekkera/Brettanomyces* populations in wines. *J. Appl. Microbiol.* **2009**, *106*, 1743–1751. [CrossRef] [PubMed]

21. Mendoza, L.; Farías, M.E. Improvement of wine organoleptic characteristics by non-*Saccharomyces* yeasts. *Appl. Microbiol.* **2010**, *2*, 908–919.

22. Ciani, M.; Comitini, F.; Mannazzu, I.; Domizio, P. Controlled mixed culture fermentation: A new perspective on the use of non-*Saccharomyces* yeasts in winemaking. *FEMS Yeast Res.* **2010**, *10*, 123–133. [CrossRef]

23. Contreras, A.; Curtin, C.; Varela, C. Yeast population dynamics reveal a potential 'collaboration' between *Metschnikowia pulcherrima* and *Saccharomyces uvarum* for the production of reduced alcohol wines during Shiraz fermentation. *Appl. Microbiol. Biotechnol.* **2015**, *99*, 1885–1895. [CrossRef]

24. Maturano, Y.P.; Mestre, M.V.; Kuchen, B.; Toro, M.E.; Mercado, L.A.; Vazquez, F.; Combina, M. Optimization of fermentation-relevant factors: A strategy to reduce ethanol in red wine by sequential culture of native yeasts. *Int. J. Food Microbiol.* **2019**, *289*, 40–48. [CrossRef]

25. Andorra, I.; Monteiro, M.; Esteve-Zarzoso, B.; Albergaria, H.; Mas, A. Analysis and direct quantification of *Saccharomyces cerevisiae* and *Hanseniaspora guilliermondii* populations during alcoholic fermentation by fluorescence in situ hybridization, flow cytometry and quantitative PCR. *Food Microbiol.* **2011**, *28*, 1483–1491. [CrossRef] [PubMed]

26. Morales, M.L.; Fierro-Risco, J.; Ríos-Reina, R.; Ubeda, C.; Paneque, P. Influence of *Saccharomyces cerevisiae* and *Lachancea thermotolerans* co-inoculation on volatile profile in fermentations of a must with a high sugar content. *Food Chem.* **2019**, *276*, 427–435. [CrossRef] [PubMed]

27. Benito, S. The impact of *Torulaspora delbrueckii* yeast in winemaking. *Appl. Microbiol. Biotechnol.* **2018**, *102*, 3081–3094. [CrossRef] [PubMed]

28. Benito, S. The impacts of *Lachancea thermotolerans* yeast strains on winemaking. *Appl. Microbiol. Biotechnol.* **2018**, *102*, 6775–6790. [CrossRef] [PubMed]

29. Benito, S.; Hofmann, T.; Laier, M.; Lochbühler, B.; Schüttler, A.; Ebert, K.; Fritsch, S.; Röcker, J.; Rauhut, D. Effect on quality and composition of Riesling wines fermented by sequential inoculation with non-*Saccharomyces* and *Saccharomyces cerevisiae*. *Eur. Food Res. Technol.* **2015**, *241*, 707–717. [CrossRef]

30. Schneider, A.; Gerbi, V.; Redoglia, M. A Rapid HPLC Method for Separation and Determination of Major Organic Acids in Grape Musts and Wines. *Am. J. Enol. Vitic.* **1987**, *38*, 151–155.

31. Semmler, H.; Sponholz, W.-R.; Rauhut, D. *Standard Operating Procedure (SOP) for the Analysis of Organic Acids in Wines by HPLC*; Hochschule Geisenheim University: Geisenheim, Germany, 2017.

32. Baumgartner, D.; Bill, R.; Roth, I. Traubenmostanalyse mit Hilfe der. FTIR-Spektroskopie. *Schweizerische Zeitschrift Obs. Weinbau* **2001**, *2*, 46–48.

33. Patz, C.D.; David, A.; Thente, K.; Kuerbel, P.; Dietrich, H. Wine analysis with FTIR spectrometry. *Wein-Wiss. Vitic. Enol. Sci.* **1999**, *54*, 80–87.

34. Rapp, A.; Yavas, I.; Hastrich, H. Einfache und schnelle Anreicherung ("Kaltronmethode") von Aromastoffen des Weines und deren quantitative Bestimmung mittels Kapillargaschromatographie. *Dtsch. Leb.* **1994**, *90*, 171–174.

35. Fritsch, S.; Brezina, S.; Ebert, K.; Rauhut, D. *Standard Operating Procedure (SOP) for the Analysis of the Fermentation Bouquet in Wines*; Department of Microbiology Biochemistry Hochschule Geisenheim University: Geisenheim, Germany, 2017.

36. Fritsch, S.; Ebert, K.; Brand, M.; Rauhut, D. *Standard Operating Procedure (SOP) for the Analysis of Terpens and Norisoprenoids in Wines*; Department of Microbiology Biochemistry Hochschule Geisenheim University: Geisenheim, Germany, 2017.

37. Rauhut, D.; Beisert, B.; Berres, M.; Gawron-Scibek, M.; Kürbel, H. Pulse flame photometric detection: An innovative technique to analyse volatile sulfur compounds in wine and other beverages. In *State-of-the-Art in Flavour Chemistry and Biology*; Hofmann, T., Rothe, M., Schieberle, P., Eds.; Deutsche Forschungsanstalt für Lebensmittelchemie: Garching, Germany, 2005; pp. 363–367. ISBN 3-00-015809-X.

38. Rauhut, D.; Beisert, B. *Standard Operating Procedure (SOP) for the Analysis of Low Volatile Sulfur Compounds Using Headspace and GC-PFPD*; Department of Microbiology Biochemistry Hochschule Geisenheim University: Geisenheim, Germany, 2017.

39. Taillandier, P.; Lai, Q.P.; Julien-Ortiz, A.; Brandam, C. Interactions between *Torulaspora delbrueckii* and *Saccharomyces cerevisiae* in wine fermentation: Influence of inoculation and nitrogen content. *World J. Microbiol. Biotechnol.* **2014**, *30*, 1959–1967. [CrossRef]

40. Gutiérrez, A.R.; Santamaría, P.; Epifanio, S.; Garijo, P.; López, R. Ecology of spontaneous fermentation in one winery during 5 consecutive years. *Lett. Appl. Microbiol.* **1999**, *29*, 411–415. [CrossRef]

41. Nissen, P.; Neilsen, D.; Arneborg, N. The relative glucose uptake abilities of non-*Saccharomyces* yeasts play a role in their coexistence with *Saccharomyces cerevisiae* in mixed cultures. *Appl. Microbiol. Biotechnol.* **2004**, *64*, 543–550. [CrossRef]

42. Stratford, M.; Anslow, P.A. Comparison of the inhibitory action on *Saccharomyces cerevisiae* of weak- acid preservatives, uncouplers, and medium-chain fatty acids. *FEMS Microbiol. Lett.* **1996**, *142*, 53–58. [CrossRef] [PubMed]

43. Stanley, G.A.; Douglas, N.G.; Every, E.J.; Tzanatos, T.; Pamment, N.B. Inhibition and stimulation of yeast growth by acetaldehyde. *Biotechnol. Lett.* **1993**, *15*, 1199–1204. [CrossRef]

44. Palfree, R.G.E.; Bussey, H. Yeast Killer Toxin: Purification and Characterisation of the Protein Toxin from *Saccharomyces cerevisiae*. *Eur. J. Biochem.* **1979**, *93*, 487–493. [CrossRef] [PubMed]

45. Santos, A.; San Mauro, M.; Bravo, E.; Marquina, D. PMKT2, a new killer toxin from Pichia membranifaciens, and its promising biotechnological properties for control of the spoilage yeast *Brettanomyces bruxellensis*. *Microbiology* **2009**, *155*, 624–634. [CrossRef]

46. Belda, I.; Ruiz, J.; Alonso, A.; Marquina, D.; Santos, A. The biology of *Pichia membranifaciens* killer toxins. *Toxins* **2017**, *9*, 112. [CrossRef] [PubMed]

47. Albergaria, H.; Francisco, D.; Gori, K.; Arneborg, N.; Gírio, F. *Saccharomyces cerevisiae* CCMI 885 secretes peptides that inhibit the growth of some non-*Saccharomyces* wine-related strains. *Appl. Microbiol. Biotechnol.* **2010**, *86*, 965–972. [CrossRef] [PubMed]

48. Holzberg, I.; Finn, R.K.; Steinkraus, K.H. A kinetic study of the alcoholic fermentation of grape juice. *Biotechnol. Bioeng.* **1967**, *9*, 413–427. [CrossRef]

49. Comitini, F.; Gobbi, M.; Domizio, P.; Romani, C.; Lencioni, L.; Mannazzu, I.; Ciani, M. Selected non-*Saccharomyces* wine yeasts in controlled multistarter fermentations with *Saccharomyces cerevisiae*. *Food Microbiol.* **2011**, *28*, 873–882. [CrossRef]

50. Azzolini, M.; Tosi, E.; Lorenzini, M.; Finato, F.; Zapparoli, G. Contribution to the aroma of white wines by controlled *Torulaspora delbrueckii* cultures in association with *Saccharomyces cerevisiae*. *World J. Microbiol. Biotechnol.* **2015**, *31*, 277–293. [CrossRef]

51. Sun, S.Y.; Gong, H.S.; Jiang, X.M.; Zhao, Y.P. Selected non-*Saccharomyces* wine yeasts in controlled multistarter fermentations with *Saccharomyces cerevisiae* on alcoholic fermentation behaviour and wine aroma of cherry wines. *Food Microbiol.* **2014**, *44*, 15–23. [CrossRef] [PubMed]

52. Mallet, S.; Arellano, M.; Boulet, J.C.; Couderc, F. Determination of tartaric acid in solid wine residues by capillary electrophoresis and indirect UV detection. *J. Chromatogr. A* **1999**, *853*, 181–184. [CrossRef]

53. Gao, C.; Fleet, G.H. Degradation of malic and tartaric acids by high density cell suspensions of wine yeasts. *Food Microbiol.* **1995**, *12*, 65–71. [CrossRef]

54. Seo, S.-H.; Rhee, C.-H.; Park, H.-D. Degradation of malic acid by *Issatchenkia orientalis* KMBL 5774, an acidophilic yeast strain isolated from Korean grape wine pomace. *J. Microbiol.* **2007**, *45*, 521–527. [PubMed]

55. Kim, D.H.; Hong, Y.A.; Park, H.D. Co-fermentation of grape must by *Issatchenkia orientalis* and *Saccharomyces cerevisiae* reduces the malic acid content in wine. *Biotechnol. Lett.* **2008**, *30*, 1633–1688. [CrossRef] [PubMed]

56. Hong, S.K.; Lee, H.J.; Park, H.J.; Hong, Y.A.; Rhee, I.K.; Lee, W.H.; Choi, S.W.; Lee, O.S.; Park, H.D. Degradation of malic acid in wine by immobilized *Issatchenkia orientalis* cells with oriental oak charcoal and alginate. *Lett. Appl. Microbiol.* **2010**, *50*, 522–529. [CrossRef] [PubMed]

57. Benito, S.; Palomero, F.; Gálvez, L.; Morata, A.; Calderón, F.; Palmero, D.; Suárez-Lepe, J.A. Quality and composition of red wine fermented with *Schizosaccharomyces pombe* as sole fermentative yeast, and in mixed and sequential fermentations with *Saccharomyces cerevisiae*. *Food Technol. Biotechnol.* **2014**, *52*, 376.

58. Kapsopoulou, K.; Kapaklis, A.; Spyropoulos, H. Growth and fermentation characteristics of a strain of the wine yeast *Kluyveromyces thermotolerans* isolated in Greece. *World J. Microbiol. Biotechnol.* **2005**, *21*, 1599–1602. [CrossRef]

59. Gobbi, M.; Comitini, F.; Domizio, P.; Romani, C.; Lencioni, L.; Mannazzu, I.; Ciani, M. *Lachancea thermotolerans* and *Saccharomyces cerevisiae* in simultaneous and sequential co-fermentation: A strategy to enhance acidity and improve the overall quality of wine. *Food Microbiol.* **2013**, *33*, 271–281. [CrossRef]

60. Benito, Á.; Calderón, F.; Benito, S. Combined use of *S. pombe* and *L. thermotolerans* in winemaking. Beneficial effects determined through the study of wines' analytical characteristics. *Molecules* **2016**, *21*, 1744. [CrossRef] [PubMed]

61. Escribano, R.; González-Arenzana, L.; Portu, J.; Garijo, P.; López-Alfaro, I.; López, R.; Santamaría, P.; Gutiérrez, A.R. Wine aromatic compound production and fermentative behaviour within different non-*Saccharomyces* species and clones. *J. Appl. Microbiol.* **2018**, *124*, 1521–1531. [CrossRef]

62. Escribano-Viana, R.; González-Arenzana, L.; Portu, J.; Garijo, P.; López-Alfaro, I.; López, R.; Santamaría, P.; Gutiérrez, A.R. Wine aroma evolution throughout alcoholic fermentation sequentially inoculated with non-*Saccharomyces/Saccharomyces* yeasts. *Food Res. Int.* **2018**, *112*, 17–24. [CrossRef] [PubMed]

63. Bartowsky, E.J.; Henschke, P.A. Acetic acid bacteria spoilage of bottled red wine—A review. *Int. J. Food Microbiol.* **2008**, *125*, 60–70. [CrossRef] [PubMed]

64. Sadoudi, M.; Tourdot-Maréchal, R.; Rousseaux, S.; Steyer, D.; Gallardo-Chacón, J.J.; Ballester, J.; Vichi, S.; Guérin-Schneider, R.; Caixach, J.; Alexandre, H. Yeast-yeast interactions revealed by aromatic profile analysis of Sauvignon Blanc wine fermented by single or co-culture of non-*Saccharomyces* and *Saccharomyces* yeasts. *Food Microbiol.* **2012**, *32*, 243–253. [CrossRef] [PubMed]

65. Contreras, A.; Hidalgo, C.; Henschke, P.A.; Chambers, P.J.; Curtin, C.; Varela, C. Evaluation of non-*Saccharomyces* yeasts for the reduction of alcohol content in wine. *Appl. Environ. Microbiol.* **2014**, *80*, 1670–1678. [CrossRef] [PubMed]

66. Vilela, A. *Lachancea thermotolerans*, the Non-*Saccharomyces* Yeast that Reduces the Volatile Acidity of Wines. *Fermentation* **2018**, *4*, 56. [CrossRef]

67. Porter, T.J.; Divol, B.; Setati, M.E. *Lachancea* yeast species: Origin, biochemical characteristics and oenological significance. *Food Res. Int.* **2019**, *119*, 378–389. [CrossRef]

68. Jolly, N.P.; Varela, C.; Pretorius, I.S. Not your ordinary yeast: Non-*Saccharomyces* yeasts in wine production uncovered. *FEMS Yeast Res.* **2014**, *14*, 215–237. [CrossRef]

69. García, M.; Esteve-Zarzoso, B.; Arroyo, T. Non-*Saccharomyces* Yeasts: Biotechnological Role for Wine Production. In *Grape and Wine Biotechnology*; InTech: London, UK, 2016.

70. du Plessis, H.; du Toit, M.; Nieuwoudt, H.; van der Rijst, M.; Kidd, M.; Jolly, N. Effect of *Saccharomyces*, Non-*Saccharomyces* Yeasts and Malolactic Fermentation Strategies on Fermentation Kinetics and Flavor of Shiraz Wines. *Fermentation* **2017**, *3*, 64. [CrossRef]

71. García, M.; Esteve-Zarzoso, B.; Cabellos, J.; Arroyo, T. Advances in the Study of *Candida stellata*. *Fermentation* **2018**, *4*, 74. [CrossRef]

72. Zohre, D.E.; Erten, H. The influence of *Kloeckera apiculata* and *Candida pulcherrima* yeasts on wine fermentation. *Process Biochem.* **2002**, *38*, 319–324. [CrossRef]

73. Clemente-Jimenez, J.M.; Mingorance-Cazorla, L.; Martínez-Rodríguez, S.; Las Heras-Vázquez, F.J.; Rodríguez-Vico, F. Molecular characterization and oenological properties of wine yeasts isolated during spontaneous fermentation of six varieties of grape must. *Food Microbiol.* **2004**, *21*, 149–155. [CrossRef]

74. Varela, C.; Sengler, F.; Solomon, M.; Curtin, C. Volatile flavour profile of reduced alcohol wines fermented with the non-conventional yeast species *Metschnikowia pulcherrima* and *Saccharomyces uvarum*. *Food Chem.* **2016**, *29*, 57–64. [CrossRef] [PubMed]

75. Belda, I.; Navascués, E.; Marquina, D.; Santos, A.; Calderon, F.; Benito, S. Dynamic analysis of physiological properties of *Torulaspora delbrueckii* in wine fermentations and its incidence on wine quality. *Appl. Microbiol. Biotechnol.* **2015**, *99*, 1911–1922. [CrossRef] [PubMed]

76. Amaya-Delgado, L.; Herrera-López, E.J.; Arrizon, J.; Arellano-Plaza, M.; Gschaedler, A. Performance evaluation of *Pichia kluyveri*, *Kluyveromyces marxianus* and *Saccharomyces cerevisiae* in industrial tequila fermentation. *World J. Microbiol. Biotechnol.* **2013**, *29*, 875–881. [CrossRef]

77. Lu, Y.; Huang, D.; Lee, P.R.; Liu, S.Q. Assessment of volatile and non-volatile compounds in durian wines fermented with four commercial non-*Saccharomyces* yeasts. *J. Sci. Food Agric.* **2016**, *96*, 1511–1521. [CrossRef]

78. Fleet, G.H. Yeast interactions and wine flavour. *Int. J. Food Microbiol.* **2003**, *86*, 11–22. [CrossRef]

79. Ruiz, J.; Kiene, F.; Belda, I.; Fracassetti, D.; Marquina, D.; Navascués, E.; Calderón, F.; Benito, A.; Rauhut, D.; Benito, S. Effects on varietal aromas during wine making: A review of the impact of varietal aromas on the flavor of wine. *Appl. Microbiol. Biotechnol.* **2019**, 1–26. [CrossRef]

80. Pereira, C.S.M.; Silva, V.M.T.M.; Rodrigues, A.E. Ethyl lactate as a solvent: Properties, applications and production processes—A review. *Green Chem.* **2011**, *13*, 2658–2671. [CrossRef]

81. Mestres, M.; Busto, O.; Guasch, J. Analysis of organic sulfur compounds in wine aroma. *J. Chromatogr. A* **2000**, *881*, 569–581. [CrossRef]

82. Jiranek, V.; Langridge, P.; Henschke, P.A. Regulation of hydrogen sulfide liberation in wine-producing *Saccharomyces cerevisiae* strains by assimilable nitrogen. *Appl. Environ. Microbiol.* **1995**, *61*, 461–467.
83. Mendes-Ferreira, A.; Mendes-Faia, A.; Leão, C. Survey of hydrogen sulphide production by wine yeasts. *J. Food Prot.* **2002**, *65*, 1033–1037. [CrossRef]
84. Renault, P.; Miot-Sertier, C.; Marullo, P.; Hernández-Orte, P.; Lagarrigue, L.; Lonvaud-Funel, A.; Bely, M. Genetic characterization and phenotypic variability in *Torulaspora delbrueckii* species: Potential applications in the wine industry. *Int. J. Food Microbiol.* **2009**, *134*, 201–210. [CrossRef]

 *fermentation*

Article

Modulation of Wine Flavor using *Hanseniaspora uvarum* in Combination with Different *Saccharomyces cerevisiae*, Lactic Acid Bacteria Strains and Malolactic Fermentation Strategies

Heinrich Du Plessis [1,2,*], **Maret Du Toit** [2], **Hélène Nieuwoudt** [2], **Marieta Van der Rijst** [3], **Justin Hoff** [1] **and Neil Jolly** [1]

[1]   Post-Harvest and Agro-Processing Technologies, Agricultural Research Council (ARC) Infruitec-Nietvoorbij, Private Bag X5026, Stellenbosch 7599, South Africa

[2]   Institute for Wine Biotechnology & Department of Viticulture and Oenology, Stellenbosch University, Private Bag X1, Matieland 7602, South Africa

[3]   ARC Biometry, Private Bag X5026, Stellenbosch 7599, South Africa

*   Correspondence: dplessishe@arc.agric.za; Tel.: +27-021-809-3063

Received: 14 June 2019; Accepted: 5 July 2019; Published: 9 July 2019

Abstract: *Hanseniaspora uvarum* is one of the predominant non-*Saccharomyces* yeast species found on grapes and in juice, but its effect on lactic acid bacteria (LAB) growth and wine flavor has not been extensively studied. Therefore, the interaction between *H. uvarum*, two *Saccharomyces cerevisiae* yeast strains, two LAB species (*Lactobacillus plantarum* and *Oenococcus oeni*) in combination with two malolactic fermentation (MLF) strategies was investigated in Shiraz wine production trials. The evolution of the different microorganisms was monitored, non-volatile and volatile compounds were measured, and the wines were subjected to sensory evaluation. Wines produced with *H. uvarum* in combination with *S. cerevisiae* completed MLF in a shorter period than wines produced with only *S. cerevisiae*. Sequential MLF wines scored higher for fresh vegetative and spicy aroma than wines where MLF was induced as a simultaneous inoculation. Wines produced with *H. uvarum* had more body than wines produced with only *S. cerevisiae*. The induction of MLF using *L. plantarum* also resulted in wines with higher scores for body. *H. uvarum* can be used to reduce the duration of MLF, enhance fresh vegetative aroma and improve the body of a wine.

Keywords: lactic acid bacteria; yeasts; chemical analyses; volatile compounds; sensory evaluation; shiraz

---

1. Introduction

The contribution of yeasts to wine composition and quality is well-known [1,2]. The *Saccharomyces* yeasts drive alcoholic fermentation by converting the grape sugar to alcohol, carbon dioxide, and other compounds affecting the wine aroma and taste [3,4]. The other group of yeasts important to winemaking are the non-*Saccharomyces* yeasts, also known as "wild yeast", which have different oenological characteristics to *S. cerevisiae* and can be used to improve wine quality through enhanced wine aroma and complexity [2,5]. Non-*Saccharomyces* yeast species such as *Hanseniaspora uvarum* (*Kloeckera apiculata*), frequently found on grapes and in grape must, are known to dominate the initial phases of spontaneous fermentations [6–8]. Certain *H. uvarum* strains can produce high levels of acetic acid and ethyl acetate, although there is great variability among strains [9–11]. It has also been reported that *H. uvarum* can produce increased levels of desirable compounds such as esters, higher alcohols, and carbonyl compounds [2,11,12]. Mendoza et al. [13] and Tristezza et al. [11] showed that mixed culture fermentations of *H. uvarum* and *S. cerevisiae* could be used to enhance wine aroma and quality.

Another process that plays an important role with regard to wine flavor and quality is malolactic fermentation (MLF), which decreases acidity by converting L-malic acid to L-lactic acid and CO_2. Malolactic fermentation can affect wine flavor by modifying the concentrations of aroma impact compounds such as diacetyl, esters, higher alcohols, and volatile acids [14,15]. Previously, *Oenococcus oeni* has been the lactic acid bacteria (LAB) of choice as a MLF starter [16], but recently more *Lactobacillus plantarum* starters have become available. *L. plantarum* produces a broader range of extracellular enzymes than *O. oeni*, which enhances flavor development [17–19]. Different MLF inoculation strategies, i.e., simultaneous inoculation (at the start of alcoholic fermentation) and sequential inoculation (after alcoholic fermentation), have been shown to affect the flavor profiles of wines [20–22].

A better understanding of how wine production methods can be manipulated to enhance wine attributes, such as aroma, flavor, body, or mouth-feel, is important for the production of a targeted wine style [23]. Du Plessis et al. [24] reported that the MLF strategy had a greater impact on the chemical and sensory profiles of Shiraz wines than yeast combinations. Only one *S. cerevisiae* strain and one LAB species were used in that study, therefore we wanted to investigate whether the same trend would be observed if different *S. cerevisiae* strains and LAB species were used. The *H. uvarum* strain in that study was shown to be compatible with MLF, had potential to enhance wine flavor and is one of the non-*Saccharomyces* yeast species frequently found on grapes and in must. The aims were to investigate the interactions between *H. uvarum*, two commercial *S. cerevisiae* strains, two LAB species (*L. plantarum* and *O. oeni*) and three MLF strategies, as well as to determine how these interactions affect shiraz wine composition and flavor.

2. Materials and Methods

2.1. Cultivation and Enumeration of Micro-Organisms

The selected yeast and LAB strains used are listed in Table 1. Similar culturing conditions and procedures were followed as described in Du Plessis et al. [24]. Eight hundred milliliters of the *H. uvarum* yeast culture was inoculated, at a concentration of ~1×10^6 cells/mL, into the Shiraz juice and skin mixture (30 kg). Commercial *S. cerevisiae* yeast and LAB cultures (*O. oeni* and *L. plantarum*) were inoculated according to the manufacturers' recommendations.

Table 1. Yeasts and lactic acid bacteria used in this study.

Reference Code	Species Name	Source
Sc1	*Saccharomyces cerevisiae*	VIN 13, commercial strain, Anchor Wine Yeast, South Africa
Sc2	*Saccharomyces cerevisiae*	NT 202, commercial strain, Anchor Wine Yeast
Hu	*Hanseniaspora uvarum*	Y0858, natural isolate, ARC Infruitec-Nietvoorbij culture collection
LAB1	*Oenococcus oeni*	Viniflora® oenos, commercial malolactic fermentation starter, Chr. Hansen A/S, Denmark
LAB2	*Lactobacillus plantarum*	Enoferm V22, commercial malolactic fermentation starter, Lallemand Inc., France

Non-*Saccharomyces* and total yeast counts for the shiraz juice and wine were obtained by plating out on Wallerstein Laboratory(WL) medium and Lysine medium, respectively (Biolab, Merck, South Africa). Bacterial counts were obtained by plating out on de Man, Rogosa and Sharpe (MRS) agar (Biolab, Merck) supplemented with 25% (v/v) grape juice and 100 mg/L Natamycin (Danisco A/S, Denmark). Yeasts were grown aerobically for 2–3 days at 28 °C, while bacteria were cultivated under facultative anaerobic conditions at the same temperature for 2–7 days. The respective control wines produced with Sc1 and Sc2, which only received a *S. cerevisiae* inoculum, were used to determine the levels of naturally occurring non-*Saccharomyces* yeasts during fermentation [24]. The naturally occurring *Saccharomyces* yeasts were only determined on day 0 and 1, and counts were obtained from the treatments without a *S. cerevisiae* inoculum, i.e., *H. uvarum* treatments. The development of the naturally occurring LAB during fermentation was monitored by sampling the treatments that were not inoculated with LAB and the sequential MLF treatments until day 5, because the commercial LAB cultures were added to induce sequential MLF after that.

2.2. Wine Production

Shiraz grapes were obtained from the Nietvoorbij research farm (Stellenbosch, South Africa). A standardized small-scale (20 L) winemaking procedure was followed as described by Du Plessis et al. [24]. The treatments that were applied are listed in Table 2. Four different yeast treatments (Sc1, Sc2, Hu + Sc1 and Hu + Sc2) were investigated. Each yeast treatment was used in combination with LAB1 and LAB2 (*O. oeni* and *L. plantarum*), respectively. Additionally, the two MLF strategies were applied: (1) Simultaneous inoculation of LAB (hereafter referred to as simultaneous MLF) and (2) sequential inoculation of LAB (hereafter referred to as sequential MLF). Wines that did not undergo MLF will be referred to as non-MLF wines, while wines that underwent MLF will be referred to as MLF wines. Sixty wines were produced, which included 20 different treatments with three replicates each.

The *S. cerevisiae* strains, Sc1 and Sc2, were inoculated on day 0 for the control treatments. *H. uvarum* was inoculated on day 0, and Sc1 and Sc2, respectively, were inoculated after 24 h (day 1) for all the mixed yeast fermentations. The LAB in the simultaneous MLF treatments was added 25 hours after the initial yeast inoculations on day 0. Fermentations were carried out at *ca.* 24 °C and after completion of the alcoholic fermentation, the sequential MLF treatments were inoculated with LAB1 or LAB2. All treatments were racked, fined, cold stabilized, and bottled as described by Minnaar et al. [25]. After bottling, all wines were stored at 15 °C until required for evaluation.

2.3. Juice and Wine Analysis

The following parameters of the grape must were measured, i.e., sugar (°Balling), free and total SO_2 (Ripper method), pH and titratable acidity (Mettler titrator) analyses as described in the South African Wine Laboratories Association (SAWLA) Manual [26]. Alcoholic fermentation was monitored using an OenoFoss™ Fourier transform infrared (FTIR) spectrometer (FOSS Analytical A/S, Denmark) and was considered to be complete when the residual sugar concentrations were below 4 g/L. The progression of MLF was monitored with the OenoFoss™ FTIR spectrometer until the malic acid levels were below 0.2 g/L, the point where MLF was considered to be complete. Oenological parameters of the wines, such as residual sugar (glucose + fructose), pH, malic and lactic acids, total acidity (TA), alcohol, volatile acidity (VA), and glycerol, were determined with a WineScan™ FT120 instrument (FOSS Analytical A/S) as described by Du Plessis et al. [24] and Louw et al. [27]. The method described by Louw et al. [27] using gas chromatography with a flame ionization detector (GC-FID) was applied to analyze the volatile compounds in the wines.

Table 2. Oenological parameters and duration of malolactic fermentation (MLF) of shiraz juice [1] and wines produced with *Saccharomyces cerevisiae* (Sc1 or Sc2) only or in combination with *Hanseniaspora uvarum* (Hu), two lactic acid bacteria strains (LAB1 or LAB2) and two MLF strategies (simultaneous or sequential inoculation). Values are means of three replicate fermentations, and the standard deviations are also shown.

Treatment	Residual Sugar (g/L)	pH	Volatile Acidity (g/L)	Total Acidity (g/L)	Malic Acid (g/L)	Lactic Acid (g/L)	Alcohol (% v/v)	Glycerol (g/L)	MLF Duration (Days)
Sc1	1.70 cdefg[3] ± 0.22	3.61 g ± 0.00	0.24 k ± 0.01	6.16 a ± 0.07	2.81 a ± 0.08	<0.20 j ± 0.00	13.65 bcdef ± 0.02	10.62 fg ± 0.03	No MLF
Sc1 + LAB1 sim MLF [2]	1.60 fg ± 0.22	3.75 cd ± 0.01	0.38 hi ± 0.01	5.36 defg ± 0.03	<0.20 d ± 0.00	1.33 fg ± 0.01	13.80 abcde ± 0.05	11.00 bcd ± 0.06	10
Sc1 + LAB1 seq MLF	1.69 defg ± 0.16	3.80 ab ± 0.02	0.45 def ± 0.01	5.39 cde ± 0.04	<0.20 d ± 0.00	1.35 ef ± 0.04	13.99 a ± 0.09	11.13 b ± 0.01	34
Sc1 + LAB2 sim MLF	1.63 efg ± 0.11	3.81 a ± 0.01	0.46 de ± 0.01	5.43 cd ± 0.03	<0.20 d ± 0.00	1.39 de ± 0.03	13.92 ab ± 0.02	11.13 b ± 0.02	34
Sc1 + LAB2 seq MLF	1.47 g ± 0.18	3.77 bc ± 0.01	0.43 efg ± 0.01	5.50 c ± 0.05	<0.20 d ± 0.00	1.40 de ± 0.02	13.87 abc ± 0.11	11.09 bc ± 0.09	34
Hu+Sc1	1.81 abcdef ± 0.07	3.60 g ± 0.00	0.29 j ± 0.01	6.17 a ± 0.02	1.82 c ± 0.11	0.24 ij ± 0.03	13.52 efg ± 0.04	10.27 h ± 0.06	No MLF
Hu + Sc1 + LAB1 sim MLF	1.97 abc ± 0.27	3.70 f ± 0.01	0.41 gh ± 0.01	5.35 defg ± 0.03	<0.20 d ± 0.00	1.50 a ± 0.02	13.61 cdefg ± 0.06	10.75 efg ± 0.08	10
Hu + Sc1 + LAB1 seq MLF	1.87 abcde ± 0.19	3.75 cd ± 0.02	0.44 ef ± 0.02	5.27 fg ± 0.02	<0.20 d ± 0.00	1.40 cde ± 0.04	13.66 bcdef ± 0.08	10.77 defg ± 0.04	22
Hu + Sc1 + LAB2 sim MLF	1.93 abcd ± 0.14	3.73 de ± 0.02	0.41 gh ± 0.01	5.29 efg ± 0.02	<0.20 d ± 0.00	1.42 bcd ± 0.03	13.61 cdefg ± 0.01	10.82 defg ± 0.02	19
Hu + Sc1 + LAB2 seq MLF	1.74 bcdef ± 0.32	3.76 cd ± 0.03	0.43 fg ± 0.02	5.25 g ± 0.05	<0.20 d ± 0.00	1.46 abc ± 0.02	13.66 bcdef ± 0.10	10.86 cdef ± 0.04	22
Sc2	1.78 abcdef ± 0.23	3.58 g ± 0.03	0.35 l ± 0.01	6.04 b ± 0.21	2.11 b ± 0.43	<0.20 j ± 0.00	13.65 bcdef ± 0.64	11.11 b ± 0.46	No MLF
Sc2 + LAB1 sim MLF	2.03 a ± 0.20	3.68 f ± 0.02	0.51 b ± 0.03	5.36 defg ± 0.04	<0.20 d ± 0.00	1.24 h ± 0.01	13.87 abc ± 0.08	11.70 a ± 0.09	10
Sc2 + LAB1 seq MLF	1.98 ab ± 0.10	3.74 cde ± 0.01	0.51 b ± 0.02	5.26 fg ± 0.04	<0.20 d ± 0.00	1.22 h ± 0.08	13.99 a ± 0.11	11.74 a ± 0.05	22
Sc2 + LAB2 sim MLF	1.99 ab ± 0.24	3.68 f ± 0.02	0.48 cd ± 0.02	5.35 defg ± 0.02	<0.20 d ± 0.00	1.19 h ± 0.02	13.95 a ± 0.12	11.60 a ± 0.17	20
Sc2 + LAB2 seq MLF	1.95 abcd ± 0.08	3.68 f ± 0.02	0.50 bc ± 0.03	5.34 defg ± 0.02	<0.20 d ± 0.00	1.2 h ± 0.00	13.82 abcd ± 0.23	11.73 a ± 0.12	22
Hu+Sc2 no MLF	1.91 abcd ± 0.25	3.58 g ± 0.01	0.43 efg ± 0.01	6.08 ab ± 0.05	1.69 c ± 0.00	0.26 l ± 0.02	13.60 cdefg ± 0.06	10.60 g ± 0.05	No MLF
Hu+Sc2 + LAB1 sim MLF	1.94 abcd ± 0.09	3.71 ef ± 0.04	0.57 a ± 0.03	5.37 def ± 0.12	<0.20 d ± 0.00	1.49 a ± 0.07	13.35 g ± 0.11	11.16 b ± 0.07	10
Hu+Sc2 + LAB1 seq MLF	1.88 abcde ± 0.07	3.71 ef ± 0.03	0.58 a ± 0.01	5.26 fg ± 0.06	<0.20 d ± 0.00	1.37 def ± 0.04	13.47 fg ± 0.05	10.97 bcde ± 0.12	18
Hu+Sc2 + LAB2 sim MLF	1.82 abcdef ± 0.16	3.76 ad ± 0.03	0.57 a ± 0.01	5.26 fg ± 0.06	<0.20 d ± 0.00	1.47 ab ± 0.06	13.55 defg ± 0.12	11.19 b ± 0.02	15
Hu+Sc2 + LAB2 seq MLF	1.93 abcd ± 0.20	3.71 ef ± 0.02	0.53 b ± 0.05	5.27 fg ± 0.09	<0.20 d ± 0.00	1.30 g ± 0.03	13.56 defg ± 0.04	10.99 bcde ± 0.33	18

[1] Juice analysis: Balling = 23.0°B, pH = 3.57, total acidity = 7.43 g/L, malic acid = 3.1 g/L, free SO_2 = 4 mg/L, and total SO_2 = 16 mg/L. [2] LAB1: *Oenococcus oeni*, LAB2: *Lactobacillus plantarum*, simultaneous (sim) and sequential (seq) MLF. [3] Values in the same column followed by the same letter do not differ significantly ($p \leq 0.05$).

2.4. Sensory Evaluation of Wines

A panel consisting of 22 experienced wine judges (13 men and nine women, aged 22 to 50 years) evaluated the wines four months after bottling. The same panelists and procedures were used as described by Du Plessis et al. [24]. The panelists were asked to rate the intensity of the aroma and taste descriptors on a 100 mm unstructured line scale. The intensity of aroma descriptors: Berry, fruity, fresh vegetative, cooked vegetative, floral, sweet associated, and spicy were rated from undetectable to prominent, while the taste descriptors were rated from low to high for acid balance, thin to full for body (mouth-feel) and undetectable to prominent for astringency and bitterness. The descriptors were scored by measuring where the mark was made on the line and expressing the value as a percentage. Each judge had a separate tasting booth and *ca.* 30 mL of the wine sample were presented in a randomized order in a standard wine glass, labeled with a three digit code. Research Randomizer (Version 4.0, http://randomizer.org) was used to generate the three digit code and to randomize the order in which the wines were presented to each panelist.

2.5. Data and Statistical Analysis

Chemical and sensory data were tested for deviation from normality by the Shapiro–Wilk test and then subjected to analysis of variance (ANOVA) using the general linear means procedure of SAS version 9.4 (SAS Institute Inc., Cary, North Carolina, USA). Fisher's least significant difference (LSD) values were calculated at the 5% probability level ($p = 0.05$) to facilitate comparison between treatment means. Principal component analysis (PCA) was performed using XLSTAT software (Version 18.07.39157, Addinsoft, New York, USA) to examine the correlation between treatments and the volatile chemical variables.

2.6. Verification of H. uvarum Implantations

Yeasts were isolated from juice and wine (day 2) samples to verify successful implantation. From WL plates with a colony count of 30 to 300, five colonies were selected randomly per replicate. Subsequently, yeast DNA was extracted using the method described by Lõoke et al. [28]. Yeast identification to the species level was carried out by amplification of the 5.8S-internal transcribed spacer (ITS) ribosomal region, using primers, ITS1 and ITS4, followed by enzyme restriction with *CfoI*, as described by Esteve-Zarzoso et al. [29]. Restriction profiles of the isolates were compared to those of known yeast species. Successful implantation of the *H. uvarum* strain was verified with random amplified polymorphic DNA (RAPD), using primer 1283 and conditions described by Pfliegler et al. [30]. Amplification products (ITS-RFLP and RAPD) were separated on 2% agarose gels, and banding patterns were visualized on a Bio-Rad image analyzer, following staining with 0.01% (*v/v*) ethidium bromide (Bio-Rad Laboratories, Inc., USA).

3. Results and Discussion

3.1. Yeast Development

The naturally occurring *Saccharomyces* and non-*Saccharomyces* yeast populations in the shiraz juice were *ca.* 4.2×10^5 and 4.1×10^5 colony forming units/mL (CFU/mL), respectively (Figure 1). The naturally occurring non-*Saccharomyces* yeast populations decreased notably on day 1, in treatments inoculated with the commercial *S. cerevisiae* yeasts, before increasing again on day 2. Thereafter the naturally occurring non-*Saccharomyces* yeast populations remained at levels of *ca.* 1×10^5 CFU/mL in wines fermented with Sc2 or decreased to *ca.* 1×10^4 CFU/mL in wines inoculated with Sc1. *S. cerevisiae* strain Sc1 had a negative effect on the growth of naturally occurring non-*Saccharomyces* yeasts because after five days of the Sc1 treatment, the non-*Saccharomyces* yeast levels were lower than for wines fermented with Sc2.

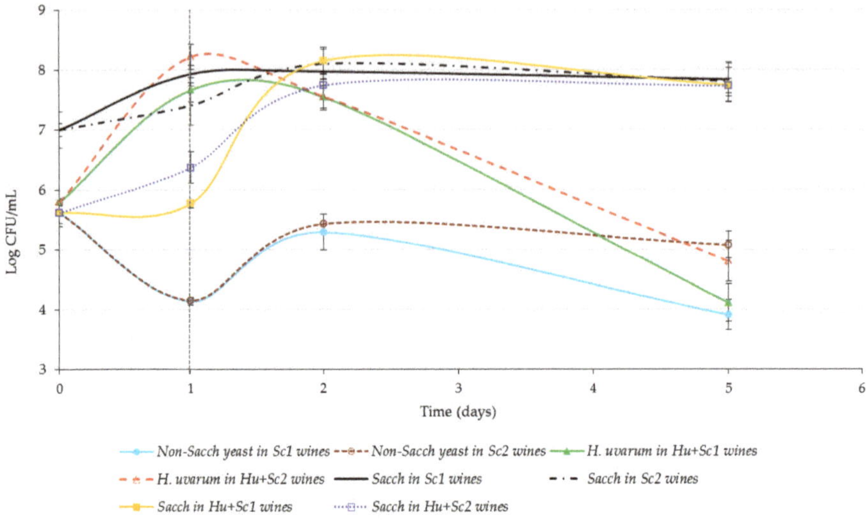

Figure 1. Cell counts (colony forming units/milliliters, CFU/mL) of naturally occurring and inoculated *Saccharomyces cerevisiae* (Sacch), naturally occurring non-*Saccharomyces* (Non-Sacch), and inoculated *Hanseniaspora uvarum* (*H. uvarum*) yeasts during alcoholic fermentation. The dashed vertical line at day 1 indicates when commercial *S. cerevisiae* yeasts were added. Abbreviations: Sc1 = commercial *S. cerevisiae* strain 1, Sc2 = commercial *S. cerevisiae* strain 2, Hu = inoculated *H. uvarum* yeasts. Values are means of three replicates and error bars indicate standard deviation.

Initial yeast counts of the wines inoculated with *H. uvarum* were just below 1×10^6 CFU/mL, but increased to levels >10 million CFU/mL after 24 h. However, this trend changed after inoculation of commercial *S. cerevisiae* yeasts (day 1, Figure 1), which resulted in the decrease of *H. uvarum* numbers. The same trend was found with regard to the inhibitory activity of Sc1 on non-*Saccharomyces* yeast viability. At the end of alcoholic fermentation, inoculated and naturally occurring non-*Saccharomyces* yeast populations were at a similar level.

The naturally occurring *Saccharomyces* yeast populations were present at moderately high numbers (4×10^5 CFU/mL), which increased after 24 h, but the inoculated *H. uvarum* yeasts were present at higher numbers (8×10^7 to 1×10^8 CFU/mL). However, both aforementioned populations were dominated by the inoculated *S. cerevisiae* yeasts, following their addition after 24 h. These results indicate that the inoculated *S. cerevisiae* strains were responsible for completing the alcoholic fermentations. However, the inoculated *H. uvarum* populations were present at high levels (10^7 to 10^8 CFU/mL) and long enough to potentially make a contribution to wine flavor. A similar trend was observed by Du Plessis et al. [24].

3.2. Yeast Verification

A selection of yeast colonies 2 was identified by amplification of the ITS-5.8S region, followed by subsequent restriction analysis. Isolate profiles were compared to profiles of known yeast species. The dominant non-*Saccharomyces* yeasts isolated from the Hu + Sc1 and Hu + Sc2 wines were identified as *H. uvarum*. DNA of these isolates were subsequently amplified using primer 1283 and the products were compared to the reference *H. uvarum* strain (Table 1). All wine isolates had similar banding patterns as the *H. uvarum* reference strain (Figure 2), indicating successful implantation. The banding patterns of *H. uvarum* juice isolates (naturally occurring strains) differed from the *H. uvarum* reference strain and were not detected in any of the implanted wines during the first two days of alcoholic fermentation. These results indicate that the inoculated *H. uvarum* dominated the naturally occurring *H. uvarum* population.

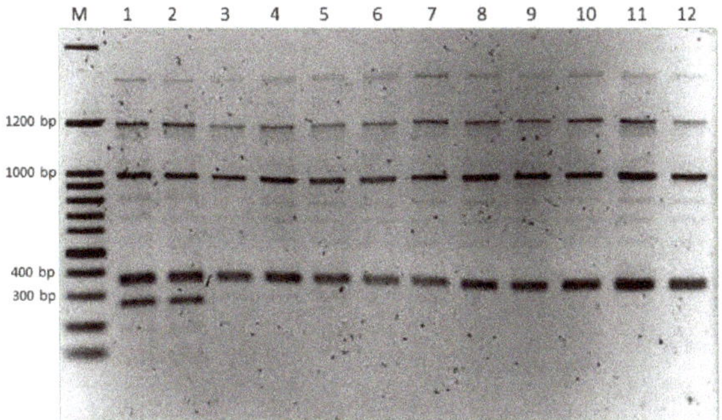

Figure 2. Random amplified polymorphic DNA products of selected *Hanseniaspora uvarum* isolates from shiraz wines produced with *Saccharomyces cerevisiae* Sc1 or Sc2 in combination with *H. uvarum*. M: 100 bp DNA ladder, lane1: *H. uvarum* strain isolated from shiraz juice, lane 2: *H. uvarum* strain isolated from shiraz juice, lane 3: *H. uvarum* reference used for implantations, lane 4 to 12: dominant non-*Saccharomyces* yeasts isolated from wines inoculated with *H. uvarum* and *S. cerevisiae*.

3.3. LAB Development and Progression of MLF

The growth and development of the naturally occurring and inoculated LAB are shown in Figure 3. The naturally occurring LAB were present at ~3.5 × 10⁴ CFU/mL in the Shiraz grape must and decreased during alcoholic fermentation in most of the treatments, with the increase in numbers at the end (day 5). This is also the typical winemaking scenario [4,31]. Individually, the numbers of naturally occurring LAB varied notably in wines, fermented with the selected yeast combinations. Based on the LAB counts from day 2 to 5, Sc1 had a greater inhibitory effect on LAB growth (decreased from 3.5 × 10⁴ to 8.8 × 10² CFU/mL) than Sc2 or *H. uvarum* in combination with Sc1 or Sc2 (decreased from 3.5 × 10⁴ to 1.8 × 10³ CFU/mL). This is in agreement with findings of Du Plessis et al. [24].

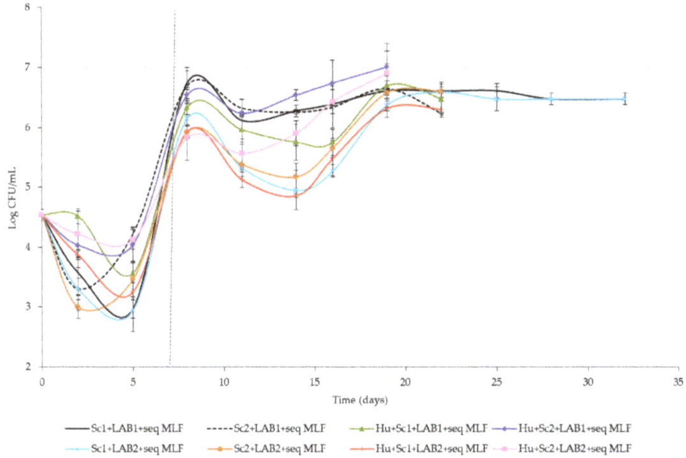

Figure 3. Cell counts (colony forming units per milliliters, CFU/mL) of the naturally occurring and sequentially inoculated lactic acid bacteria (LAB) in shiraz juice and wine produced with *Saccharomyces cerevisiae* (Sc1 or Sc2) on its own or in combination with *Hanseniaspora uvarum* (Hu) and two LAB species (LAB1 or LAB2). The dashed vertical line at day 7 indicates inoculation of the commercial LAB for sequential malolactic fermentation (seq MLF). Values are means of three replicate fermentations and error bars indicate standard deviation.

The alcoholic fermentation was completed after five days and the commercial LAB were inoculated on day 7 to induce sequential MLF in the selected treatments. The addition of commercial LAB resulted in an expected increase of LAB numbers from ~1×10^3–10^4 to >7×10^5 CFU/mL (Figure 3). No notable delays in MLF was observed in sequentially inoculated wines (Table 2), despite inoculated LAB1 (*O. oeni*) and LAB2 (*L. plantarum*) counts decreasing from 5.0×10^6 to 4.5×10^5 CFU/mL, and 6.8 to 1.9×10^5 CFU/mL, respectively (Figure 3). Wines produced with Hu + Sc1 + LAB1 and Hu + Sc2 + LAB2 completed MLF in the shortest time (18 days), while wines produced with Sc1 + LAB1 and Sc1 + LAB2 took the longest to complete MLF (34 days). The delay in MLF of the Sc1+LAB2 wines can be correlated to lower LAB numbers (<1×10^6 CFU/mL), but the trend was not observed for Sc1 + LAB1 wines, which contained high LAB numbers (>1×10^6 CFU/mL) throughout MLF (Figure 3).

The development of LAB that were inoculated at the same time as the yeasts are shown in Figure 4. LAB1 (*O. oeni*) numbers remained above 1×10^6 CFU/mL and completed MLF within 10 days (Table 2), while LAB2 (*L. plantarum*) numbers decreased to below 1×10^6 CFU/mL, before increasing again, which resulted in the MLF taking longer to complete. For the wines inoculated with LAB2, the Hu + Sc2 treatment completed MLF within 15 days, while the Sc1 + LAB2 treatment took 34 days to complete MLF. There was a negative interaction between Sc1 and LAB2. The inhibition of LAB2 growth might be due to the depletion of essential nutrients needed for LAB growth or the production of toxic metabolites.

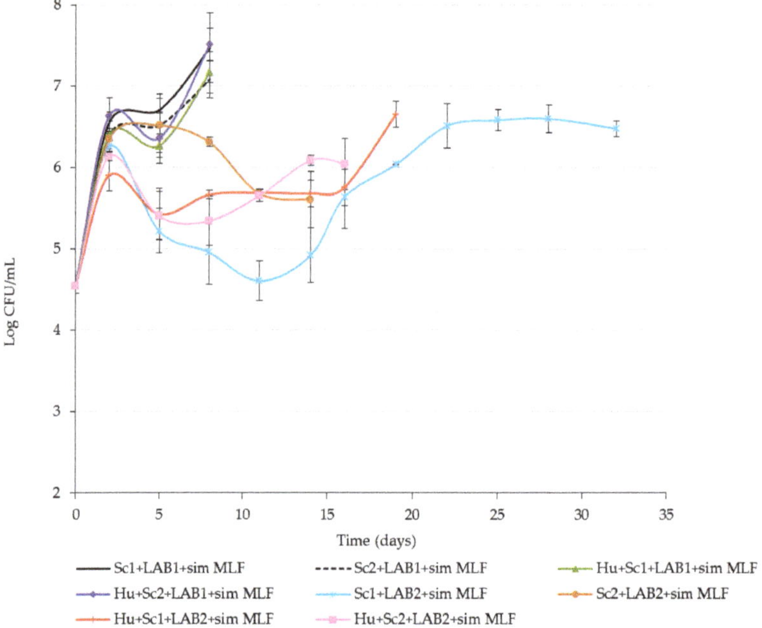

Figure 4. Cell counts (colony forming units per milliliters, CFU/mL) of the naturally occurring and inoculated lactic acid bacteria (LAB) in shiraz juice and wine produced with *Saccharomyces cerevisiae* (Sc1 or Sc2) on its own or in combination with *Hanseniaspora uvarum* and two LAB species (LAB1 or LAB2). Malolactic fermentation induced as a simultaneous inoculation (sim MLF). Values are means of three replicate fermentations and error bars indicate standard deviation.

In general, *O. oeni* is known to be better suited to harsh conditions found in wine than *L. plantarum*, which explains why LAB1 performed better than LAB2. Overall, simultaneous MLF completed in a shorter time than sequential MLF. This trend is in agreement with findings of other researchers [20,32].

3.4. Standard Oenological Parameters

The interaction between yeast combinations, LAB strains and MLF strategies had a significant effect ($p \leq 0.05$) on pH, VA, malic acid, and lactic acid concentrations of wines (Table S1). In addition, yeast combination also had a significant effect on alcohol and glycerol concentrations of wines, while the interaction between LAB strain and MLF strategy had a significant impact on TA, and glycerol concentrations.

3.4.1. Wines without MLF

All wines were fermented to dryness and contained residual sugar levels of less than 4 g/L (Table 2). Alcohol concentrations of wines produced with *H. uvarum* in combination with Sc1 and Sc2 were slightly lower than those produced with only Sc1 or Sc2. This trend is in agreement with findings of Mendoza et al. [13,33]. Wines produced with only *S. cerevisiae* yeasts contained significantly higher glycerol concentrations than wines produced with the *H. uvarum* and *S. cerevisiae* combinations. Mendoza et al. [13] reported similar findings, but Liu et al. [34] reported the contrary, which indicates that this is not a species trait, but rather strain dependent.

None of the treatments produced excessively high concentrations of VA (>0.7 g/L). However, VA concentrations in wines produced with *H. uvarum* in combination with Sc1 (0.29 g/L) and Sc2 (0.43 g/L) were slightly higher than wines produced with Sc1 (0.24 g/L) and Sc2 (0.35 g/L) on their own. This is in agreement with the findings of Mendoza et al. [13] and also confirmed reports that some *H. uvarum* (*K. apiculata*) strains can produce lower VA levels comparable to those of *S. cerevisiae* [11,35,36]. Malic acid concentrations in wines produced with *H. uvarum* in combination with Sc1 (1.82 g/L) and Sc2 (1.69 g/L) were significantly lower than wines produced with Sc1 (2.81 g/L) and Sc2 (2.11 g/L) on their own. The ability of this *H. uvarum* strain to degrade malic acid has been reported by Du Plessis et al. [24,37].

3.4.2. Wines That Underwent MLF

In most cases, non-MLF wines contained lower alcohol levels than MLF wines (Table 2). These findings are contrary to those of Mendoza et al. [13] and Abrahamse and Bartowsky [20], but in agreement with results of Izquierdo-Cañas et al. [37] and Du Plessis et al. [24]. The differences reported might be due to the LAB strain used or LAB and yeast interactions. In general, the alcohol levels were lower for simultaneous MLF wines than for sequential MLF wines, which are in agreement with the findings of Mendoza et al. [13] and Abrahamse and Bartowsky [20], but contrary to the findings of Izquierdo-Cañas et al. [32] and Tristezza et al. [22]. MLF wines had significantly higher glycerol levels than non-MLF wines. In most cases, simultaneous MLF wines contained slightly lower glycerol levels than sequential MLF wines.

Overall, MLF wines contained significantly higher VA values (0.38 to 0.58 g/L) than non-MLF wines (0.24 to 0.43 g/L). Similar results have been reported by Mendoza et al. [13] and Izquierdo-Cañas et al. [37]. Most of the simultaneous MLF wines had slightly lower VA levels than sequential MLF wines, which is similar to reports of Tristezza et al. [22].

The conversion of malic acid to lactic acid resulted in a significant decrease in the total acidity levels of the MLF wines, with the expected increase in the pH of those wines. In most cases, simultaneous MLF wines had slightly higher total acidity levels than sequential MLF wines, which is similar to the findings of Mendoza et al. [13].

3.5. Multivariate Data Analysis of Wines

Principal component analysis (PCA) was used to investigate the association among yeast combinations, LAB strain, MLF strategy, and volatile composition of shiraz wines (Figure 5). The first two principal components explain 65% of the variance in the data (PC1 = 38.78% and PC2 = 26.27%). Four distinct clusters (indicated by different colors) can be observed, i.e., Hu + Sc1 non-MLF and MLF wines (top right quadrant), Hu + Sc2 non-MLF and MLF wines (top left quadrant), Sc2 non-MLF and MLF wines (bottom left quadrant), and Sc1 non-MLF and MLF wines (bottom right quadrant). Results clearly show that the yeast combinations had a significant effect on volatile chemical composition of the wines (Table S2 and Figure 5). The distribution of the data points within the aforementioned four clusters shows that there was some within-group variation. This within-group variation is due to the LAB strain or MLF strategy that was applied. These results indicate that yeast combination has the greatest impact on the chemical composition, but LAB strain and MLF strategy also have a significant effect ($p \leq 0.05$) on the chemical composition of the wines (Table S2).

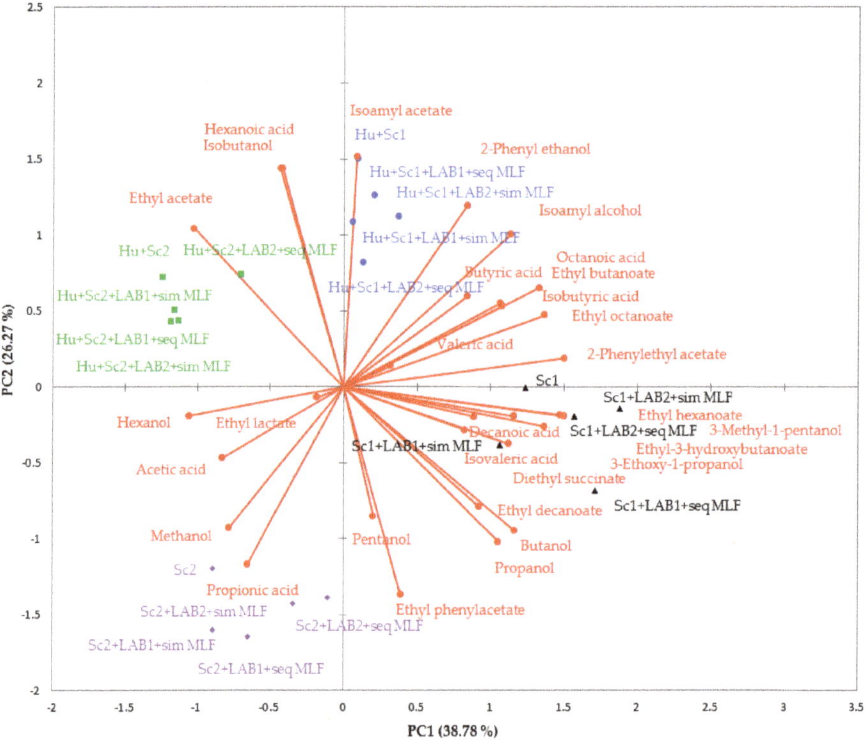

Figure 5. Principal component biplot of volatile compounds of shiraz wines produced with *Saccharomyces cerevisiae* (Sc1 and Sc2) in combination with *Hanseniaspora uvarum*, two lactic bacteria strains (LAB1 and LAB2), and two malolactic fermentation (MLF) strategies (simultaneous or sequential inoculation). Mean values of three replicate fermentations. Abbreviations: LAB1 = *Oenococcus oeni*, LAB2 = *Lactobacillus plantarum*, sim = simultaneous MLF, and seq = sequential MLF.

Based on the contribution and squared cosines of the variables, the main compounds responsible for differentiating among wines produced with the selected yeast combinations, LAB strain and MLF strategies were, isoamyl acetate, ethyl hexanoate, ethyl octanoate, ethyl-3-hydroxybutanoate, ethyl phenylacetate, 2-phenyl acetate, isobutanol, 3-methyl-1-pentanol, hexanoic acid, and octanoic acid (Figure 5).

All wines produced with Sc1 show a positive correlation with 2-phenylethyl acetate, 3-methyl-pentanol, ethyl hexanoate, decanoic acid, ethyl-3-hydroxybutanoate, 3-ethoxy-1-propanol, isovaleric acid, diethyl succinate, ethyl decanoate, butanol, and propanol. The aforementioned wines were negatively correlated with ethyl acetate.

The Sc2 wines show a positive correlation with methanol, propionic acid, pentanol, and ethyl phenylacetate, and a negative correlation with isoamyl acetate, 2-phenyl ethanol, isoamyl alcohol, octanoic acid, isobutyric acid, and ethyl butanoate.

All wines produced with Hu + Sc1 show a positive correlation with isoamyl acetate, 2-phenyl ethanol, isoamyl alcohol, and butyric acid, and a negative correlation with propionic acid, methanol, and acetic acid. Octanoic acid, ethyl butanoate, isobutyric acid, valeric acid, and ethyl octanoate show a positive correlation with wines produced with Sc1 only and those produced with *H. uvarum* in combination with Sc1.

The Hu + Sc2 wines show a positive correlation with ethyl acetate and are negatively correlated with ethyl decanoate, butanol, propanol, diethyl succinate, isovaleric acid, decanoic acid, and ethyl-3-hydroxybutanoate. Isobutanol and hexanoic acid show a positive correlation with wines produced with *H. uvarum* in combination with Sc1 and Sc2. This indicates that these compounds are linked to the growth and metabolism of the *H. uvarum* strain.

3.6. Sensory Evaluation

Sensory evaluation results indicate how yeast selection, LAB combination, and MLF strategy can impact the volatile composition and sensory profiles of wines. ANOVA of the sensory data show that the interactions among the selected yeast combinations, LAB strains and even MLF strategies had a significant ($p \leq 0.05$) impact on fresh vegetative, cooked vegetative, spicy, and floral aromas (Table 3). Yeast treatment had a significant effect on fresh vegetative and spicy aroma, as well as body and astringency of the wines. Wines produced with the selected LAB strains and MLF strategies were significantly different with regard to berry, fruity, sweet associated, and spicy aroma, as well as acidity and body. Only the sensory attributes that showed significant differences for at least two of the treatment interactions will be discussed in detail (Table 3).

Table 3. Probability (p) values [1] of shiraz wines produced with different yeast treatments and malolactic fermentation (MLF) strategies. Probability (p) values [1] of the sensory descriptors of shiraz wines produced with *Saccharomyces cerevisiae* (Sc1 or Sc2) only, or in combination with *Hanseniaspora uvarum* (Hu), two lactic acid bacteria strains (LAB 1 or LAB2), and two MLF strategies (simultaneous or sequential inoculation).

Descriptor	Treatment		
	Yeast	LAB Strain × MLF Strategy	Yeast × LAB Strain × MLF Strategy
Berry	0.3042	0.0004	0.8400
Fruity	0.7647	0.0191	0.9095
Sweet associated	0.4417	0.0023	0.5761
Fresh vegetative	0.0001	0.1245	0.0418
Cooked vegetative	0.5094	0.2079	0.0420
Spicy	0.0165	0.0009	0.0548
Floral	0.0602	0.5104	0.0159
Acid balance	0.0905	0.0001	0.3488
Body	0.0001	0.0020	0.1454
Astringency	0.0010	0.0876	0.1182
Bitterness	0.7069	0.2683	0.0800

[1] Probability (p) values ≤ 0.05 indicate significant differences between treatments.

3.6.1. Fresh Vegetative Aroma

Non-MLF wines produced with *S. cerevisiae* only (Sc1 and Sc2) scored lower for fresh vegetative aroma than non-MLF wines produced with *H. uvarum* in combination with the two *S. cerevisiae* strains (Figure 6). Sequential MLF wines scored higher for fresh vegetative aroma than simultaneous MLF and non-MLF wines. Of all the different treatments, Hu + Sc1 + LAB2 seq MLF wines scored the highest (35.27%) for fresh vegetative aroma (Table S3). The Hu + Sc1 combination consistently produced wines with high fresh vegetative aroma scores and this was observed for non-MLF and MLF. The opposite trend was found for wines produced with Sc2. These results indicate that this Hu + Sc1 combination can be used to enhance the fresh vegetative character in wines where this attribute is lacking or to produce a wine style with a predominant fresh vegetative flavor profile. On the other hand, if a wine with low fresh vegetative character is preferred, the use of a yeast strain, such as Sc2 is recommended.

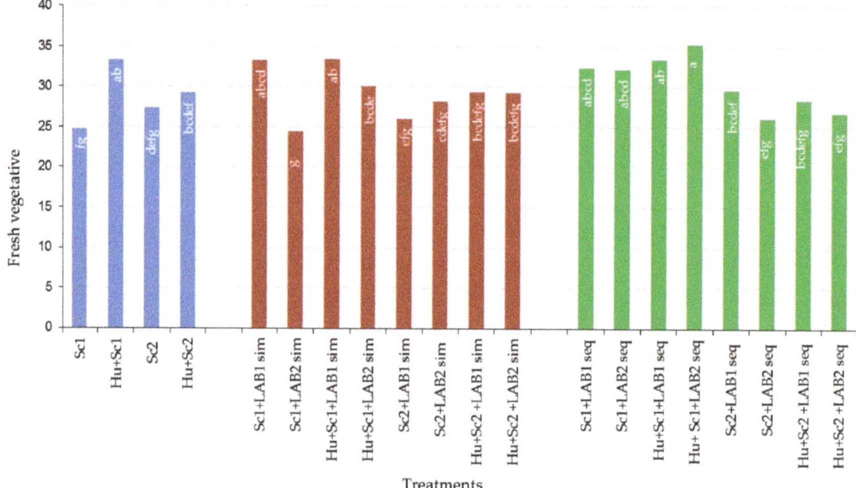

Figure 6. Percentage (%) of fresh vegetative aroma of shiraz wines produced with *Saccharomyces cerevisiae* (Sc1 and Sc2) in combination with *Hanseniaspora uvarum*, two lactic bacteria strains (LAB1 and LAB2), and two malolactic fermentation (MLF) strategies (simultaneous or sequential inoculation). Abbreviations: LAB1 = *Oenococcus oeni*, LAB2 = *Lactobacillus plantarum*, sim = simultaneous MLF, and seq = sequential MLF. The letters inside the bars refer to differences among the treatments and treatments that have the same letter/s do not differ significantly ($p \leq 0.05$).

Differences in fresh vegetative aroma scores were observed for wines produced with the two LAB strains, and were also affected by MLF strategy applied. In most cases, wines inoculated with LAB1 scored higher for vegetative aroma than wines inoculated with LAB2. Therefore to increase fresh vegetative flavor in Shiraz wines *O. oeni* should be used to induce MLF, but to reduce the fresh vegetative flavor, *L. plantarum* is recommended.

3.6.2. Spicy

Non-MLF wines produced with Sc2 scored the highest for spicy aroma (32.71%; Figure 7 and Table S3). Overall, sequential MLF wines scored higher for spicy aroma than simultaneous MLF and non-MLF wines (Figure 7). Of all the various treatments, sequential MLF wines produced with Hu + Sc1 + LAB2 scored the highest for spicy aroma. Differences in spicy aroma scores were found for wines produced with the two LAB strains, and were affected by yeast combination as well as the MLF strategy. Therefore to increase spicy flavor in wine MLF should be induced as a sequential inoculation.

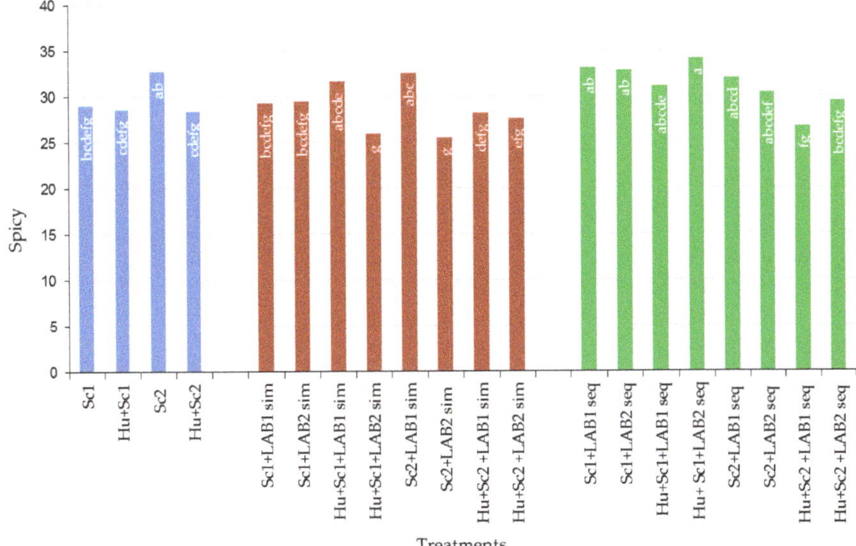

Figure 7. Percentage (%) of spicy aroma of shiraz wines produced with *Saccharomyces cerevisiae* (Sc1 and Sc2) in combination with *Hanseniaspora uvarum*, two lactic bacteria strains (LAB1 and LAB2), and two malolactic fermentation (MLF) strategies (simultaneous or sequential inoculation). Abbreviations: LAB1 = *Oenococcus oeni*, LAB2 = *Lactobacillus plantarum*, sim = simultaneous MLF, and seq = sequential MLF. The letters inside the bars refer to differences among the treatments and treatments that have the same letter/s do not differ significantly ($p \leq 0.05$).

3.6.3. Body

The non-MLF wines produced with Sc1 and Sc2 scored lower for the taste descriptor, body (mouth-feel) than those where *H. uvarum* was used (Figure 8). MLF wines scored higher for body than non-MLF wines. MLF wines produced with Sc1 scored slightly higher for body than those inoculated with Sc2. MLF wines produced with LAB2 scored higher for body than those inoculated LAB1. It is noteworthy that the relative scores for body varied according to the yeast combination used. Winemakers can manipulate the body (mouth-feel) of wines by applying the aforementioned combinations to achieve the wine style they prefer. To increase the body of a wine, *H. uvarum* in combination with *S. cerevisiae* should be used and MLF should be induced using LAB2 (*L. plantarum*).

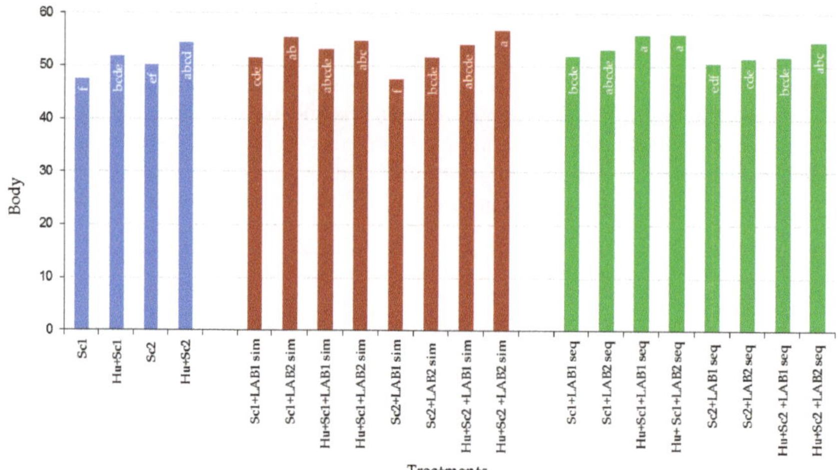

Figure 8. Percentage (%) of body of shiraz wines produced with *Saccharomyces cerevisiae* (Sc1 and Sc2) in combination with *Hanseniaspora uvarum*, two lactic bacteria strains (LAB1 and LAB2), and two malolactic fermentation (MLF) strategies (simultaneous or sequential inoculation). Abbreviations: LAB1 = *Oenococcus oeni*, LAB2 = *Lactobacillus plantarum*, sim = simultaneous MLF, and seq = sequential MLF. The letters inside the bars refer to differences among the treatments and treatments that have the same letter/s do not differ significantly ($p \leq 0.05$).

3.7. Overall Effects

Chemical and sensory results support our opinion that the selected *H. uvarum* strain contributed positively to wine flavor. None of the treatment combinations produced off-flavors. Wines produced with *H. uvarum* in combination with Sc1 and Sc2 were different to wines produced with the Sc1 or Sc2 on their own. These results show that *H. uvarum* can be used to reduce the duration of MLF and to change the style or flavor profile of a wine. Wines where yeast and LAB were added as a simultaneous inoculation, reduced MLF duration and the flavor profiles differed from those that were sequentially inoculated. Notable differences were also observed between wines inoculated with LAB1 and LAB2 with regard to their flavor profiles, which supports the concept of *L. plantarum* playing a greater role in the future of MLF as envisaged by Du Toit et al. [38]. The yeast treatments, LAB strains and MLF strategies had a significant effect on the standard chemical parameters and volatile composition of the wines, and these differences in chemical composition translated to perceivable sensory differences.

4. Conclusions

H. uvarum had a positive effect on the growth of inoculated and naturally occurring LAB, which resulted in shorter MLF periods for wines. Allowing the naturally occurring yeast population to develop for at least 24 hours may be beneficial to winemakers that want MLF to proceed quickly and successfully. Wines produced with the selected yeasts, LAB strains and MLF strategies differed with regard to fermentation kinetics, chemical composition, and sensory properties. Yeast treatment had a greater effect on the volatile chemical composition of the wines than LAB strain or MLF strategy, but LAB strain and MLF strategy also had a significant impact. The sensory differences between non-MLF and MLF wines were as significant as wines produced with different yeast strains. *H. uvarum* in combination with *O. oeni* as a sequential inoculation can be used to increase vegetative aroma of shiraz wines. The spicy flavor can be increased by inducing MLF as a sequential inoculation and increased body can be achieved by using *H. uvarum* in combination with *S. cerevisiae* to conduct the alcoholic fermentation and *L. plantarum* to induce MLF. The flavor profile of shiraz wines can be enhanced by using different yeasts, LAB strains, MLF strategies, or a combination of the aforementioned options.

Supplementary Materials: Supplementary materials can be found at http://www.mdpi.com/2311-5637/5/3/64/s1.

Author Contributions: H.d.P., M.d.T. and N.J. conceived and designed the experiments; H.d.P. and J.H. performed the experiments; H.N. assisted with FTIR and GC analyses and interpretation of data; M.v.d.R. contributed to the statistical analysis and interpretation of the data; H.d.P., M.d.T., H.N. and N.J. contributed to writing the paper.

Acknowledgments: The authors thank the ARC, Winetech and the National Research Foundation of South Africa (THRIP programme; grant numbers UID 71526 and 90103) for funding. The opinions, findings and conclusions expressed in this publication are those of the authors. The National Research Foundation accepts no liability in this regard. Mses P. Adonis, C. du Plessis, D. Blaauw, R. Louw and L. Isaacs, as well as Messrs H. Jumat, M. Mewa Ngongang and J. Boonzaier, are thanked for technical assistance.

Conflicts of Interest: The authors declare no conflict of interest.

References

1. Swiegers, J.H.; Bartowsky, E.J.; Henschke, P.A.; Pretorius, I.S. Yeast and bacterial modulation of wine aroma and flavour. *Aust. J. Grape Wine Res.* **2005**, *11*, 139–173. [CrossRef]
2. Jolly, N.P.; Valera, C.; Pretorius, I.S. Not your ordinary yeast: Non-*Saccharomyces* yeasts in wine production uncovered. *FEMS Yeast Res.* **2014**, *14*, 215–237. [CrossRef] [PubMed]
3. Fleet, G.H. Yeast interactions and wine flavour. *Int. J. Food Microbiol.* **2003**, *86*, 11–22. [CrossRef]
4. Ribéreau-Gayon, P.; Dubourdieu, D.; Donéche, B.; Lonvaud, A. *Handbook of Enology, Volume 1: The Microbiology of Wine and Vinifications*, 2nd ed.; John Wiley & Sons Ltd.: London, UK, 2006.
5. Hranilovic, A.; Li, S.; Boss, P.K.; Bindon, K.; Ristic, R.; Grbin, P.R.; Van der Westhuizen, T.; Jiranek, V. Chemical and sensory profiling of Shiraz wines co-fermented with commercial non-*Saccharomyces* inocula. *Aust. J. Grape Wine Res.* **2018**, *24*, 166–180. [CrossRef]
6. Romano, P.; Capece, A.; Jespersen, L. Taxonomic and ecological diversity of food and beverage yeasts. In *Yeasts in Food and Beverages*; Querol, A., Fleet, G., Eds.; Springer: Berlin, Germany, 2006; pp. 13–54.
7. Capozzi, V.; Garofalo, C.; Chiriatti, M.A.; Grieco, F.; Spano, G. Microbial terroir and food innovation: The case of yeast biodiversity in wine. *Microbiol. Res.* **2015**, *181*, 75–83. [CrossRef]
8. Ye, D.Q.; Sun, Y.; Song, Y.Y.; Ma, X.L.; Ren, X.N.; Gong, X.; Liu, Y.L. Diversity and identification of yeasts isolated from tumultuous stage of spontaneous table grape fermentations in central China. *S. Afr. J. Enol. Vitic.* **2019**, *40*, 1–7. [CrossRef]
9. Romano, P.; Fiore, C.; Paraggio, M.; Caruso, M.; Capece, A. Function of yeast species and strains in wine flavour. *Int. J. Food Microbiol.* **2003**, *86*, 169–180. [CrossRef]
10. De Benedictis, M.; Bleve, G.; Grieco, F.; Tristezza, M.; Tufariello, M. An optimized procedure for the enological selection of non-*Saccharomyces* starter cultures. *Antonie Leeuwenhoek* **2011**, *99*, 189–200. [CrossRef]
11. Tristezza, M.; Tufariello, M.; Capozzi, V.; Spano, G.; Mita, G.; Grieco, F. The oenological potential of *Hanseniaspora uvarum* in simultaneous and sequential co-fermentation with *Saccharomyces cerevisiae* for industrial wine production. *Front. Microbiol.* **2016**, *7*, 1–14. [CrossRef]
12. Moreira, N.; Mendes, F.; de Pinho, R.G.; Hogg, T.; Vasconcelos, I. Heavy sulphur compounds, higher alcohols and esters production profile of *Hanseniaspora uvarum* and *Hanseniaspora guilliermondii* grown as pure and mixed cultures in grape must. *Int. J. Food Microbiol.* **2008**, *124*, 231–238. [CrossRef]
13. Mendoza, L.M.; Merín, M.G.; Morata, V.I.; Farías, M.E. Characterization of wines produced by mixed culture of autochthonous yeasts and *Oenococcus oeni* from the northwest region of Argentina. *J. Ind. Microbiol. Biotechnol.* **2011**, *38*, 1777–1785. [CrossRef] [PubMed]
14. Bartowsky, E.J.; Costello, P.; Henschke, P.A. Management of malolactic fermentation—Wine flavour manipulation. *Aust. N. Z. Grapegrow. Winemak.* **2002**, *461*, 7–8 and 10–12.
15. Lerm, E.; Engelbrecht, L.; Du Toit, M. Malolactic fermentation: The ABC's of MLF. *S. Afr. J. Enol. Vitic.* **2010**, *31*, 186–212. [CrossRef]
16. Dimopoulou, M.; Raffenne, J.; Claisse, O.; Miot-Sertier, C.; Iturmendi, N.; Moine, V.; Coulon, J.; Dols-Lafargue, M. *Oenococcus oeni* exopolysaccharide biosynthesis, a tool to improve malolactic starter performance. *Front. Microbiol.* **2018**, *9*, 1276. [CrossRef] [PubMed]
17. Grimaldi, A.; Bartowsky, E.; Jiranek, V. Screening of *Lactobacillus* spp. and *Pediococcus* spp. for glycosidase activities that are important in oenology. *J. Appl. Microbiol.* **2005**, *99*, 1061–1069. [CrossRef] [PubMed]
18. Mtshali, P.S.; Divol, B.; Van Rensburg, P.; du Toit, M. Genetic screening of wine-related enzymes in *Lactobacillus* species isolated from South African wines. *J. Appl. Microbiol.* **2010**, *108*, 1389–1397. [CrossRef]

19. Brizuela, N.; Tymczyszyn, E.E.; Semorile, L.C.; La Hens, D.V.; Delfederico, L.; Hollmann, A.; Bravo-Ferrada, B. *Lactobacillus plantarum* as a malolactic starter culture in winemaking: A new (old) player? *Electron. J. Biotechnol.* **2019**, *38*, 10–19. [CrossRef]

20. Abrahamse, C.; Bartowsky, E. Timing of malolactic fermentation inoculation in Shiraz grape must and wine: influence on chemical composition. *World J. Microbiol. Biot.* **2012**, *28*, 255–265. [CrossRef]

21. Bartowsky, E.J.; Costello, P.J.; Chambers, P.J. Emerging trends in the application of malolactic fermentation. *Aust. J. Grape Wine Res.* **2015**, *21*, 663–669. [CrossRef]

22. Tristezza, M.; di Feo, L.; Tufariello, M.; Grieco, F.; Capozzi, V.; Spano, G.; Mita, G. Simultaneous inoculation of yeasts and lactic acid bacteria: Effects on fermentation dynamics and chemical composition of Negroamaro wine. *LWT-Food Sci. Technol.* **2016**, *66*, 406–412. [CrossRef]

23. Lesschaeve, I. Sensory evaluation of wine and commercial realities: Review of current practices and perspectives. *Am. J. Enol. Vitic.* **2007**, *58*, 252–258.

24. Du Plessis, H.; du Toit, M.; Nieuwoudt, H.; van der Rijst, M.; Kidd, M.; Jolly, N. Effect of *Saccharomyces*, non-*Saccharomyces* yeasts and malolactic fermentation strategies on fermentation kinetics and flavor of Shiraz wines. *Fermentation* **2017**, *3*, 64. [CrossRef]

25. Minnaar, P.P.; Ntushelo, N.; Ngqumba, Z.; Van Breda, V.; Jolly, N.P. Effect of *Torulaspora delbrueckii* yeast on the anthocyanin and flavanol concentrations of Cabernet franc and Pinotage wines. *S. Afr. J. Enol. Vitic.* **2015**, *36*, 50–58. [CrossRef]

26. Anonymous. Methods of analyses for wine laboratories. Compiled by the South African Wine Laboratories Association. South African Society for Viticulture and Oenology, P.O. Box 2092, Dennesig, Stellenbosch 7601, South Africa. 2003. Available online: http://www.sasev.org (accessed on 8 July 2019).

27. Louw, L.; Roux, K.; Tredoux, A.; Tomic, O.; Naes, T.; Nieuwoudt, H.H.; Van Rensburg, P. Characterisation of selected South African cultivar wines using FTMIR spectroscopy, gas chromatography and multivariate data analysis. *J. Agric. Food. Chem.* **2009**, *57*, 2623–2632. [CrossRef] [PubMed]

28. Lõoke, M.; Kristjuhan, K.; Kristjuhan, A. Extraction of genomic DNA from yeasts for PCR-based applications. *BioTechniques* **2011**, *50*, 325–328. [CrossRef] [PubMed]

29. Esteve-Zarzoso, B.; Belloch, C.; Uruburu, F.; Querol, A. Identification of yeasts by RFLP analysis of the 5.8S rRNA gene and the two ribosomal internal transcribed spacers. *Int. J. Syst. Bacteriol.* **1999**, *49*, 329–337. [CrossRef] [PubMed]

30. Pfliegler, W.P.; Horváth, E.; Kállai, Z.; Sipiczki, M. Diversity of *Candida zemplinina* isolates inferred from RAPD, micro/minisatellite and physiological analysis. *Microbiol. Res.* **2014**, *169*, 402–410. [CrossRef] [PubMed]

31. Costantini, A.; García-Moruno, E.; Moreno-Arribas, M.V. Biochemical transformations produced by malolactic fermentation. In *Wine Chemistry and Biochemistry*; Moreno-Arribas, M.V., Polo, M.C., Eds.; Springer: Berlin, Germany, 2009; pp. 27–57.

32. Izquierdo-Cañas, P.M.; Romero, E.G.; Pérez-Martín, F.; Seseña, S.; Palop, M.L. Sequential inoculation versus co-inoculation in Cabernet Franc wine fermentation. *Food Sci. Technol. Int.* **2015**, *21*, 203–212. [CrossRef] [PubMed]

33. Mendoza, L.M.; Manca de Nadra, M.C.; Farías, M.E. Kinetics and metabolic behaviour of a composite culture of *Kloeckera apiculata* and *Saccharomyces cerevisiae* wine related strains. *Biotechnol. Lett.* **2007**, *29*, 1057–1063. [CrossRef]

34. Liu, P.T.; Lu, L.; Duan, C.Q.; Yan, G.L. The contribution of indigenous non-*Saccharomyces* wine yeast to improved aromatic quality of Cabernet Sauvignon wines by spontaneous fermentation. *LWT-Food Sci. Technol.* **2016**, *71*, 356–363. [CrossRef]

35. Ciani, M.; Beco, L.; Comitini, F. Fermentation behaviour and metabolic interactions of multistarter wine yeast fermentations. *Int. J. Food Microbiol.* **2006**, *108*, 239–245. [CrossRef] [PubMed]

36. Whitener, M.E.B.; Stanstrup, J.; Carlin, S.; Divol, B.; Du Toit, M.; Vrhovsek, U. Effect of non-*Saccharomyces* yeasts on the volatile chemical profile of Shiraz wine. *Aust. J. Grape Wine Res.* **2017**, *23*, 179–192. [CrossRef]

37. Izquierdo-Cañas, P.M.; Mena-Morales, A.; García-Romero, E. Malolactic fermentation before or during wine aging in barrels. *LWT-Food Sci. Technol.* **2016**, *66*, 468–474. [CrossRef]

38. Du Toit, M.; Engelbrecht, L.; Lerm, E.; Krieger-Weber, S. *Lactobacillus*: The next generation of malolactic fermentation starter cultures—an overview. *Food Bioproc. Tech.* **2010**, *4*, 876–906. [CrossRef]

 *fermentation*

Article

Impact on Sensory and Aromatic Profile of Low Ethanol Malbec Wines Fermented by Sequential Culture of *Hanseniaspora uvarum* and *Saccharomyces cerevisiae* Native Yeasts

María Victoria Mestre [1,2,†], Yolanda Paola Maturano [1,2,*,†], Candelaria Gallardo [1,2], Mariana Combina [2,3], Laura Mercado [3], María Eugenia Toro [1], Francisco Carrau [4], Fabio Vazquez [1] and Eduardo Dellacassa [4]

[1] Instituto de Biotecnología, UNSJ, Av. San Martín 1109 (O), San Juan 5400, Argentina
[2] Consejo Nacional de Investigaciones Científicas y Tecnológicas (CONICET), Godoy Cruz 2290 C1425FQB CABA, Argentina
[3] Estación Experimental Agropecuaria Luján de Cuyo (INTA), Av. San Martin 3853 Luján de Cuyo, Mendoza M5507, Argentina
[4] Laboratorio de Biotecnología de Aromas, Facultad de Química, UdelaR. Av. Gral. Flores 2124, Montevideo 11800, Uruguay
*   Correspondence: paolamaturano@yahoo.com.ar Tel.: +54-0264-4211700/+54-0264-4619625
†   These authors contributed equally.

Received: 25 June 2019; Accepted: 8 July 2019; Published: 11 July 2019

Abstract: It is well known that high ethanol levels in wines adversely affect the perception of new wine consumers. Moreover, numerous issues, such as civil restrictions, health risk and trade barriers, are associated with high ethanol concentrations. Several strategies have been proposed to produce wines with lower alcoholic content, one simple and inexpensive approach being the use of new wine native yeasts with less efficiency in sugar to ethanol conversion. Nevertheless, it is also necessary that these yeasts do not impair the quality of wine. In this work, we tested the effect of sequential culture between *Hanseniaspora uvarum* BHu9 and *Saccharomyces cerevisiae* BSc114 on ethanol production. Then, the wines produced were analyzed by GC-MS and tested by a sensorial panel. Co-culture had a positive impact on ethanol reduction and sensory profile when compared to the *S. cerevisiae* monoculture. Wines with lower alcohol content were related to fruity aroma; moreover, color intensity was associated. The wines obtained with *S. cerevisiae* BSc114 in pure conditions were described by parameters linked with high ethanol levels, such as hotness and astringency. Moreover, floral profile was related to this treatment. Based on these findings, this work provides a contribution to answer the current consumers' preferences and addresses the main challenges faced by the enological industry.

Keywords: low-ethanol wines; sequential culture; *Hanseniaspora uvarum* yeast; aromatic/sensorial profiles

---

1. Introduction

Well-structured and full-body wines have become the preferences of many new wine consumers. In order to obtain these characteristics, it is necessary to ensure optimal phenolic maturity of grapes, which requires longer grape ripening times [1]. However, in the context of global warming, this practice results in a significant increase in the berry sugar content at the moment of harvesting, and consequently higher alcohol levels in the wine [2]. Numerous issues are associated with high ethanol levels in wine such as consumers' rejection, civil restrictions, health risk, and trade barriers [1,3,4]. The sensorial quality of wines is also significantly affected because of an increase in the perception of bitterness, sweetness, astringency and hotness, and masking of volatile aromatic compounds [5,6]. In this context, different technological solutions have been evaluated: harvest of unripe berries, increase in crop load, shading bunches, choosing proper irrigation techniques, and modulation of source–sink relationships by removing leaves or topping shoots [7–10]. Other authors have tried partial dealcoholization with physical methods [11–13].

More recently, microbiological solutions have been proposed by using selected non-*Saccharomyces* and *Saccharomyces cerevisiae* yeast strains in simultaneous or sequential fermentations [4,14–16]. The use of non-*Saccharomyces* yeasts has become a common trend in the main wine regions, particularly because of their effects on the composition, flavor and color of the wine [17,18]. In addition to the aforementioned effects, this yeast group is also known to be less efficient in the production of ethanol from consumed sugars when compared with *S. cerevisiae* yeasts [19].

Hanseniaspora genera as a whole and particularly *Hanseniaspora uvarum* species are non-*Saccharomyces* yeasts commonly encountered at high concentrations on the grape surface and throughout the fermentation process [20]. Recently, 28 *H. uvarum* isolates were evaluated by our research group and they demonstrated interesting enological characteristics such us: ability to grow at high sugar, ethanol and SO_2 contents; to produce high concentrations of glycerol; low acetic acid and hydrogen sulfide levels; and the release of proteolytic enzymes [21]. Moreover, it is important to highlight that *H. uvarum* was also found to be a potential candidate to produce less ethanol because it requires more than 19 g/L of consumed sugar to produce 1% *v/v* of ethanol [21]. In a more recent study, a selected *H. uvarum* yeast strain was assessed in sequential inoculations with *S. cerevisiae* yeasts under optimized fermentation conditions [22]. The authors found that the ethanol levels were significantly reduced compared with fermentations carried out with *S. cerevisiae* monocultures. Nevertheless, and in order to achieve holistic knowledge, the aim of the present work was to assess the aromatic impact of an optimized inoculum of *H. uvarum/S. cerevisiae* yeasts in fresh must and compare the findings with a *S. cerevisiae* monoculture. It is also relevant to establish how ethanol reduction affects sensorial perception. The results would allow the design of a comprehensive microbiological strategy in order to answer the current consumers' preferences and address the main challenges faced by the enological industry.

2. Materials and Methods

2.1. Microorganisms

Hanseniaspora uvarum BHu9 and *Saccharomyces cerevisiae* BSc114 were used in the present study. Both strains of yeasts were previously selected based on their oxidative and fermentative metabolism in order to obtain reduced ethanol wines [21]. Strains were obtained from the Culture Collection of Autochthonous Microorganisms (Institute of Biotechnology, School of Engineering—UNSJ, San Juan, Argentina) and preserved at −80 °C until use.

2.2. Yeast Inoculum Preparation

Each strain was grown on YEPD agar for 48 h and the biomasses were transferred to YEPD broth (130 rpm during 4 h) [22]. Then, strains were transferred to grape must (13° Brix, pH 3.8) supplemented with 0.1% yeast extract and 0.4% peptone, and incubated at 25 °C during 24 h under aerobic conditions (130 rpm). YEPD broth was used for pre-adaptation in order to reduce the lag-stage in the grape must, which allows strains to grow immediately exponentially in the grape juice [22]. Once the pre-adaptation process had finished, cells were counted with an improved Neubauer chamber.

2.3. Grapes and Vineyard Location

All experiments were carried out using *Vitis vinifera* L. cv. Malbec grapes harvested during the 2017 vintage from a vineyard located in Cañada Honda, San Juan, Argentina (31°58′34″S 68°32′52″W) at 610 m altitude. Grapes were manually destemmed and mixed to obtain a homogeneous solution. The composition of the fresh juice was as follows: sugar (glucose and fructose), 238.2 g/L; pH, 3.8; titratable acidity, 5.3 g/L; and yeast assimilable nitrogen (YAN), 175 mg/L. Then, 5-L vessels equipped with a Muller valve were filled with juice (3 L) and supplemented with 50 ppm of free SO_2 before fermentation. The Muller valve was filled with a solution of 50% sulfuric acid and 50% sterile water distilled. Vinifications were performed in triplicate.

2.4. Inoculation and Winemaking

Lab-scale fermentations were conducted under optimized factors previously determined by Maturano et al. [22]. Treatment 1 (T1): *H. uvarum* BHu9 was inoculated (T0) at a concentration of 5×10^6 cells/mL, and 48 h later, 2×10^6 cells/mL of *S. cerevisiae* BSc114 were sequentially inoculated. In parallel, a single culture of 2×10^6 cells/mL of *S. cerevisiae* yeasts was inoculated at T0 as control treatment (TC). Both fermentations were performed at 25 ± 1 °C under static conditions. Musts were supplemented with nitrogen by adding 20 mg/L of $(NH_4)_2HPO_4$ twice: after 48 h and in the middle of the fermentation (when 5% weight loss was verified). Nitrogen supplement was established based on nitrogen uptake previously determined with selected yeasts (data not shown). Punch down was carried out every 24 h in order to keep acceptable dissolved oxygen levels throughout the process. The fermentation progress was evaluated by the weight loss caused by CO2 production and vessels were weighed every 24 h.

Samples were collected periodically and viable cell counts were determined by plating onto Wallerstein Laboratory Nutrient (WLN) Agar medium (Oxoid, Hampshire, UK). Dilutions of 10^{-3}, 10^{-4} and 10^{-5} were spread onto WLN agar medium and incubated for 7 days at 28 °C. Green colonies (*H. uvarum* BHu9) and creamy colonies (*S. cerevisiae* BSc114) were differentiated and counted [23].

After the sugar was completely consumed, 50 mg/L of free SO_2 was added. The wines were chemically stabilized, filtered, bottled, and conserved at 16 ± 1 °C until sensorial analysis. Samples of 50 mL were stored at −20 °C in order to carry out volatile composition analysis.

2.5. Chemical Analysis

Glycerol, residual sugars, total acidity and acetic, malic, lactic, and tartaric acid were measured periodically using an ALPHA FT-IR Wine Analyzer (Bruker Optik Gmbh, Ettlingen, Germany). Ethanol concentration was determined according to the OIV OENO 379-2009 ES official method. The pH was measured with a multi-parameter Adwa (AD1030 PHM_MES_6362).

2.6. Sensorial Analysis

After 4 months of bottle stabilization, wines were evaluated by descriptive analysis according to Lawless and Heymann [24]. A well-trained panel carried out the evaluation of 13 sensorial attributes: three color/appearance descriptors (color intensity, red and brown color), five aroma descriptors (mineral note, frutal, floral, chili pepper, and toasted) and five taste parameters (acidity, sweetness, astringency, hotness, and bitterness). The intensity of each attribute was assessed using a structured scale from 0 to 5, where 0 indicates that the descriptor was not perceived and values between 1 and 5 indicate that the intensity of the descriptors was very low to very high. The panel consisted of seven individuals (five males and two females between 35 and 50 years old) from the Wine Sensorial Analysis Department (Instituto Nacional de Vitivinicultura, Mendoza, Argentina). Vinifications were tasted blindly and in duplicate from a constant volume of 30 mL at room temperature.

2.7. Free aromatic Analyses

2.7.1. Solid Phase Extraction (SPE)

The extraction of aroma compounds was performed by adsorption and the molecules were separate elutions from an Isolute ENV+ cartridge (IST Ltd., Mid Glamorgan, UK) packed with 1 g of the highly cross-linked styrene divinylbenzene (SDVB) polymer according to Boido et al. [25] with some modifications.

2.7.2. GC-MS Analyses

GC-MS analyses were conducted using a Shimadzu QP 2020 (Shimadzu Corporation, Kyoto, Japan) mass spectrometer. A Carbowax 20 M capillary column (Agilent Technologies, Walt and Jennings Scientific, Wilmington, DE, USA) (30m × 0.25mm × 0.25μm film thickness) was used. The experimental conditions were as follows: The initial column temperature was 40 °C (8 min), which was then increased to 180 °C (3 °C/min) and then increased again to 250 °C (20 min) at 20 °C/min; injector temperature, 250 °C; injection mode, split; split ratio, 1:30; volume injected, 1.0 μL; carrier gas H2, 30 kPa; energy 70 eV. The wine aroma components were identified by comparison of their linear retention indices (LRI) determined with a homologous series of n-alkanes (C9–C26), with those from pure standards or reported in the literature according to their elution order with Carbowax 20 M [26–28]. Comparison of mass spectral fragmentation patterns with those stored in databases was also performed. GC-MS instrumental procedures using 1-heptanol as an internal standard were applied for quantitative purposes. GC-MS analyses were carried out with two samples of each wine.

2.8. Odor Activity Value (OAV) and Relative Odor Contributions (ROCs)

The contribution of each volatile compound was quantitatively evaluated using Odor Activity Values (OAVs). The OAV was obtained by dividing the mean concentration of each volatile compound by its odor threshold value in a hydroalcoholic solution [29]. The volatile compounds contribute to wine aroma when its concentration in wine is above the perception threshold, therefore, the OAV value is above 1. In this study, the threshold values were obtained from information available in the literature. Moreover, the identified compounds were classified according to aromatic descriptors and grouped in seven aromatic series which were classified according to the associated descriptor: 1, solvent; 2, sweet; 3, herbaceous; 4, floral; 5, fruity; 6, fatty; and 7, toasted.

From the volatile compounds that presented OAV > 1, the relative odor contribution (ROC) was calculated. The relative odor contribution (ROC) represents the percentage of contribution of a particular aroma compound and this was determined as the ratio between the OAV of the respective compound and the total OAV of each wine ((individual OAV/$\sum$OAV) * 100) [30].

2.9. Statistical Analysis

Chemical data and population analysis were expressed as the means ± standard deviation from three repetitions and aromatic analysis as the means of two repetitions. One-way ANOVA was used to evaluate differences between treatments. Statistical analysis was performed using the InfoStat professional version (Cordoba, Argentina, 2016).

3. Results

The current study assessed the contribution of *H. uvarum* BHu9 and *S. cerevisiae* BSc114 yeasts to the ethanol content and sensorial and aromatic impact on wine.

3.1. Fermentative Kinetics and Population Dynamics

In the present study, fermentative kinetics are represented by sugar consumption and CO2 release in both fermentations: BHu9/BSc114 (T1) and BSc114 (TC) (Figure 1). Both treatments completed alcoholic fermentation after 8 days. During the first 24 h, both vinifications showed a similar sugar consumption, but from day 2 until day 6, T1 exhibited a slower fermentation rate than TC ($p < 0.05$). During day 7 and 8, sugar consumption was not significantly different ($p > 0.05$), and at the end of the process, both treatments behaved similarly.

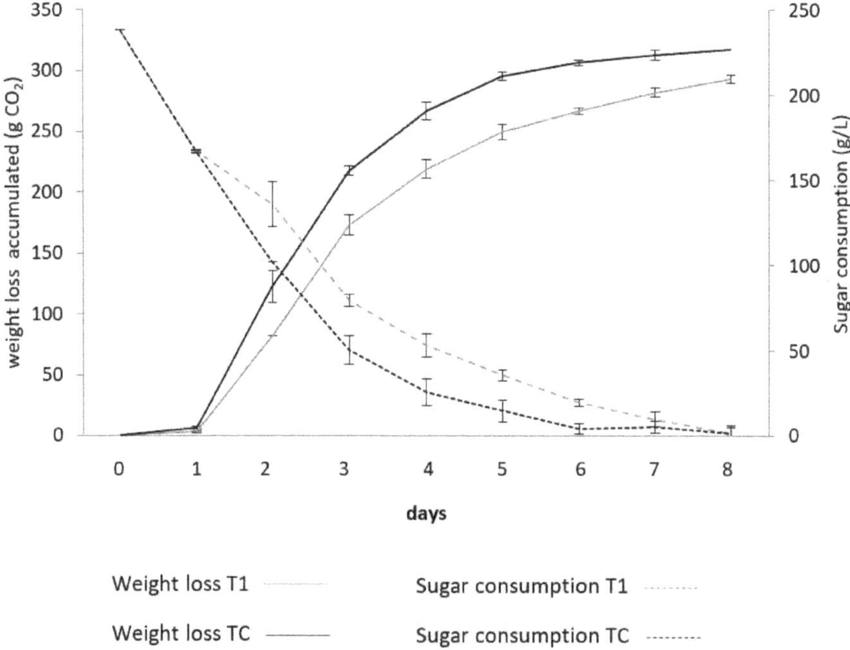

Figure 1. Release of CO2 (g) and sugar consumption in T1 (BHu9/BSc114) and TC (BSc114 control).

Like the sugar consumption, CO2 release showed a similar behavior for both treatments during the first 24 h. From day 2 until the end, T1 demonstrated a lower rate than TC (BSc114) ($p < 0.05$). Total CO2 production was 293.33 ± 17 g and 320 ± 10 g for the BHu9/BSc114 co-culture and *S. cerevisiae* monoculture, respectively (Figure 1).

The population dynamics of T1 (*H. uvarum* BHu9/*S. cerevisiae* BSc114) and TC (*S. cerevisiae* BSc114) are shown in Figure 2. *H. uvarum* BHu9 population increased during the early stages reaching a maximum of 8.18×10^7 cells/mL on day three. During the first 48 h, (previous *S. cerevisiae* inoculation) BHu9 consumed 102.54 g/L of sugar with an ethanol production of 3.49% *v/v*. Therefore, when BSc114 was inoculated (after 48 h), the available sugar concentration was 135 g/L. In co-inoculation trials, *H. uvarum* BHu9 maintained its population up to day 4, after which the concentrations were undetectable with the technique applied in this study. Hence, *H. uvarum* BHu9 and *S. cerevisiae* BSc114 coexisted only during 2 days. During this coexistence period, the sugar consumption was 51.74 g/L, which means that the sugar consumption by both strains was less than the consumption by BHu9 before *S. cerevisiae* inoculation, and the ethanol production at this stage was 5.93% *v/v*. At the final fermentation stage (day 5 to 8), *S. cerevisiae* BSc114 consumed 74.67 g/L of sugar, and the average ethanol production was 3.22% *v/v*. The dynamic population of *S. cerevisiae* in T1 presented an increase in the number of cells from 2×10^6 cells/mL to 1.82×10^8 cells/mL on day 7, whereas the maximum population achieved by BSc114 (TC, control) was 1.9×10^8 cells/mL on day 6.

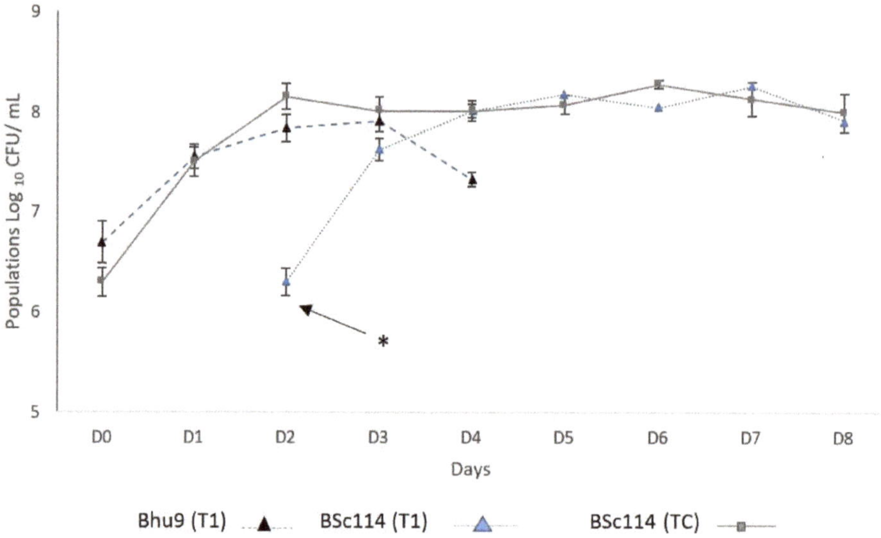

Figure 2. Dynamic populations of T1 (BHu9/BSc114) and TC (BSc114). * indicate the inoculation moment of BSc114 in T1 after 48 h.

3.2. Enological Parameters

The analyses of the main chemical parameters at the end of the fermentations are summarized in Table 1. Both treatments finished with sugar concentrations below 1.8 g/L, indicating that the fermentations had been successfully finished. Ethanol concentration in T1 was significantly lower (12.63 ± 0.05% *v/v*) than TC (13.15 ± 0.28% *v/v*), representing an average reduction of 0.52% *v/v*. Likewise, pH values were lowest in wines produced with co-cultures, but tartaric acid was higher compared to control (TC) wines. No significant differences were observed for acetic, malic and lactic acid, or for glycerol and total acidity under the experimental conditions.

Table 1. Principal chemical parameters in wines obtained from BHu9/Bcs114 co-inoculation and BSc114 control.

Chemical Compounds	T1	TC
Ethanol (%*v/v*)	12.63 ± 0.05	13.5 ± 0.28 (*)
Acetic acid (g/L)	0.56 ± 0.02	0.49 ± 0.04
Lactic acid (g/L)	0.5 ± 0.01	0.47 ± 0.02
Malic acid (g/L)	2.85 ± 0.21	2.85 ± 0.07
Tartaric acid (g/L)	1.31 ± 0.08	1 ± 0.01 (*)
Glycerol (g/L)	10 ± 0.71	9.3 ± 0.57
pH	3.43 ± 0.00	3.49 ± 0.01 (*)
Total acidity (g/L)	5.95 ± 0.07	5.80 ± 0.14
Residual sugar (g/L)	1.55 ± 0.64	1.70 ± 0.28

REFERENCES: (*) indicate significant differences between treatments at $p < 0.05$.

3.3. Aromatic Composition

Volatile products of the fermented musts were quantified by SPE-GC-MS according to Boido et al. (2003). Table 2 shows volatile compounds and their concentrations, odorant descriptors, perception thresholds, odorant activity values (OAVs), and aromatic series found in the Malbec wines analyzed. A total of 38 volatile compounds were identified and quantified, and classified into four groups: esters (ethyl and acetate esters), higher alcohols, fatty acids, and lactones.

Table 2. Aromatic description of wines obtained (mean and SD of two measure).

Compounds (µg/L)	Treatments T1	TC	$p < 0.05$	Descriptor	Threshold Perception (µg/L)	Ref	OAV T1	OAV TC	Aromatic Serie
Ethyl esters									
Ethyl hexanoate	600 ± 15	200 ± 83	*	fruity, apple	14	1	42.85	14.28	5
Ethyl octanoate	872 ± 58	369 ± 187	*	pineapple, pear	5	1	174.4	73.8	5
Ethyl-3-hydroxybutanoate	212 ± 53	174 ± 14		grape, caramel	67,000	2	0.003	0.002	2
Ethyl decanoate	1140 ± 40	590 ± 81	*	floral	200	1	5.7	2.95	4
Ethyl dodecanoate	672 ± 52	nd		leaf, fruity	1500	1	0.448	-	5
Ethyl tetradecanoate	31 ± 1	121 ± 22	*	waxy	2000	1	0.015	0.060	6
Ethyl palmitate	345 ± 23	224 ± 14	*	waxy	1500	1	0.23	0.149	6
Ethyl succinate	127 ± 0.011	145 ± 6	*	ripe melon	1,000,000	1	0.0001	0.0001	5
Ethyl lactate	603 ± 21	409 ± 63	*	strawberrry	14,000	1	0.043	0.029	5
Σ Ethyl esters	**4600**	**2232**							
Acetate esters									
Isoamyl acetate	3210 ± 18	2892 ± 191	*	banana	30	2	107	96.4	5
Hexyl acetate	24 ± 24	297 ± 11	*	red fruit	1500	1	0.03	0.443	5
Σ acetate esters	**3234**	**3189**							
TOTAL ESTERS	**7491 (1.77%)**	**5421 (1.03%)**							
Higher Alcohols									
2-Methyl-1-propanol	18,480 ± 1620	12,990 ± 299	*	solvent	7000	3	2.64	1.85	1
1-Butanol	444 ± 1	692 ± 72	*	solvent	9000	3	0.049	0.076	1
3-Methyl-1-butanol	318,400 ± 12,300	371,000 ± 16,140	*	burned, alcohol	30,000	3	10.61	12.36	1
1-Pentanol	40 ± 2	nd		fruity, balsmic	4000	3	0.01	-	5
4-Methyl-1-pentanol	59 ± 8	60 ± 18		almond	50,000	3	0.001	0.001	2
3-Methyl-1-pentanol	214 ± 18	220 ± 10		herbaceous	50,000	3	0.004	0.044	3
1-Hexanol	1080 ± 142	852 ± 82		grass, green leaf	2500	4	0.432	0.340	3
trans-3-Hexenol	40 ± 0	nd		herbaceous, land	400	1	0.1	-	3
3-Ethoxy-1-propanol	150 ± 11	65 ± 3	*	ripe pear	100	1	1.5	0.65	5
cis-3-Hexenol	54 ± 8	57 ± 8		cutted grass	400	1	0.135	0.014	6
2-Ethyl hexanol	256 ± 12	255 ± 17		rose, citrus	8000	1	0.032	0.031	4
2,3-Butanediol	233 ± 15	340 ± 13	*	butter	120,000	1	0.001	0.002	6
Furfurol	102 ± 19	190 ± 20	*	floral	5000	4	0.02	0.038	4
3-(Methylthio)-1-propanol	819 ± 41	1450 ± 38	*	cooked vegetal	1000	5	0.819	1.45	3
Benzyl alcohol	610 ± 50	11 ± 4	*	caramelo, cítrico	10,000	4	0.061	0.0001	2
2-Phenylethyl alcohol	53,885 ± 3012	86,072 ± 731	*	rose	14,000	4	3.848	6.148	2
Tyrosol	17,110 ± 895	23,071 ± 3245	*	honey	-	-			2
Tryptophol	1910 ± 98	1780 ± 98	*		-	-			

Table 2. *Cont.*

Compounds (µg/L)	Treatments T1	TC	$p < 0.05$	Descriptor	Threshold Perception (µg/L)	Ref	OAV T1	OAV TC	Aromatic Serie
Σ Higher alcohols	414,357 (98.14%)	515,754 (98.45%)							
Fatty acids									
Acetic acid	39 ± 6	nd		vinegar	200	1	0.195	-	6
Isobutanoic acid	181 ± 21	396 ± 9	*	butter, cheese	8100	3	0.022	0.048	6
Butanoic acid	56 ± 3	295 ± 29	*	fatty, rancid	1000	3	0.056	0.295	6
Hexanoic acid	176 ± 53	231 ± 42	*	cheese, sudor	3000	3	0.058	0.077	6
Octanoic acid	235 ± 50	824 ± 9	*	rancid butter	3000	3	0.078	0.276	6
Decanoic acid	94 ± 9	773 ± 310		fatty, rancid	10,000	3	0.009	0.007	6
Dodecanoic acid	49 ± 1	43 ± 1		fatty, rancid	10,000	3	0.005	0.004	6
Σ Acids	83 (0.019%)	2562 (0.49%)							
Lactones									
gamma-Valerolactone	39 ± 1	49 ± 3	*	sweet, cocconut	10	7	3.9	4.9	2
gamma-Butyrolactone	236 ± 8	163 ± 26	*	caramel	35	7	6.74	4.65	2
Σ Lactones	275 (0.06%)	212 (0.04%)							
Σ compounds (µg/L)	422,206	523,858							

References: (*) indicate significant differences between treatments. Grey cells indicate compounds with OAV > 1. References: [1] Tao and Zang (2010), [2] Moreno et al. (2005), [3] Rychlik et al. (1996), [4] Leffingwell & Associates (2009), [5] Burdock (2016), [6] Culleré et al. (2004), [7] Lopez de Lerma and Peinado (2011).

Alcohols formed the most abundant group of volatile compounds, followed by esters, fatty acids and lactones. Higher alcohols represented 98.14 and 98.45% of the total aroma content in T1 and TC wines, respectively, while esters and fatty acids constituted 1.77–1.03% and 0.019–0.49% in T1 and TC wines, respectively (Table 2).

Overall, the total concentration of higher alcohols and fatty acids was higher in control treatment TC, fermented with *S. cerevisiae* BSc114, than in wines produced by the sequential fermentation of *H. uvarum* BHu9/*S. cerevisiae* BSc114 (T1). In contrast, esters and lactones (γ-butyrolactone and γ-valerolactone) content was higher in T1 than in TC. These compounds represented 1.77 and 1.03 % in esters, in T1 and TC respectively. The lactones proportions were 0.06 % and 0.04 % in T1 and TC respectively. (Table 2).

Some compounds such as ethyl hexanoate, ethyl decanoate, 3-ethoxy-1-propanol, isoamyl acetate, and γ-butyrolactone were detected at higher concentrations in T1 than in TC ($p < 0.05$). In contrast, *S. cerevisiae* BSc114 fermentation showed higher concentrations of ethyl octanoate, 3-methyl-1-butanol, 3-(methyl thio)-1-propanol, 2-phenylethanol, and γ-valerolactone compared to wines obtained with BHu9/BSc114 ($p < 0.05$) (Table 2).

Compounds such us ethyl hexanoate (fruity, apple), ethyl octanoate (pineapple, pear), isoamyl acetate (banana), γ-butyrolactone (caramel, coconut), and γ-valerolactone (sweet, coconut) showed OAVs > 1 in both treatments. Comparing pure with mixed, fermented 3-ethoxy-1-propanol (ripe pear) exhibited an OAV > 1 only in T1, and 3-(methyl thio)-1-propanol (cooked vegetables) showed an OAV > 1 in wine fermented by the *S. cerevisiae* Bc114 monoculture (Table 2).

Table 3 presents compounds with an OAV > 1 and their relative odor contribution (ROC). When considering the ester contribution to the odorant composition, ethyl octanoate greatly contributed to wines obtained with BHu9/BSc114 (49.35%), while isoamyl acetate was the main contributor to the control treatment fermented with pure BSc114 control. The higher alcohols that demonstrated major contributions to wines in both treatments were 3-methyl-1-butanol and 2-phenylethanol, but their relative odor contributions were higher in TC wines.

Table 3. Compounds with an OAV > 1 and their relative odor contribution (ROC) in T1 and TC wines.

Compounds	T1		TC		Aromatic Serie
	OAV	ROC (%)	OAV	ROC (%)	
Ethyl hexanoate	42.85	12.12	14.28	6.60	5 Fruity
Ethyl octanoate	174.40	49.35	73.8	34.09	5 Fruity
Ethyl decano ate	1.03	0.29	0	0.00	4 Floral
Isoamyl acetate	107.00	30.28	96.40	44.53	5 Fruity
2-Methyl-1-propanol	2.64	3.72	1.85	6.51	1 Solvent
3-Methyl-1-butanol	10.61	3.00	12.36	5.71	1 Solvent
3-Ethoxy-1-propanol	1.50	0.42	<1	-	5 Fruity
3-(Methylthio)-1-propanol	<1	-	1.45	0.67	3 Herbaceous
2-Phenylethanol	5.38	1.52	8.61	3.98	4 Floral
gamma-Valerolactone	3.90	1.10	4.90	2.26	2 Sweet
gamma-Butirolactone	6.742	1.91	4.66	2.15	2 Sweet

Figure 3 shows the aromatic profile of the analyzed wines based on the sum of the components with an OAV > 1 and ROC values according to each descriptor. Wines fermented with BSc114 were related to the aromatic "floral", "solvent", "herbaceous" and "sweet" families, while co-culture fermented wines were characterized by "frutal" descriptors.

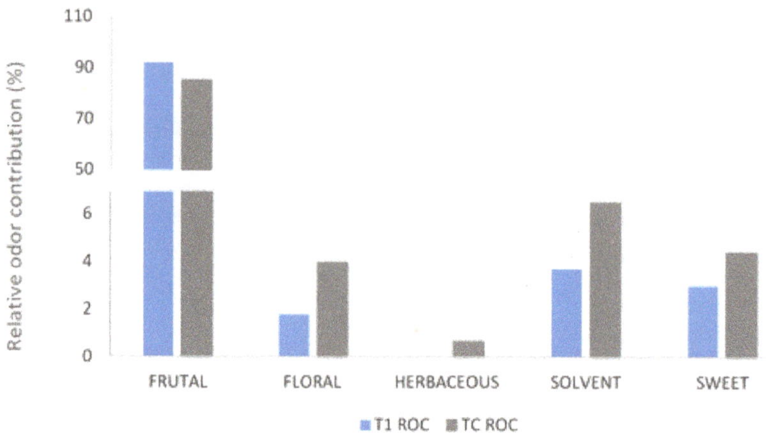

Figure 3. Aromatic profile of wines produced by BHu9/BSc114 (T1) and BSc114 (TC control).

3.4. Sensorial Analysis

Figure 4 shows the sensorial analyses of the wines obtained. Wines fermented with *H. uvarum* BHu9/*S. cerevisiae* BSc114 (T1) could be defined as fruity ($p < 0.05$). These results are in agreement with the aromatic profile obtained with ROC analysis. Another parameter that significantly affected reduced ethanol wines (T1) was color intensity. Wines produced with *S. cerevisiae* BSc114 (TC) were more related to the floral descriptor, which is in agreement with the ROC results due to 2-phenylethanol concentration; moreover, these wines were associated with astringency and hotness mouthfeel.

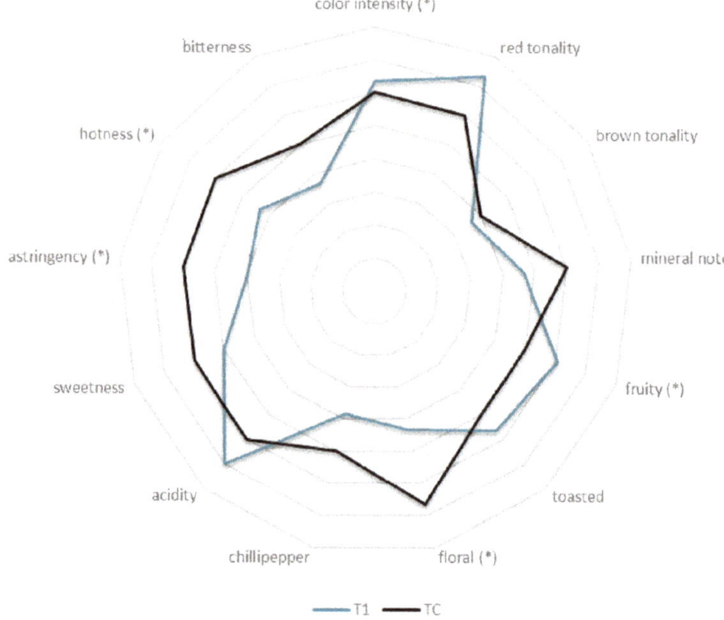

Figure 4. Sensory analysis of wines obtained from BHu9/BSc114 (T1) and BSc114 (TC). (*) Difference significant at 95% confidence level.

4. Discussion

The current wine market requires wines with lower ethanol concentrations and complex flavor and color perception. Several microbiological strategies have been proposed in order to obtain these characteristics. The present study intends to verify the behavior of selected yeasts regarding ethanol production and the sensorial impact on the wine quality.

Several studies have reported reduced ethanol levels with sequential inoculations of non-*Saccharomyces* and *Saccharomyces* yeasts and under different winemaking conditions [4,16,31]. It is well known that *H. uvarum* is the most representative yeast species found on grape surfaces showing prevalence during early stages of spontaneous alcoholic fermentation [32]. This yeast has several characteristics that could be used to reduce ethanol content in wines [21]. In the current study, inoculation of *H. uvarum* BHu9 prior to inoculation of *S. cerevisiae* BSc114 demonstrated a sugar consumption of 35.7 g/L for 1% of ethanol produced. In contrast, *S. cerevisiae* control (TC) used 17.5 g /L of sugar for 1% of ethanol produced. It is reported that *S. cerevisiae* yeast uses 16.83 to 17 g/L on average [1]. The decrease in ethanol production can sometimes be explained by an increase in glycerol and acetic acid. However, in the present study, both glycerol and acetic acid did not show significant differences. Sugars were probably partially consumed through the oxidative pathway to produce biomass and other products.

It must be highlighted that when both strains remained together, sugar consumption was lower than in the *H. uvarum* monoculture (prior to *S. cerevisiae* inoculation). There is evidence that presence of non-*Saccharomyces* yeasts in the early stages of fermentation could affect the metabolic activity of *S. cerevisiae*, probably encouraging a competition for nutrients [33–35]. For example, Bisson et al. [36] demonstrated that *K. apiculata* consumed thiamine and other micronutrients, generating inefficiency in the metabolic development of *S. cerevisiae*. Another study established that immobilized *Starmerella bombicola* cells in a mixed fermentation affected decarboxylase and alcohol dehydrogenase levels of *S. cerevisiae* [37]. Recently, the research of Petitgonnet et al. [38] demonstrated that sequential culture between *Lachaceae thermotholerans* and *S. cerevisiae* provokes a negative interaction between the two species to the detriment of *S. cerevisiae*, due to a cell–cell contact mechanism and essential nutrients uptake. When *H. uvarum* and *S. cerevisiae* are mixed inoculated, the cultivability of *H. uvarum* is significantly affected; however, the final ethanol concentrations are lower compared to the pure culture of *S. cerevisiae* [39]. Nevertheless, to answer the results of this work, further studies should be carried out to fully understand the interactions between *H. uvarum* and *S. cerevisiae* strains employed in the present study.

With respect to the aromatic composition, many studies have shown that non-*Saccharomyces* yeasts such as *Candida, Debaryomyces, Pichia, Hansenula,* and *Hanseniaspora,* that display oxidative metabolism and/or are weakly fermentative produced higher ester levels than a single *S. cerevisiae* culture [40]. In accordance, the total ester concentrations in wines produced by *H. uvarum/S. cerevisiae* co-cultures were superior to that of wines produced by control treatment. The co-inoculation showed higher levels of ethyl hexanoate, ethyl octanoate and ethyl decanoate which allowed the "fruity" aromatic profile of the wines.

Fusel alcohol production was higher in wine fermented with a monoculture of *S. cerevisiae*. Fusel alcohol production is related to amino acid production by yeasts, which varies according to genera, species and strain [41]. *S. cerevisiae* yeasts have been reported to produce higher quantities of these compounds compared with certain non-*Saccharomyces* yeasts [42]. The aromatic series that best describes the TC profile is "floral", which is associated with 2-phenylethanol levels. As was expected, sensorial analysis of TC wine related it to floral descriptors ($p < 0.05$).

It is well known that ethanol significantly affects the sensorial perception of wines [43]; for example, it decreases the perception of higher alcohols and aldehydes and shows a similar effect for ethyl esters [44]. According to our results, wines obtained from a BSc114 (TC) monoculture could be associated with attributes such us astringency, bitterness, hotness, and sweetness, which is in agreement with the results by Tilloy et al. [45]. The authors found that high ethanol levels enhanced the perception of the abovementioned attributes. In contrast, wines with lower ethanol levels obtained with *H. uvarum/S. cerevisiae* in the present study were related to red fruit by the sensorial panel ($p < 0.05$). Although the control treatment presented elevated concentrations of chemical compounds which are commonly related to fruity descriptors, it has been reported that high ethanol contents can mask certain flavor-related volatile compounds like those related to fruity and floral profiles [46].

To our knowledge, this is the first time that a *H. uvarum* species, submitted to a previous selection process, has been proposed to carry out sequential fermentations with *S. cerevisiae* under optimized conditions to reduce ethanol in wines. The results obtained in the present study have demonstrated the impact of this co-culture on the ethanol concentration and the chemical aromatic composition; and, in addition, it has evidenced that ethanol levels affect sensorial perception. Therefore, the present study could be considered an additional step to a successful change in the wine industry to face current consumers' demands. It is possible, however, that more research is necessary in order to fully understand the impact of this co-culture on a major production scale.

Author Contributions: Investigation, M.V.M.; Methodology, M.V.M., Y.P.M., C.G. and E.D.; Project administration, Y.P.M., M.E.T., F.V. and E.D.; Supervision, F.C.; Writing—original draft, M.V.M., F.V. and E.D.; Writing—review & editing, M.V.M., Y.P.M., M.C., L.M., M.E.T. and F.C.

Funding: This research was funded by [UNSJ-SECITI] grant number [3635-2015-2017] and [UNSJ-CICITCA] grant number [1531-2016].

Conflicts of Interest: The authors declare no conflict of interest.

References

1. Zamora, F. Dealcoholised Wines and Low-Alcohol Wines. In *Wine Safety, Consumer Preference, and Human Health*; Springer: Cham, Switzerland, 2016; pp. 163–182.
2. De Orduna, R.M. Climate change associated effects on grape and wine quality and production. *Food Res. Int.* **2010**, *43*, 1844–1855. [CrossRef]
3. Meillon, S.; Dugas, V.; Urbano, C.; Schlich, P. Preference and acceptability of partially dealcoholized white and red wines by consumers and professionals. *Am. J. Enol. Vitic.* **2010**, *61*, 42–52.
4. Contreras, A.; Hidalgo, C.; Henschke, P.A.; Chambers, P.J.; Curtin, C.; Varela, C. Evaluation of non-Saccharomyces yeasts for the reduction of alcohol content in wine. *Appl. Environ. Microbiol.* **2014**, *80*, 1670–1678. [CrossRef] [PubMed]
5. Fischer, U.; Noble, A.C. The effect of ethanol, catechin concentration, and pH on sourness and bitterness of wine. *Am. J. Enol. Vitic.* **1994**, *45*, 6–10.
6. King, E.S.; Heymann, H. The effect of reduced alcohol on the sensory profiles and consumer preferences of white wine. *J. Sens. Stud.* **2014**, *29*, 33–42. [CrossRef]
7. Kliewer, W.M.; Dokoozlian, N.K. Leaf area/crop weight ratios of grapevines: Influence on fruit composition and wine quality. *Am. J. Enol. Vitic.* **2005**, *56*, 170–181.
8. Walker, R.R.; Blackmore, D.H.; Clingeleffer, P.R.; Tarr, C.R. Rootstock effects on salt tolerance of irrigated field-grown grapevines (*Vitis vinifera* L. cv. Sultana). 3. Fresh fruit composition and dried grape quality. *Aust. J. Grape Wine Res.* **2007**, *13*, 130–141. [CrossRef]
9. Chorti, E.; Guidoni, S.; Ferrandino, A.; Novello, V. Effect of different cluster sunlight exposure levels on ripening and anthocyanin accumulation in Nebbiolo grapes. *Am. J. Enol. Vitic.* **2010**, *61*, 23–30.
10. Palliotti, A.; Panara, F.; Silvestroni, O.; Lanari, V.; Sabbatini, P.; Howell, G.S.; Gatti, M.; Poni, S. Influence of mechanical postveraison leaf removal apical to the cluster zone on delay of fruit ripening in Sangiovese (*Vitis vinifera* L.) grapevines. *Aust. J. Grape Wine Res.* **2013**, *19*, 369–377. [CrossRef]
11. Schmidtke, L.M.; Blackman, J.W.; Agboola, S.O. Production technologies for reduced alcoholic wines. *J. Food Sci.* **2012**, *77*, R25–R41. [CrossRef]
12. Catarino, M.; Mendes, A. Dealcoholizing wine by membrane separation processes. *Innov. Food Sci. Emerg.* **2011**, *12*, 330–337. [CrossRef]
13. Belisario-Sanchez, Y.Y.; Taboada-Rodriguez, A.; Marin-Iniesta, F.; Lopez-Gomez, A. Dealcoholized wines by spinning cone column distillation: Phenolic compounds and antioxidant activity measured by the 1, 1-diphenyl-2-picrylhydrazyl method. *J. Agric. Food Chem.* **2009**, *57*, 6770–6778. [CrossRef] [PubMed]

14. Quirós, M.; Rojas, V.; Gonzalez, R.; Morales, P. Selection of non-Saccharomyces yeast strains for reducing alcohol levels in wine by sugar respiration. *Int. J. Food Microbiol.* **2014**, *181*, 85–91. [CrossRef] [PubMed]

15. Varela, C.; Dry, P.R.; Kutyna, D.R.; Francis, I.L.; Henschke, P.A.; Curtin, C.D.; Chambers, P.J. Strategies for reducing alcohol concentration in wine. *Aust. J. Grape Wine Res.* **2015**, *21*, 670–679. [CrossRef]

16. Englezos, V.; Rantsiou, K.; Cravero, F.; Torchio, F.; Ortiz-Julien, A.; Gerbi, V.; Rolle, L.; Cocolin, L. Starmerella bacillaris and Saccharomyces cerevisiae mixed fermentations to reduce ethanol content in wine. *Appl. Microbiol. Biotechnol.* **2016**, *100*, 5515–5526. [CrossRef] [PubMed]

17. Jolly, N.P.; Varela, C.; Pretorius, I.S. Not your ordinary yeast: Non-Saccharomyces yeasts in wine production uncovered. *FEMS Yeast Res.* **2014**, *14*, 215–237. [CrossRef]

18. Varela, C.; Sengler, F.; Solomon, M.; Curtin, C. Volatile flavour profile of reduced alcohol wines fermented with the non-conventional yeast species *Metschnikowia pulcherrima* and *Saccharomyces uvarum*. *Food Chem.* **2016**, *209*, 57–64. [CrossRef]

19. Gonzalez, R.; Quirós, M.; Morales, P. Yeast respiration of sugars by non-Saccharomyces yeast species: A promising and barely explored approach to lowering alcohol content of wines. *Trends Food Sci. Technol.* **2013**, *29*, 55–61. [CrossRef]

20. Martin, V.; Valera, M.; Medina, K.; Boido, E.; Carrau, F. Oenological impact of the Hanseniaspora/Kloeckera yeast genus on wines—A review. *Fermentation* **2018**, *4*, 76. [CrossRef]

21. Mestre Furlani, M.V.; Maturano, Y.P.; Combina, M.; Mercado, L.A.; Toro, M.E.; Vazquez, F. Selection of non-Saccharomyces yeasts to be used in grape musts with high alcoholic potential: A strategy to obtain wines with reduced ethanol content. *FEMS Yeast Res.* **2017**, *17*. [CrossRef]

22. Maturano, Y.P.; Mestre, M.V.; Kuchen, B.; Toro, M.E.; Mercado, L.A.; Vazquez, F.; Combina, M. Optimization of fermentation-relevant factors: A strategy to reduce ethanol in red wine by sequential culture of native yeasts. *Int. J. Food Microbiol.* **2019**, *289*, 40–48. [CrossRef] [PubMed]

23. Pallmann, C.L.; Brown, J.A.; Olineka, T.L.; Cocolin, L.; Mills, D.A.; Bisson, L.F. Use of WL medium to profile native flora fermentations. *Am. J. Enol. Vitic.* **2001**, *52*, 198–203.

24. Lawless, H.T.; Heymann, H. Physiological and psychological foundations of sensory function. In *Sensory Evaluation of Food*; Springer: New York, NY, USA, 2010; pp. 19–56.

25. Boido, E.; Lloret, A.; Medina, K.; Fariña, L.; Carrau, F.; Versini, G.; Dellacassa, E. Aroma composition of *Vitis vinifera* cv. Tannat: The typical red wine from Uruguay. *J. Agric. Food Chem.* **2003**, *51*, 5408–5413. [CrossRef] [PubMed]

26. Mondello, L. *Mass Spectra of Flavors and Fragrances of Natural and Synthetic Compounds*, 3rd ed.; MS Wil GmbH: Zurich, Switzerland, 2015.

27. Adams, R.P. *Identification of Essential Oil Components by Gas Chromatography/Mass Spectrometry*; Allured publishing corporation: Carol Stream, IL, USA, 2007; Volume 456.

28. McLafferty, F.W.; Stauffer, D.B. *The Wiley/NBS Registry of Mass Spectral Data*; Wiley and Sons: New York, NY, USA, 1991.

29. Hellin, P.; Manso, A.; Flores, P.; Fenoll, J. Evolution of aroma and phenolic compounds during ripening of superior seedless grapes. *J. Agric. Food Chem.* **2010**, *58*, 6334–6340. [CrossRef] [PubMed]

30. Welke, J.E.; Zanus, M.; Lazzarotto, M.; Zini, C.A. Quantitative analysis of headspace volatile compounds using comprehensive two-dimensional gas chromatography and their contribution to the aroma of Chardonnay wine. *Food Res. Int.* **2014**, *59*, 85–99. [CrossRef]

31. Tristezza, M.; Tufariello, M.; Capozzi, V.; Spano, G.; Mita, G.; Grieco, F. The oenological potential of *Hanseniaspora uvarum* in simultaneous and sequential co-fermentation with *Saccharomyces cerevisiae* for industrial wine production. *Front. Microbiol.* **2016**, *7*, 670. [CrossRef] [PubMed]

32. Fleet, G.H.; Heard, G.M. Yeasts-growth during fermentation. In *Wine Microbiology and Biotechnology*; Fleet, G.H., Harwood Academic, Eds.; CRC Press: Chur, Switzerland, 1993; pp. 27–54.

33. Mendoza, L.M.; de Nadra, M.C.M.; Farías, M.E. Kinetics and metabolic behavior of a composite culture of *Kloeckera apiculata* and Saccharomyces cerevisiae wine related strains. *Biotechnol. Lett.* **2007**, *29*, 1057–1063. [CrossRef] [PubMed]

34. Domizio, P.; Romani, C.; Lencioni, L.; Comitini, F.; Gobbi, M.; Mannazzu, I.; Ciani, M. Outlining a future for non-Saccharomyces yeasts: Selection of putative spoilage wine strains to be used in association with Saccharomyces cerevisiae for grape juice fermentation. *Int. J. Food Microbiol.* **2011**, *147*, 170–180. [CrossRef]

35. Suzzi, G.; Arfelli, G.; Schirone, M.; Corsetti, A.; Perpetuini, G.; Tofalo, R. Effect of grape indigenous Saccharomyces cerevisiae strains on Montepulciano d'Abruzzo red wine quality. *Food Res. Int.* **2012**, *46*, 22–29. [CrossRef]
36. Bisson, L.F. Stuck and sluggish fermentations. *Am. J. Enol. Vitic.* **1999**, *50*, 107–119.
37. Milanovic, V.; Ciani, M.; Oro, L.; Comitini, F. Starmerella bombicola influences the metabolism of Saccharomyces cerevisiae at pyruvate decarboxylase and alcohol dehydrogenase level during mixed wine fermentation. *Microb. Cell Factories* **2012**, *11*, 18. [CrossRef] [PubMed]
38. Petitgonnet, C.; Klein, G.L.; Roullier-Gall, C.; Schmitt-Kopplin, P.; Quintanilla-Casas, B.; Vichi, S.; Diane, J.D.; Alexandre, H. Influence of cell-cell contact between *L. thermotolerans* and *S. cerevisiae* on yeast interactions and the exo-metabolome. *Food Microbiol.* **2019**, *83*, 122–133. [CrossRef] [PubMed]
39. Wang, C.; Mas, A.; Esteve-Zarzoso, B. Interaction between Hanseniaspora uvarum and Saccharomyces cerevisiae during alcoholic fermentation. *Int. J. Food Microb.* **2015**, *206*, 67–74. [CrossRef] [PubMed]
40. Erten, H.; Campbell, I. The production of low-alcohol wines by aerobic yeasts. *J. Inst. Brew.* **1953**, *59*, 207–215. [CrossRef]
41. Giudici, P.; Romano, P.; Zambonelli, C. A biometric study of higher alcohol production in Saccharomyces cerevisiae. *Can. J. Microbiol.* **1990**, *36*, 61–64. [CrossRef] [PubMed]
42. Mateos, J.R.; Pérez-Nevado, F.; Fernández, M.R. Influence of Saccharomyces cerevisiae yeast strain on the major volatile compounds of wine. *Enzym. Microb. Technol.* **2006**, *40*, 151–157. [CrossRef]
43. Guth, H.; Sies, A. Flavour of wines: Towards an understanding by reconstitution experiments and an analysis of ethanol's effect on odour activity of key compounds. In Proceedings of the 11th Australian Wine Industry Technical Conference, Glen Osmond, SA, Australia, 7–11 October 2002; pp. 128–139.
44. Conner, J.M.; Birkmyre, L.; Paterson, A.; Piggott, J.R. Headspace concentrations of ethyl esters at different alcoholic strengths. *J. Sci. Food Agric.* **1998**, *77*, 121–126. [CrossRef]
45. Tilloy, V.; Ortiz-Julien, A.; Dequin, S. Reduction of ethanol yield and improvement of glycerol formation by adaptive evolution of the wine yeast *Saccharomyces cerevisiae* under hyperosmotic conditions. *Appl. Environ. Microbiol.* **2014**, *80*, 2623–2632. [CrossRef]
46. Escudero, A.; Gogorza, B.; Melus, M.A.; Ortin, N.; Cacho, J.; Ferreira, V. Characterization of the aroma of a wine from Maccabeo. Key role played by compounds with low odor activity values. *J. Agric. Food Chem.* **2004**, *52*, 3516–3524. [CrossRef]

 *fermentation*

Article

Influence of Native *Saccharomyces cerevisiae* Strains from D.O. "Vinos de Madrid" in the Volatile Profile of White Wines

Margarita García [1],*[image_ref], Braulio Esteve-Zarzoso [2][image_ref], Julia Crespo [1], Juan Mariano Cabellos [1] and Teresa Arroyo [1][image_ref]

[1]   Department of Food and Agricultural Science, IMIDRA, 28805 Alcalá de Henares, Spain;
      julia.crespo.garcia@madrid.org (J.C.); juan.cabellos@madrid.org (J.M.C.); teresa.arroyo@madrid.org (T.A.)
[2]   Department of Chemistry and Biotechnology, Rovira i Virgili University, 43007 Tarragona, Spain;
      braulio.esteve@urv.cat
*   Correspondence: margarita_garcia_garcia@madrid.org; Tel.: +34-918-879-382

Received: 11 August 2019; Accepted: 28 October 2019; Published: 30 October 2019

Abstract: Yeasts during alcoholic fermentation form a vast number of volatile compounds that significantly influence wine character and quality. It is well known that the capacity to form aromatic compounds is dependent on the yeast strain. Thus, the use of native yeast strains, besides promoting biodiversity, encourages the conservation of regional sensory properties. In this work, we studied the volatile profile of Malvar wines fermented with 102 *Saccharomyces cerevisiae* yeast strains, isolated from vineyards and cellars belonging to the D.O. "Vinos de Madrid". The wines elaborated with different *S. cerevisiae* showed a good classification by cellar of origin. Additionally, seven sensory descriptors have helped to classify the wines depending on their predominant aromatic character. Twenty-nine *Saccharomyces* strains, belonging to five of six cellars in the study, were characterized by producing wines with a fruity/sweet character. Floral, solvent, and herbaceous descriptors are more related to wines elaborated with *Saccharomyces* strains from organic cellars A, E, and F. Based on these findings, winemakers may use their best native *S. cerevisiae* strains, which add personality to their wine. Therefore, this study contributes to promoting the use of native *Saccharomyces* yeasts in winemaking.

Keywords: native yeast; *Saccharomyces cerevisiae*; aroma; Malvar (*Vitis vinifera* L. cv.); white wine

---

1. Introduction

Yeasts contribute to wine aroma by several mechanisms: firstly, by alcoholic fermentation of the grape must; secondly by the de novo biosynthesis of volatile compounds; and lastly, by the transformation of neutral grape compounds into flavor-active components [1,2]. Among fermentation-derived volatiles are esters, higher alcohols, and volatile acids, as well as varietal compounds, i.e., thiols and terpenes; all of these are the most abundant in the total wine aroma composition [3].

Aroma is one of the most influential factors on wine quality and consumers preferences, as well as the prime contributor to overall flavor perception [4,5]. Since the 90s, wine has been described as containing around 600 to 800 volatile aroma compounds arising from the grapes, from alcoholic fermentation, and from the aging process [6]. The particular importance of a specific volatile compound to wine aroma perception is related to its odor threshold value (OTV), which can be considered as the lowest concentration detected by smelling [7]. Another parameter extensively used to estimate the sensory contribution of aromatic compounds to the overall aroma of wine is the odor activity value (OAV). The OAV is obtained from the ratio between the concentration of an individual compound and its perception threshold. A volatile compound contributes to overall aroma when its concentration in wine is above the perception threshold; therefore, odorants with OAV ≥ 1 can be perceived [8,9]. Nevertheless, some authors presented evidence that compounds with low OAV values may act as significant impact odorants [10,11]. Therefore, the characterization of wine aroma compounds and their odorant profiles are currently among the research targets in winemaking [12–14]. In addition, several works have determined the aromatic series as groups of all volatile compounds with similar sensory descriptors [15–17], and a generalized OAV for each aromatic series can be calculated by adding the OAV of each aromatic series component [18].

The monitoring of fermentation is an effective method for modulating the wine aroma [19]. Typically, the use of commercial *Saccharomyces cerevisiae* starter cultures is widespread to obtain control and homogeneity of the fermentation process. In fact, their predominance reduces the risk of wine spoilage, so their dominant growth makes the development of indigenous spoilage species difficult [3,20,21]. However, the continued use of commercial yeasts has resulted in an excessive standardization of wines, regardless of their vinicultural region of origin [22]. For this reason, in recent decades, most studies have focused on using indigenous yeast strains as a way of expressing singular characters and to encourage the aromatic profiles of wines from a given region or appellation [23–25]. This relationship between wine microbiota and *terroir* has gained relevance in the wine industry [26–29]. The concept of *terroir* is linked to the natural environment, the physico-chemical characteristics of the soil, and climatic conditions in a delimited area that affect grape characteristics, so the obtained wine is also affected by this territoriality. Thus, the microbiota from a determined *terroir* is able to confer a unique quality to the wine [30].

The Denomination of Origin (D.O.) "Vinos de Madrid", created in 1990, is located in the center of Spain and covers an area of 8390 ha. This D.O. comprises 46 wineries in three regions: Arganda (27 wineries), Navalcarnero (5 wineries), and San Martín de Valdeiglesias (14 wineries). Recently, the new region of El Molar has become part of this D.O. The climate of this region is Mediterranean continental, with temperatures ranging from −8 °C minimum in winter to a maximum of 41 °C in summer [31]. The annual rainfall ranges between 460 and 660 mm. Winemakers in this region base their production on the cultivation of the vine varieties Airen and Malvar (white), and Garnacha and Tempranillo (red) (*Vitis vinifera* L. cv.). Malvar is an autochthonous cultivar for this D.O., while Airen, Garnacha, and Tempranillo have major extensions all over the Iberian Peninsula.

In the last few years, our research activity has been directed to the exploitation of native microbiota potential to enhance the quality of regional Malvar wines. In the present investigation, small volume fermentations were carried out with 101 autochthonous *S. cerevisiae* strains isolated from vineyards and cellars of D.O. "Vinos de Madrid" and compared with a control of *S. cerevisiae* CLI889 previously isolated and exhibiting good oenological aptitudes [32,33]. This work seeks to study the impact of *S. cerevisiae* strains isolated from their oenological region on the volatile composition of Malvar wines, providing an opportunity for wineries to elaborate products with their own typicity.

2. Materials and Methods

2.1. Yeast Strains, Origin, and Vinification Procedure

A total of 101 native *S. cerevisiae* yeast strains have been used for wine elaboration in this study. These strains were isolated from six vineyards and commercial cellars (A-F) belonging to the D.O. "Vinos de Madrid" as stated by Tello et al. [34]. The location of cellars is shown in Figure 1. As published by Tello et al. [34] and García et al. [35], four of the wineries (A, D, E, and F) use an organic system of wine production, in contrast to cellars B and C, that utilize a conventional production system. In wineries A, E, and F, the fermentation was spontaneous, and different commercial *S. cerevisiae* strains induced the fermentation in cellars B, C, and D. One autochthonous strain, *S. cerevisiae* CLI 889 from the IMIDRA collection, selected by our group for Airen white wine elaboration, was used as a control [32,33]. This strain has been deposited in the Spanish Type Culture Collection (CECT 13145).

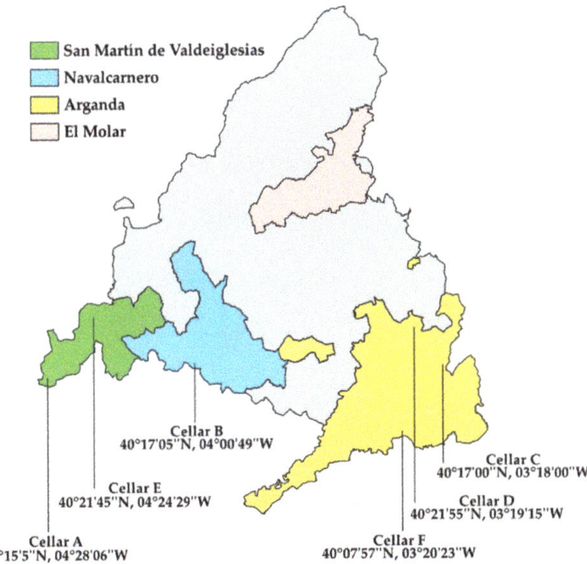

Figure 1. Map of D.O. "Vinos de Madrid". Different regions and location of cellars (A-F) included in this study are given in this figure.

The different genotypes of *S. cerevisiae* were identified by microsatellite multiplex PCR analysis using the highly polymorphic loci SC8132X, YOR267C, and SCPTSY7 [36]. The size of the fragments was determined by automatic electrophoresis with an ABI 3130 Genetic Analyzer (Applied Biosystems, Tres Cantos, Madrid, Spain), whose results were published by Tello et al. [34].

Grapes from the Malvar cultivar (*Vitis vinifera* L. cv.) were hand-collected from IMIDRA's experimental vineyard located in the Madrid winegrowing region, Spain (40°31' N, 3°17' W and 610 m altitude) during the 2010 vintage at commercial maturity. The must was clarified at 4 °C by pectolityc enzymes (Enozym Altair, Agrovin, Spain) (0.01 g/L) and stored at −20 °C until needed. In order to carry out the study under the same conditions, the grape must was adjusted to 200 g/L of reducing sugars; then, the pH value was 3.2, total acidity (as g/L of tartaric acid) was 5.0, and there was 165 mg/L of yeast assimilable nitrogen (YAN). The fermentations were performed in sterile flasks with 100 mL of pasteurized Malvar must with constant agitation (150 rpm) under anaerobic conditions. Each *S. cerevisiae* strain was inoculated in grape must at a concentration of 10^6 cells/mL, from a culture grown for 48 h in YPD liquid medium at 28 °C. The fermentation was performed at 20 °C in a JP Selecta™ incubator (Abrera, Barcelona, Spain), and the alcoholic fermentation kinetics was controlled daily by weight loss. When its value was constant for two consecutive days, the fermentation process was considered complete, and clarified wine samples were frozen in order to carry out volatile composition analyses. All experiments were performed in duplicate.

2.2. Volatile Fraction Analysis

Quantification of major volatile compounds was undertaken by gas chromatography coupled to flame ionization detector (GC-FID) (Agilent Technologies, Santa Clara, CA, USA). The column was DB-Wax column (60 m × 0.32 mm × 0.5 µm film thickness) from J&W Scientific (Folsom, CA, USA). The oven temperature program was: 40 °C for 5 min, then increased at 3 °C/min up to 200 °C. Helium was used as carrier gas at 2 mL/min. Two µL of aroma extract were injected at 250 °C in splitless mode. Total run time was 75 min per sample. The extraction and analysis methodologies of volatile compounds were done following the procedures proposed by Ortega et al. [37]. Analyses were carried out in duplicate.

2.3. Statistical Analysis

The odor activity values of all volatile compounds have been statistically analyzed to study how the use of different *S. cerevisiae* strains affects the aromatic profile of wines. Thus, a discriminant analysis was carried out to determine the impact of wine aroma profiles on their classification by cellar. In addition, a principal component analysis (PCA) was elaborated to link wines produced by *Saccharomyces* strains with the seven aromatic descriptors to describe the volatile profiles of Malvar young wines. Both statistical analyses were performed using SPSS Statistics 25 (SPSS Inc., Chicago, IL, USA).

3. Results and Discussion

Most of the native strains were capable of completing the vinification (residual sugars below 5 g/L), although there are differences in the time required, ranging between 8 to 18 days. Only 13% of *Saccharomyces* strains did not complete the alcoholic fermentation, which stopped at 5–59 g/L residual sugar.

Volatile acidity expressed as acetic acid (g/L) content affects the wine quality. In general, most of the elaborated wines contained moderate levels of acetic acid, i.e., between 0.23–0.70 g/L, except for the wines elaborated with the strains G8 (1.14) and G16 (0.98) from Cellar A, G462 (1.08) and G493 (0.90) from Cellar E, and G529 (1.37) from Cellar F. The legal limit is 1.2 g/L of acetic acid under European legislation [38]. However, acetic acid may provide an unpleasant vinegar aroma and an undesirable acidic taste to wine at concentrations above 0.8 g/L [39].

3.1. Aromatic Profile of Wines Elaborated with Different S. cerevisiae Strains

Table 1 shows the major volatile compounds quantified in Malvar white wines. This table also contains the odor threshold values and descriptors for each aromatic compound. Moreover, each compound was attributed to one or more aromatic series depending on its principal sensory description: solvent, sweet, herbaceous, floral, fruity, microbiological, and fatty. These seven classes of sensory descriptors were employed to link odorous compounds with similar sensory descriptors into classes (aromatic series) [16,18,40,41] and give an organoleptic profile of wines elaborated with the different *S. cerevisiae* native strains. Moreover, the contribution of each volatile compound to each series can be determined. This procedure, which is based on more objective criteria than other existing alternatives, allows for the connection of quantitative information obtained from chemical analysis to sensory perceptions in order to achieve an aroma profile for the wine [16].

In order to analyze the aroma composition of wines, the OAVs were calculated for each of the 31 volatile compounds quantified in the wines (Table S1, Supplementary Materials). As can be seen, only isoamyl alcohol, several esters such as ethyl butyrate, ethyl isovalerate, isoamyl acetate, ethyl hexanoate, acids, i.e., isobutyric acid, isovaleric acid, hexanoic acid, and octanoic acid, and one ketone compound, i.e., diacetyl, has OAVs above the unity in all wines. The OAV for β-phenylethyl alcohol was greater than 1 in all wines elaborated with *S. cerevisiae* from D.O. "Vinos de Madrid" cellars, whereas that for the *S. cerevisiae* CLI 889 (control), this OAV value was lower than unity. In the case of 2-phenylethyl acetate, only two strains (G12, Cellar A and G507, Cellar F) did not exceed the unity. In contrast, it should be pointed out that the OAVs of 1-butanol, benzyl alcohol, ethyl-3-hydroxybutyrate, diethyl succinate, furfural, benzaldehyde, and acetoin, were below 0.1 in all cases.

Fusel alcohols (isoamyl alcohol, isobutanol, and β-phenylethyl alcohol) contribute to the wine odor of the analyzed Malvar wines. These alcohols are usually present in wines, formed as the fermentation products by yeasts. High concentrations of these volatiles (above 300 mg/L) can have a detrimental effect on wine, whereas concentrations below this value add a desirable level of complexity to the wine [2,42]. Esters are one of the most important classes of volatiles, and are responsible for the fruity and floral character in wines; their synthesis is mainly dependent on yeasts [6]. However, these compounds in excess can mask varietal aromas; for example, ethyl acetate over 90 mg/L, or 200 mg/L of total esters, can have a negative effect [42]. In our work, the total esters of samples ranged between 6.42 mg/L and 88.04 mg/L, in no case exceeding 200 mg/L (data not shown).

Table 1. Major aroma compounds quantified in wines. Odor description (ODE), Odor threshold value (OTV) (in mg/L), and assignation of compounds to different aromatic series.

Compound	ODE	OTV [1]	Aromatic Serie [2]
1-Propanol	Alcohol, ripe fruit	9 [a]	1, 5
1-Butanol	Soap, fatty, diesel	150 [a]	1
Isobutanol	Bitter, fusel, alcohol	40 [b]	1
Isoamyl alcohol	Harsh, bitter	30 [b]	1
(Z)-3-Hexen-1-ol	Lemon, fresh	0.4 [b]	3, 5
1-Hexanol	Green grass, fresh	8 [b]	3
Metionol	Garlic	1 [b]	3
Benzyl alcohol	Pleasant, soft	200 [c]	2, 4
β-Phenylethyl alcohol	Flowery, roses	14 [b]	2, 4
Ethyl butyrate	Fruity, sweet, apple	0.02 [b]	2, 5
Ethyl isovalerate	Fruity, sweet, banana	0.003 [b]	2, 5
Isoamyl acetate	Banana, sweet, fruity	0.03 [b]	2, 5
Ethyl hexanoate	Pineapple, apple	0.014 [b]	5
Ethyl-3-hydroxybutyrate	Fruity	20 [c]	5
Hexyl acetate	Fruity, green, pear	1 [d]	5
2-Phenylethyl acetate	Flowery, lilac	0.25 [b]	4
Diethyl succinate	Camphor	100 [d]	5, 6
Ethyl octanoate	Fresh, flowery, pineapple	0.58 [a]	2, 4, 5
Ethyl lactate	Lactic	154 [a]	6
Isobutyric acid	Rancid, butter, cheese	0.05 [e]	7
Butyric acid	Butter, cheese, stinky	0.173 [b]	7
Isovaleric acid	Cheese	0.033 [b]	7
Hexanoic acid	Cheese	0.42 [b]	7
Octanoic acid	Sweet, cheesy	0.5 [b]	7
Decanoic acid	Rancid, fatty	1 [b]	7
Diacetyl	Butter	0.1 [b]	7
Furfural	Bread, toasty, candy	15 [d]	6
Benzaldehyde	Sweet, candy, wood	5 [b]	2, 4
Phenylacetaldehyde	Roses	1 [f]	4
Acetoin	Butter	150 [a]	7
γ-Butyrolactone	Coconut	35 [c]	2

[1] References: a, thresholds from Etievant et al. [43]; b, thresholds from Ferreira et al. [15]; c, thresholds from Aznar et al. [44]; d, thresholds from Chaves et al. [45]; e, thresholds from Van Gemert and Nettenbreijer [46]; f, thresholds from Culleré et al. [47]. [2] 1, solvent; 2, sweet; 3, herbaceous; 4, floral; 5, fruity; 6, microbiological; 7, fatty.

The family of fatty acids has been reported to derive not only from yeasts, but from grapes as well [48], providing fruity, cheese, fatty, and rancid notes to wines [7]. Among these fatty acids, we mention the importance of isobutyric, isovaleric, hexanoic, and octanoic acids as active odorants, whose OAVs were higher than 1 in all the studied wines. A greater fatty acid proportion than other aromatic descriptors was found in samples G462 and G475 (Cellar E). Finally, we denoted the relevant content of the ketone diacetyl in the Malvar samples, showing the highest amounts in G7 (1.50 mg/L) and G502 (1.51 mg/L) from cellars A and E, respectively. Diacetyl concentrations exceeding 5–7 mg/L are considered undesirable, although depending on the style and type of wine, this compound is recognized to contribute a desirable buttery and butterscotch-like flavor at amounts around 1–4 mg/L [49].

To determine whether the volatile composition of wines is related to the cellar to which the yeast strains belong, the OAV data from the 31 compounds were submitted to discriminant statistical analysis to find the canonical parameters that explain the maximum variability between the studied wines (Figure 2). The results of this analysis showed six discriminant functions, where the first two accounted for 41.3% and 28.5% of the total variance, respectively, so the total variance explained by these two functions reached 69.8%. The wines elaborated with different *S. cerevisiae* genotypes presented a good correlation by cellar of origin. These results are in agreement with those obtained by Knight et al. [50], who revealed that there is a significant correlation between the region of isolation of *S. cerevisiae* and the aroma profile in New Zealand wines. In relation with discriminant function 1, 1-propanol, (Z)-3-hexen-1-ol, 1-hexanol, and isovaleric acid were the most significant compounds in the differentiation between wines. In the case of function 2, 1-butanol, 2-phenylethyl acetate and benzaldehyde contributed most to the discriminant model. The same analysis exhibited correct classification of 86.7% of the wines elaborated with *S. cerevisiae* native strains, according to their cellar of origin (data not shown). Figure 2 shows that the most aromatically-different wines were those elaborated with strains CLI 889 (control), G113 (Cellar B) and G114 (Cellar C). As previously indicated, the fermentation process was induced by commercial *S. cerevisiae* strains in these two cellars, and these strains were the only ones which were isolated throughout the fermentation from each winery [34]. The use of these starter yeast cultures for winemaking guarantees that the must ferments in the expected way [51]. In contrast, some authors have found that native yeasts produced wines with high concentrations of pleasant aromas and special bouquets not which are available with commercial yeast strains [23,24,52,53]. In our case, these wines showed a fruity character, highlighting ethyl isovalerate and ethyl hexanoate concentrations, while acetate ester contents were lower than those in wines elaborated with the native strains.

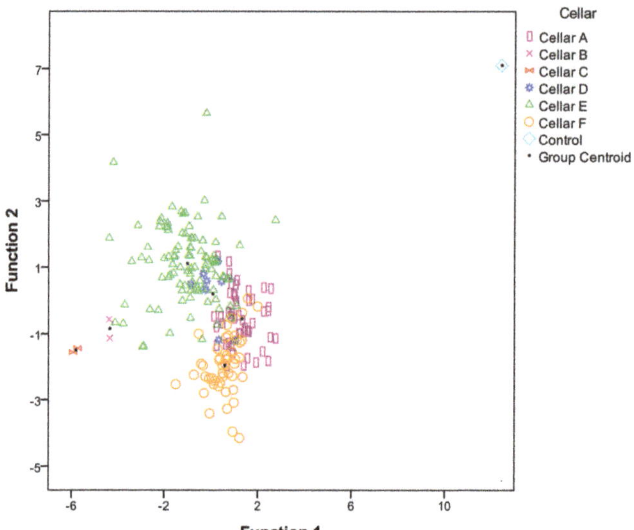

Figure 2. Application of discriminant analyses of the OAV data of volatile compounds studied in wines, classifying the samples by cellar of origin.

3.2. Principal Component Analysis

A PCA analysis was done to cluster yeast fermentations according to the aromatic descriptors (Figure 3). In reference to the seven defined classes of aromatic descriptors (Figure 3a), a generalized OAV for each class of sensory descriptor was calculated by adding up the OAVs of all compounds belonging to that class. Then, this generalized OAV calculated by wine sample was used to calculate the proportion (% OAV; Table S2, Supplementary Materials) that each aromatic descriptor represents into wines elaborated with the *Saccharomyces* strains (Figure 3b). Calculation of the aroma series by the accumulation of OAVs cannot be considered as an arithmetical addition of the odorant sensations, and the assignment of some compounds in a particular series or in several series may be questionable [40,54]. However, several authors have employed the proposed method, which groups the compounds into odorant series, since it reduces the number of variables to be interpreted and, consequently, is a valid and simple way to compare a wine's aroma character [18,41]. It can be particularly useful in many contexts where a sensorial study is not available or affordable, and a first analysis of wine aroma peculiarity is outlined [18].

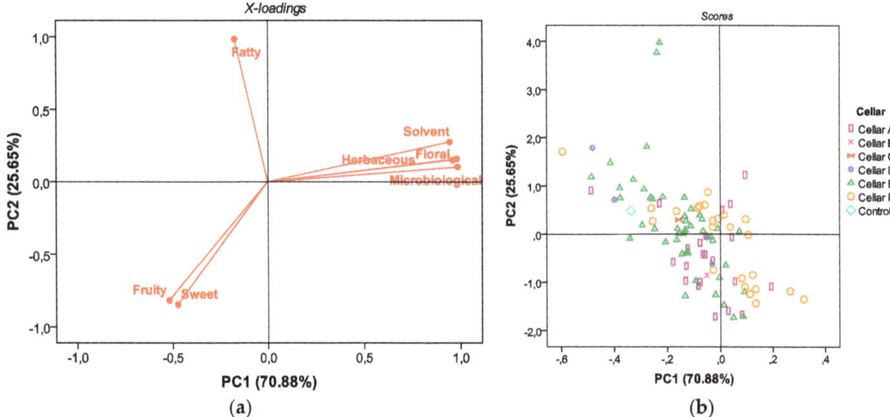

Figure 3. Results of the principal component analysis carried out on the aromatic series matrix: (a) loadings of the variables on the first and second principal components; (b) scores of the % OAV on each sensory descriptor adding up the different strains by cellar of origin in the plane formed by the first and second principal components. Values are the mean of two % OAV ratios.

The PCA explained the 96.53% of the total variance. Wine samples closely related with sweet and fruity descriptors appear in the left bottom corner of the PC plane. These two sensory descriptors are mainly determined by ester content in wines; specifically, the fruity descriptor represents the highest proportion of aroma composition in most wines (Table S2, Supplementary Materials). Five of the six cellars studied are included in this group (Figure 3b); therefore, we have not found a direct correlation between fruity/sweet descriptors and a determined area or cellar. Wine samples classified at the top right plane are more associated with compounds related to floral, solvent, microbiological, and herbaceous descriptors. However, it is worth noting that the volatile compounds comprising microbiological character (diethyl succinate, ethyl lactate, and furfural) have an OAV lower than unity (OAV < 1) in all Malvar wines (Table S1, Supplementary Materials). The floral parameter was mainly constituted by β-phenylethyl alcohol and 2-phenylethyl acetate; solvent is mostly related to isoamyl alcohol; and herbaceous is determined by metionol. In this quarter, we can find the wines elaborated with the native strains G3, G9, and G19 from Cellar A, G465 from Cellar F, and G513, G514, G515, and G518 from Cellar E. In this case, a connection point between the strains named above is that their cellars of origin utilize spontaneous fermentation and an organic system of wine production. In accordance with these results, Lorenzo et al. [55] observed that the volatile composition of wines from organic or non-organic grapes was considerably different. In particular, they concluded that the OAVs of wines from ecologically-grown grapes had more chemical and floral aromas, while the wines from conventional practices presented a fruitier character. Finally, the fatty character was nearly correlated with two samples from Cellar E (G475 and G462), due to the high proportion of fatty acids within these two samples.

Although compounds with OAV ≥ 1 are called critical compounds essential to total aroma [8], the statistical treatments of this work also considered the compounds with OAV < 1, in agreement with the theory that sub-threshold volatile compounds may contribute to wine aroma through the additive effects of compounds with a similar odor or structure [56]. In contrast, some compounds can mask the perception of others, so they remain undetected at supra-threshold concentrations [57,58]. Atanasova et al. [59] concluded that the fruity character of wine might be masked by woody components when presented at supra-threshold concentrations.

In a previous work by our group, some of these *S. cerevisiae* native strains that showed a pleasant aromatic profile were also recognized for their good fermentation abilities and for resistance to the stresses inherent to wine fermentation in warm areas [35].

4. Conclusions

The knowledge of the volatile profile of wines elaborated with different *S. cerevisiae* strains, together with their fermentation aptitudes and stress resistance, provide important information which contributes to promoting the use of these autochthonous strains in winemaking. Thus, we suggest that each winery uses their best native *S. cerevisiae* strains, which may add personality to their wines. However, more studies are necessary to know the fermentative behavior of these *Saccharomyces* strains at industrial scales. Furthermore, it could be considered an opportunity for some of these *S. cerevisiae* strains to become commercially available.

Supplementary Materials: The following are available online at http://www.mdpi.com/2311-5637/5/4/94/s1, Table S1: Odor activity value (OAV) for the aroma compounds studied in Malvar wines, Table S2: Odor activity value proportion (% OAV) of each aromatic descriptor in wines elaborated with the *Saccharomyces cerevisiae* yeast strains.

Author Contributions: M.G., T.A. and B.E.-Z. designed the experiments, analyzed the results, discussion of the results and wrote the manuscript. M.G., J.C., and J.M.C. performed experiments and analyzed results.

Funding: This work was financially supported by the project RM2010-00009-C03-01 funded by INIA.

Acknowledgments: We are grateful to E. Martin for revising the manuscript. M.G. acknowledges the IMIDRA for his grant. We thank to the wine cellars from D.O. "Vinos de Madrid" that kindly participated in this study.

Conflicts of Interest: The authors declare no conflict of interest. The funders had no role in the design of the study; in the collection, analyses, or interpretation of data; in the writing of the manuscript, and in the decision to publish the results.

References

1. Fleet, G.H. Yeast interactions and wine flavour. *Int. J. Food Microbiol.* **2003**, *86*, 11–22. [CrossRef]
2. Styger, G.; Prior, B.; Bauer, F.F. Wine flavor and aroma. *J. Ind. Microbiol. Biotechnol.* **2011**, *38*, 1145–1159. [CrossRef] [PubMed]
3. Lambrechts, M.G.; Pretorius, I.S. Yeast and its importance to wine aroma-A review. *S. Afr. J. Enol. Vitic.* **2000**, *21*, 97–129. [CrossRef]
4. Polaskova, P.; Herszage, J.; Ebeler, S.E. Wine flavor: Chemistry in a glass. *Chem. Soc. Rev.* **2008**, *37*, 2478–2489. [CrossRef] [PubMed]
5. King, E.S.; Kievit, R.L.; Curtin, C.; Swiegers, J.H.; Pretorius, I.S.; Bastian, S.E.P.; Leigh Francis, I. The effect of multiple yeasts co-inoculations on Sauvignon Blanc wine aroma composition, sensory properties and consumer preference. *Food Chem.* **2010**, *122*, 618–626. [CrossRef]
6. Rapp, A. Natural flavours of wine: Correlation between instrumental analysis and sensory perception. *Anal. Chem.* **1990**, *337*, 777–785. [CrossRef]
7. Rocha, S.M.; Rodrigues, F.; Coutinho, P.; Delgadillo, I.; Coimbra, M.A. Volatile composition of Baga red wine: Assessment of the identification of the would-be impact odourants. *Anal. Chim. Acta* **2004**, *513*, 257–262. [CrossRef]
8. Guth, H. Quantitation and sensory studies of character impact odorants of different white wine varieties. *J. Agric. Food Chem.* **1997**, *43*, 3027–3032. [CrossRef]
9. Gil, M.; Cabellos, J.M.; Arroyo, T.; Prodanov, M. Characterization of the volatile fraction of young wines from the Denomination of Origin "Vinos de Madrid" (Spain). *Anal. Chim. Acta* **2006**, *563*, 145–153. [CrossRef]

10. Ryan, D.; Prenzler, P.D.; Saliba, A.J.; Scollary, G.R. The significance of low impact odorants in global odour perception. *Trends Food Sci. Technol.* **2008**, *19*, 383–389. [CrossRef]

11. Escudero, A.; Campo, E.; Fariña, L.; Cacho, J.; Ferreira, V. Analytical characterization of the aroma of five premium red wines. Insights into the role of odor families and the concept of fruitiness of wines. *J. Agric. Food Chem.* **2007**, *55*, 4501–4510. [CrossRef] [PubMed]

12. Robinson, A.L.; Boss, P.K.; Solomon, P.S.; Trengove, R.D.; Heymann, H.; Ebeler, S.E. Origins of grape and wine aroma. Part 2. Chemical and sensory analysis. *Am. J. Enol. Vitic.* **2014**, *65*, 25–42. [CrossRef]

13. Mina, M.; Tsaltas, D. Contribution of Yeast in Wine Aroma and Flavour. In *Yeast—Industrial Applications*; Morata, A., Loira, I., Eds.; IntechOpen: London, UK, 2017; pp. 117–134. [CrossRef]

14. Kutyna, D.R.; Borneman, A.R. Heterologous production of flavour and aroma compounds in *Saccharomyces cerevisiae*. *Genes* **2018**, *9*, 326. [CrossRef] [PubMed]

15. Ferreira, V.; López, R.; Cacho, J.F. Quantitative determination of the odorants of young red wines from different grape varieties. *J. Sci. Food Agric.* **2000**, *80*, 1659–1667. [CrossRef]

16. Peinado, R.A.; Moreno, J.; Bueno, J.E.; Moreno, J.A.; Mauricio, J.C. Comparative study of aromatic compounds in two young white wines subjected to pre-fermentative cryomaceration. *Food Chem.* **2004**, *84*, 585–590. [CrossRef]

17. Mestre, M.V.; Maturano, Y.P.; Gallardo, C.; Combina, M.; Mercado, L.; Toro, M.E.; Carrau, F.; Vazquez, F.; Dellacassa, E. Impact on sensory and aromatic profile of low ethanol Malbec wines fermented by sequential culture of *Hanseniaspora uvarum* and *Saccharomyces cerevisiae* native yeasts. *Fermentation* **2019**, *5*, 65. [CrossRef]

18. Capone, S.; Tufariello, M.; Siciliano, P. Analytical characterisation of Negroamaro red wines by "Aroma Wheels". *Food Chem.* **2013**, *141*, 2906–2915. [CrossRef]

19. King, E.S.; Swiegers, J.H.; Travis, B.; Leigh Francis, I.; Bastian, S.E.P.; Pretorius, I.S. Coinoculated fermentations using *Saccharomyces* yeasts affect the volatile composition and sensory properties of *Vitis vinifera* L. cv. Sauvignon Blanc wines. *J. Agric. Food Chem.* **2008**, *56*, 10829–10837. [CrossRef]

20. Romano, P. Metabolic characteristics of wine strains during spontaneous and inoculated fermentation. *Food Technol. Biotechnol.* **1997**, *35*, 255–260.

21. Pretorius, I.S. Tailoring wine yeast for the new millennium: Novel approaches to the ancient art of winemaking. *Yeast* **2000**, *16*, 675–729. [CrossRef]

22. Swiegers, J.H.; Bartowsky, E.J.; Henschke, P.A.; Pretorius, I.S. Yeast and bacterial modulation of wine aroma and flavour. *Aust. J. Grape Wine Res.* **2005**, *11*, 139–173. [CrossRef]

23. Martini, A. Biotechnology of natural and winery-associated strains of *Saccharomyces cerevisiae*. *Int. Microbiol.* **2003**, *6*, 207–209. [CrossRef] [PubMed]

24. Callejon, R.M.; Clavijo, A.; Ortigueira, P.; Troncoso, A.M.; Paneque, P.; Morales, M.L. Volatile and sensory profile of organic red wines produced by different selected autochthonous and commercial *Saccharomyces cerevisiae* strains. *Anal. Chim. Acta* **2010**, *660*, 68–75. [CrossRef] [PubMed]

25. Tufariello, M.; Chiriatti, M.A.; Grieco, F.; Perrotta, C.; Capone, S.; Rampino, P.; Tristezza, M.; Mita, G.; Grieco, F. Influence of autochthonous *Saccharomyces cerevisiae* strains on volatile profile of Negroamaro wines. *LWT Food Sci. Technol.* **2014**, *58*, 35–48. [CrossRef]

26. Di Maio, S.; Polizzotto, G.; Di Gangi, E.; Foresta, G.; Genna, G.; Verzera, A.; Scacco, A.; Amore, G.; Oliva, D. Biodiversity of indigenous *Saccharomyces* populations from old wineries of South-Eastern Sicily (Italy): Preservation and economic potential. *PLoS ONE* **2012**, *7*, e30428. [CrossRef]

27. Tristezza, M.; Vetrano, C.; Bleve, G.; Spano, G.; Capozzi, V.; Logrieco, A.; Mita, G.; Grieco, F. Biodiversity and safety aspects of yeast strains characterized from vineyards and spontaneous fermentations in the Apulia Region, Italy. *Food Microbiol.* **2013**, *36*, 335–342. [CrossRef]

28. Capozzi, V.; Garofalo, C.; Chiriatti, M.A.; Grieco, F.; Spano, G. Microbial terroir and food innovation: The case of yeast biodiversity in wine. *Microbiol. Res.* **2015**, *181*, 75–83. [CrossRef]

29. Vaudano, E.; Quinterno, G.; Costantini, A.; Pulcini, L.; Pessione, E.; García-Moruno, E. Yeast distribution in Grignolino grapes growing in a new vineyard in Piedmont and the technological characterization of indigenous *Saccharomyces* spp. strains. *Int. J. Food Microbiol.* **2019**, *289*, 154–161. [CrossRef]

30. Csoma, H.; Zakany, N.; Capece, A.; Romano, P.; Sipiczki, M. Biological diversity of *Saccharomyces* yeasts of spontaneously fermenting wines in four wine regions: Comparative genotypic and phenotypic analysis. *Int. J. Food Microbiol.* **2010**, *140*, 239–248. [CrossRef]

31. Cordero-Bueso, G.; Esteve-Zarzoso, B.; Cabellos, J.M.; Gil-Díaz, M.; Arroyo, T. Biotechnological potential of non-*Saccharomyces* yeasts isolated during spontaneous fermentations of Malvar (*Vitis vinifera* cv. L.). *Eur. Food Res. Technol.* **2013**, *236*, 193–207. [CrossRef]

32. Arroyo, T. Estudio de la Influencia de Diferentes Tratamientos Enológicos en la Evolución de la Microbiota y en la Calidad de los Vinos Elaborados con la Variedad "Airén", en la D.O. "Vinos de Madrid". Ph.D. Thesis, University of Alcalá, Madrid, Spain, 2000.

33. Cordero-Bueso, G.; Esteve-Zarzoso, B.; Gil-Díaz, M.; García, M.; Cabellos, J.; Arroyo, T. Improvement of Malvar wine quality by use of locally-selected *Saccharomyces cerevisiae* strains. *Fermentation* **2016**, *2*, 7. [CrossRef]

34. Tello, J.; Cordero-Bueso, G.; Aporta, I.; Cabellos, J.M.; Arroyo, T. Genetic diversity in commercial wineries: Effects of the farming system and vinification management on wine yeasts. *J. Appl. Microbiol.* **2012**, *112*, 302–315. [CrossRef] [PubMed]

35. García, M.; Greetham, D.; Wimalasena, T.T.; Phister, T.G.; Cabellos, J.M.; Arroyo, T. The phenotypic characterization of yeast strains to stresses inherent to wine fermentation in warm climates. *J. Appl. Microbiol.* **2016**, *121*, 215–233. [CrossRef] [PubMed]

36. Vaudano, E.; García-Moruno, E. Discrimination of *Saccharomyces cerevisiae* wine strains using microsatellite multiplex PCR and band pattern analysis. *Food Microbiol.* **2008**, *25*, 56–64. [CrossRef] [PubMed]

37. Ortega, C.; López, R.; Cacho, J.; Ferreira, V. Fast analysis of important wine volatile compounds-Development and validation of a new method based on gas chromatographic-flame ionisation detection analysis of dichloromethane microextracts. *J. Chromatogr. A* **2001**, *923*, 205–214. [CrossRef]

38. Office Internationale de la Vigne et du Vin. *International Code of Oenological Practices*; OIV: Paris, France, 2009.

39. Bely, M.; Rinaldi, A.; Dubourdieu, D. Influence of assimilable nitrogen on volatile acidity production by *Saccharomyces cerevisiae* during high sugar fermentation. *J. Biosci. Bioeng.* **2003**, *96*, 507–512. [CrossRef]

40. Moyano, L.; Zea, L.; Moreno, J.; Medina, M. Analytical study of aromatic series in sherry wines subjected to biological aging. *J. Agric. Food Chem.* **2002**, *50*, 7356–7361. [CrossRef] [PubMed]

41. Wu, Y.; Duan, S.; Zhao, L.; Gao, Z.; Luo, M.; Song, S.; Xu, W.; Zhang, C.; Ma, C.; Wang, S. Aroma characterization based on aromatic series analysis in table grapes. *Sci. Rep.* **2016**, *6*. [CrossRef] [PubMed]

42. Belda, I.; Ruiz, J.; Esteban-Fernández, A.; Navascués, E.; Marquina, D.; Santos, A.; Moreno-Arribas, M.V. Microbial contribution to wine aroma and its intended use for wine quality improvement. *Molecules* **2017**, *22*, 189. [CrossRef]

43. Etiévant, X.P. Wine. In *Volatile Compounds in Foods and Beverages*; Maarse, H., Ed.; Marcel Dekker: New York, NY, USA, 1991; pp. 483–546.

44. Aznar, M.; López, R.; Cacho, J.; Ferreira, V. Prediction of aged red wine aroma properties from aroma chemical composition. Partial least squares regression models. *J. Agric. Food Chem.* **2003**, *51*, 2700–2707. [CrossRef]

45. Chaves, M.; Zea, L.; Moyano, L.; Medina, M. Changes in color and odorant compounds during oxidative aging of Pedro Ximenez sweet wines. *J. Agric. Food Chem.* **2007**, *55*, 3592–3598. [CrossRef] [PubMed]

46. Van Gemert, L.J.; Nettenbreijer, A.H. *Compilation of Odour Threshold Values in Air and Water*; National Institute for Water Supply/Central Institute for Nutrition and Food Research TNO: Zeist and Voorburg, The Netherlands, 1977.

47. Culleré, L.; Cacho, J.; Ferreira, V. An assessment of the role played by some oxidation-related aldehydes in wine aroma. *J. Agric. Food Chem.* **2007**, *55*, 876–881. [CrossRef] [PubMed]

48. Yunoki, K.; Tanji, M.; Murakami, Y.; Yasui, Y.; Hirose, S.; Ohnishi, M. Fatty acid compositions of commercial red wines. *Biosci. Biotechnol. Biochem.* **2004**, *68*, 2623–2626. [CrossRef] [PubMed]

49. Rankine, B.C.; Fornachon, J.C.M.; Annette Bridson, D. Diacetyl in Australian dry red table wines and its significance in wine quality. *Vitis* **1969**, *8*, 129–134.

50. Knight, S.; Klaere, S.; Fedrizzi, B.; Goddard, M.R. Regional microbial signatures positively correlate with differential wine phenotypes: Evidence for a microbial aspect to terroir. *Sci. Rep.* **2015**, *5*, 14233. [CrossRef]

51. Ribéreau-Gayon, P. New developments in wine microbiology. *Am. J. Enol. Vitic.* **1985**, *36*, 1–10.

52. Saberi, S.; Cliff, M.A.; van Vuuren, H.J.J. Impact of mixed *S. cerevisiae* strains on the production of volatiles and estimated sensory profiles of Chardonnay wines. *Food Res. Int.* **2012**, *48*, 725–735. [CrossRef]

53. Capece, A.; Romaniello, R.; Pietrafesa, R.; Romano, P. Indigenous *Saccharomyces cerevisiae* yeasts as a source of biodiversity for the selection of starters for specific fermentations. *BIO Web Conf.* **2014**, *3*, 02003. [CrossRef]

54. Noguerol-Pato, R.; González-Barreiro, C.; Cancho-Grande, B.; Martínez, M.C.; Santiago, J.L.; Simal-Gándara, J. Floral, spicy and herbaceous active odorants in Gran Negro grapes from shoulders and tips into the cluster, and comparison with Brancellao and Mouratón varieties. *Food Chem.* **2012**, *135*, 2771–2782. [CrossRef]

55. Lorenzo, C.; Garde-Cerdan, T.; Lara, J.F.; Martínez-Gil, A.M.; Pardo, F.; Salinas, M.R. Volatile composition of wines elaborated from organic and non-organic grapes. *Ferment. Technol.* **2015**, *4*. [CrossRef]

56. Francis, I.L.; Newton, J.L. Determining wine aroma from compositional data. *Aust. J. Grape Wine Res.* **2005**, *11*, 114–126. [CrossRef]

57. Dalton, P.; Doolittle, N.; Nagata, H.; Breslin, P.A.S. The merging of the senses: Integration of subthreshold taste and smell. *Nat. Neurosci.* **2000**, *3*, 431–432. [CrossRef] [PubMed]

58. Ishii, A.; Roudnitzky, N.; Béno, N.; Bensafi, M.; Hummel, T.; Rouby, C.; Thomas-Danguin, T. Synergy and masking in odor mixtures: An electrophysiological study of orthonasal vs. retronasal perception. *Chem. Senses* **2008**, *33*, 553–561. [CrossRef] [PubMed]

59. Atanasova, B.; Thomas-Danguin, T.; Langlois, D.; Nicklaus, S.; Chabanet, C.; Etiévant, P. Perception of wine fruity and woody notes: Influence of peri-threshold odorants. *Food Qual. Prefer.* **2005**, *16*, 504–510. [CrossRef]

 *fermentation*

Article

The Influence of Selected Autochthonous *Saccharomyces cerevisiae* Strains on the Physicochemical and Sensory Properties of Narince Wines

Zeynep Dilan Çelik [1,2], Hüseyin Erten [2] and Turgut Cabaroglu [2,*]

[1]   Department of Biotechnology, Cukurova University, 01330 Adana, Turkey
[2]   Department of Food Engineering, Cukurova University, 01330 Adana, Turkey
*   Correspondence: tcabar@cu.edu.tr; Tel.: +90-322-338-6997

Received: 29 June 2019; Accepted: 30 July 2019; Published: 1 August 2019

Abstract: *Vitis vinifera* cv. Narince is a Turkish native white grape variety. In this study, volatile and sensory properties of Narince wines that are produced with autochthonous *Saccharomyces cerevisiae* (*S. cerevisiae*) strains and commercial strain were compared. Autochthonous yeast strains 1044 (MG017575), 1088 (MG017577), and 1281 (MG017581) were previously isolated from spontaneous fermentations of Narince grapes. Volatile compounds formed in wines were extracted using a liquid–liquid extraction method and determined by GC-MS-FID. All yeast strains fermented Narince grape juice to dryness. The differences between the volatile profiles of the yeast strains were determined. Wines fermented with autochthonous strains 1281 and 1044 produced a higher amount of acetates and ethyl esters. While the highest concentrations of ethyl hexanoate and hexyl acetate were found in wine fermented with 1044, the highest concentrations of ethyl octanoate, ethyl decanoate, isoamyl acetate, and 2-phenylethyl acetate were found in wine fermented with strain 1281. Also, the highest contents of 2-phenyl ethanol and linalool were found in wine fermented with strain 1281. According to sensory analysis, the wine fermented with 1281 achieved the best scores in floral and fruity attributes, as well as balance and global impression. The data obtained in the present study showed that autochthonous yeast strains affect the final physicochemical composition and sensory profile of Narince wines.

Keywords: narince; autochthonous; *Saccharomyces cerevisiae*; aroma; white wine

---

1. Introduction

Wine quality is influenced, in part, by the composition of the grape juice and by the microbial communities present during the fermentation process. Aroma is one of the main characteristics that determine the quality and value of wine, especially white wines. The aroma of wine is a unique mixture of volatile compounds originating from grapes (varietal compounds), secondary products formed during the wine fermentation (fermentative compounds), and aging (post-fermentative compounds) [1–3]. Alcoholic fermentation is carried out by yeasts that convert sugars not only into ethanol and carbon dioxide but also into different secondary metabolites, such as higher alcohols, esters, and fatty acids [4,5]. The ability to produce these secondary compounds depends on the yeast species and yeast strains. Therefore, it is important to determine the dynamics of fermentation populations during fermentation, since the metabolism of yeasts has an effect on the chemical and sensory properties of the wine [4,6].

At present, commercial *S. cerevisiae* strains are widely used in winemaking, and in Turkey, most of them are imported. This practice usually guarantees fermentation control and quality of wines.

However, in some cases, the commercial inoculated *S. cerevisiae* strains cannot compete successfully with indigenous strains, and therefore, cannot dominate the fermentation as expected. Local selected strains of *S. cerevisiae*, which are better adapted to micro-area conditions of the wine production region and easily dominate the natural biota, are rather advisable as starters, and contribute to the regional characteristics of the wine. Recently, there has been an increase in the use of autochthonous or locally selected yeasts to carry out must fermentation [4,7–10]. In Turkey, Narince wine production is generally carried out by commercial *S. cerevisiae* strains imported from abroad.

Vitis vinifera cv. Narince is one of the most important native white grape varieties grown in the mid-southern Anatolia Region (Tokat and Cappadocia) of Turkey. Narince makes straw-yellow colored wines with floral notes, yellow fruit, and citrus aromas on the nose. On the palate, it produces round, medium to full-bodied wines. Because of their balanced acidity, these wines are suitable for aging and acquire a rich and complex bouquet over time [11,12].

The aim of this work was to monitor the effect of three selected autochthonous yeast cultures previously isolated from spontaneous fermentation of Narince grapes on the volatile and sensory profiles of wine samples and compare these experimental variants with a control sample produced by a commercial starter strain widely used for the production of Narince wines.

2. Materials and Methods

2.1. Yeast Strains

The autochthonous *S. cerevisiae* strains 1044, 1088, and 1281 used in this study were previously isolated from spontaneous fermentations of Narince grapes. These strains were chosen due to their good technological properties (Table 1). Commercial yeast strain X5 (Laffort, Bordeaux, France) was used as a control. Among autochthonous yeasts, technological properties of strain 1088 were previously explained by Çelik et al. [12].

Table 1. Technological properties of autochthonous *S. cerevisiae* strains previously isolated from spontaneous fermentations of Narince grapes.

Technological Properties	Strain 1044	Strain 1088	Strain 1281
Resistance to 12% (*v/v*) ethanol	**	***	**
Resistance to 200 mg/L SO_2	**	***	**
Growth at low temperature 15 °C	**	**	**
H_2S Production	3	4	2
Killer activity	+	+	+
Growth at Brix 30°	***	***	***
Foam production (15/20 °C)	F1/F2	F0/F1	F1/F2
Fermentation rate (g CO_2/L.h)	1.27 ± 0.0	0.99 ± 0.0	2.47 ± 0.2
Fermentation vigor (% h/h)	9.9 ± 0.0	10 ± 0.1	10.12 ± 0.1
Volatile acidity (g/L)	0.74 ± 0.0	0.85 ± 0.0	0.58 ± 0.0
Flocculation (%)	98	98	95
Esterase (C4)	2	1	3
Esterase Lipase (C8)	3	3	3
Ethyl acetate (mg/L)	29.9 ± 0.5	27.26 ± 0.1	24.26 ± 0.04 ± 0.2
Acetaldehyde (mg/L)	8.39 ± 0.4	18.4 ± 0.2	12.49 ± 0.05 ± 0.2
Higher alcohols (mg/L)	263.6 ± 0.6	252.96 ± 0.5	241.05 ± 0.5

Note: ** = medium growth rate; *** = high growth rate, F1: 2–4 mm, F2: 4 mm, and higher; 1 = very low activity; 2 = low activity; 3 = medium activity; 4 = high activity; 5 = very high activity, + = positive activity.

2.2. Culture Media and Chemical Standards

Yeast peptone dextrose agar (YPD) and YPD broth were purchased from Sigma Aldrich (St Louis, MO, USA) and L-lysin agar was purchased from Oxoid (Basingstoke, UK). Dichloromethane (≥99.9% purity), sodium sulfate anhydrous (99%), internal standard (4-nonanol), and a mixture of n-alkane standards ranging from C_8–C_{40} were purchased from Merck (Darmstadt, Germany). Standard volatile compounds, glucose, fructose, glycerol, tartaric acid, and lactic acid used in the study were obtained from Sigma Aldrich (St Louis, MO, USA).

2.3. Fermentations

Grapes from *Vitis vinifera* L. Narince were harvested at optimum maturity during the 2015 vintage in the commercial vineyard of Kavaklıdere (Cappadocia Region, Nevşehir, Turkey). The grape juice had the following main analytical composition: pH 3.32; initial sugar content 214 g/L; total acidity 5.62 g/L; and free amino nitrogen (FAN) content 131 mg/L. Grapes were crushed and pressed, and 50 mg/L of SO_2 were added. After pressing, the juice was allowed to settle at 10 °C for 12 h, then separated from the lees and randomly distributed into twelve 1L glass bottles. The fermentation trials were carried out in bottles containing 750 mL of Narince grape juice. Each fermentation experiment was performed in triplicate using standard protocols for white wines. Autochthonous strains 1044, 1088, and 1281 were previously grown in YPD medium at 28 °C for 24 h on an orbital shaker (rotation, 150 rpm); following this, the cells were recovered by centrifugation and washed with sterile water. Yeasts were counted by using Thoma counting chamber by light microscopy (Olympus CX22, Olympus Optical Co Ltd., Tokyo, Japan) before inoculation. The final concentration of each yeast was adjusted to 1×10^6 cells mL/L and added to the must. Control strain was added as suggested by the manufacturer. The bottles were locked with a fermentation airlock containing water and sulphuric acid to allow only CO_2 to escape from the system. All fermentations were conducted at 18 °C in a temperature-controlled room. The development of alcoholic fermentation (density and temperature) was monitored daily with a digital densimeter (Mettler Toledo, Inc., Columbus, OH, USA) until the end of alcoholic fermentation. The final wines were analyzed for residual sugars (glucose and fructose) using HPLC method, which is explained below. At the end of the alcoholic fermentation, all wines were racked off lees and 50 mg/L sulfur dioxide was added. After this, the wines were bottled and stored at 13–15 °C for 3 months until analysis.

2.4. Chemical Analysis and Microbial Enumeration

Density, alcohol, titrable acidity, pH, volatile acidity, reducing sugar, free SO_2, and total SO_2 were measured according to the methods outlined by International Organization of Vine and Wine (OIV) [13], while free amino nitrogen (FAN) was measured according to Ough and Amerine [14]. Glucose, fructose, tartaric acid, malic acid, and glycerol were quantified using HPLC LC-10A (Shimadzu, Kyoto, Japan) equipped with a refractive index detector (RID-10A) for the analysis of sugar and glycerol, and a UV/Vis detector (SPD-20A) for the analysis of organic acids monitored at 210 nm. Sugars, glycerol, and organic acids were simultaneously analyzed using an Aminex HPX-87H column (300 × 7.8 mm; Bio-Rad, Hercules, CA, USA). The column was eluted with 0.5 mM sulfuric acid at 50 °C at a flow rate of 0.5 mL/min. Before HPLC, wine samples were filtered through a membrane (0.45 μm) and passed through a C18 Sep-Pak. Quantification of glucose, fructose, tartaric acid, malic acid, and glycerol were done by external standard method [13,15].

The enumeration of culturable yeasts was performed during the first day after the initiation of fermentation, in the middle (when about 50% of total sugar was fermented), and at the end of fermentation (stabilization of the density). Samples of must and wine (1 mL) diluted in 0.1% peptone-water (decimal dilutions) were inoculated onto plates of yeast peptone dextrose YPD agar for total yeast count. Lysine agar was used for non-*Saccharomyces* yeast count, and modified YPD agar (% 10 ethanol *v/v* and 2 g/L potassium metabisulphite) was used for *S. cerevisiae* count. All agars were supplemented with chloramphenicol and sodium propionate to inhibit bacteria and filamentous fungi, respectively, and plates were incubated at 28 °C for 48 h [16].

2.5. Volatile Compounds Analysis

A liquid–liquid extraction method was used for the isolation of volatile compounds [17,18]. The extraction of volatile compounds was performed using dichloromethane. Then, 100 mL wine samples containing 40 mL of dichloromethane and 34 mg/L of 4-nonanol (5 µL, as an internal standard) were poured into a 500 mL flask, which was stirred at 4 °C, 700 rpm, for 30 min under nitrogen gas. The mixture was then centrifuged at 4 °C (9000 rpm, 15 min). After the dehydration process, using anhydrous sodium sulfate, the pooled organic extract was concentrated to a volume of 0.5 mL with a Vigreux distillation column prior to gas chromatography/mass spectrometry (GC/MS) analysis. Each sample was extracted in triplicate. The concentration of volatile compounds was quantified from the flame ionization detector FID peaks areas and the internal standard, 4-nonanol. The response factor was set to 1 for all compounds. The analytical methods for GC/MS-FID were well explained by Arslan et al. [11].

The determination of acetaldehyde and ethyl acetate was carried out by direct injection into gas chromatography using Agilent 6890 N equipped with FID. Each sample was prepared and analyzed as reported by Arslan et al. [11].

2.6. Sensory Analysis

The sensory characteristics of the final wines were evaluated according to Lawless and Heymann [19]. The sensory panel comprised 6 females and 4 males, 25–55 years of age, all belonging to the laboratory staff and having substantial experience with sensory analysis. The panelist used a 15-point scale, from 0 (no intensity) to 15 (very strong intensity). Each panelist smelled and then tasted the wines in a tasting glass to detect the intensity of the 8 attributes (floral, fruity, honey, herbaceous, acidity, persistence, balance, global impression). Sensory analysis was done in five-booth sensory panel room at 22 °C equipped with white fluorescent lighting. Wines were served (50 mL at 12 °C) in a tulip-shaped wine glasses covered by glass Petri dishes. The tasting glasses were coded with different three-digit numbers.

2.7. Statistical Analysis

The results were compared by the analysis of variance (ANOVA) using SPSS (for Windows version 16.0). Duncan's multiple-range tests were used to compare the significant differences of the mean values with $p < 0.05$. Principal component analysis (PCA) was used as a tool for screening, extraction, and compression of volatile compounds using XLStat Pro (Addisonsoft).

3. Results and Discussion

3.1. Yeast Growth and Fermentation Kinetic

The yeast counts (log CFU/mL) are shown in Figure 1. On the lysine agar no count was obtained. For this reason, only a modified agar count has been given. The initial yeast level was similar in all of the samples (6.4–6.9 Log CFU/mL). In general, a yeast population ranging from 8.0 Log CFU/mL to 8.5 Log CFU/mL was found in samples analyzed in the middle of fermentation, while values from 7.0 Log CFU/mL to 8.3 Log CFU/mL were found at the end of the fermentation.

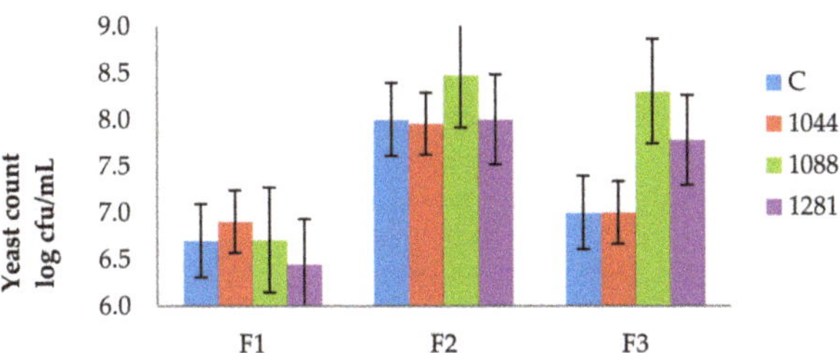

Figure 1. The growth of yeasts during fermentation. F1 is the beginning of fermentation, F2 is middle of fermentation, and F3 is end of fermentation. C represents wine inoculated with control strain, 1044 represents wine inoculated with 1044, 1088 is wine inoculated with 1088, and 1281 is wine inoculated with 1281.

Alcoholic fermentation of must started one day after inoculation in all Narince musts (Figure 2). The duration of alcoholic fermentation with control strain, autochthonous 1088, and 1281 strains were shorter (11 days) compared to the autochthonous 1044 strain (13 days). All musts were fermented to dryness.

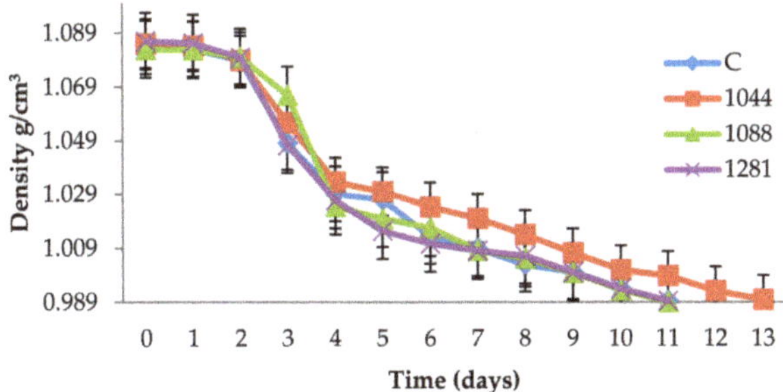

Figure 2. Daily fermentation monitoring in Narince must be fermented with different yeasts. C represents wine inoculated with control strain, 1044 is wine inoculated with 1044, 1088 is wine inoculated with 1088, and 1281 is: wine inoculated with 1281.

3.2. General Composition of Wines

The physicochemical compositions of Narince wines are summarized in Table 2. Glycerol and tartaric acid did not show significant differences among the wines made with autochthonous and commercial wine strains. Wines obtained by autochthonous strain have slightly higher ethanol strength compared to control. Acetic acid belongs to the group of volatile acids and is undesirable in wine. This acid is produced predominantly by oxidation of ethanol. However, it can also be imported into wine with grapes and small amounts of acetic acid may be produced by yeasts under anaerobic conditions [20]. Three autochthonous *S. cerevisiae* strains showed significant differences in volatile acid production. The 1088 strain produced the lowest amount of volatile acid. The concentration of residual sugar was lower than 4 g/L in all wines.

Table 2. General composition of Narince wines.

General Composition	Control	1044	1088	1281	F
Alcohol (% *v/v*)	11.40 ± 0.30 [b]	11.50 ± 0.50 [a,b]	11.65 ± 0.25 [a]	11.63 ± 0.18 [a]	*
Total acidity (g/L) **	5.95 ± 0.21 [c]	6.40 ± 0.14 [a]	6.04 ± 0.06 [b]	6.71 ± 0.10 [a]	*
pH	3.35 ± 0.22 [b]	3.35 ± 0.15 [b]	3.63 ± 0.04 [a]	3.37 ± 0.12 [b]	*
Volatile acidity (g/L) ***	0.56 ± 0.02 [b]	0.65 ± 0.06 [a]	0.41 ± 0.01 [c]	0.57 ± 0.02 [b]	*
Residual sugar (g/L)	3.10 ± 0.10 [a]	2.85 ± 0.16 [b]	2.65 ± 0.07 [c]	2.45 ± 0.02 [d]	*
Glycerol (g/L)	5.35 ± 0.34	5.40 ± 0.20	5.25 ± 0.26	5.35 ± 0.16	ns
Total SO_2 (mg/L)	43.50 ± 0.60 [b]	39.06 ± 0.65 [b]	51.00 ± 2.10 [a]	32.66 ± 1.50 [c]	*
Sugars (g/L)					
Glucose	1.10 ± 0.10 [b]	1.5 ± 0.11 [a]	1.65 ± 0.02 [a]	1.6 ± 0.12 [a]	*
Fructose	2.00 ± 1.27 [a]	1.27 ± 0.11 [b]	0.95 ± 0.10 [c]	0.8 ± 0.02 [c]	*
Organic acids (g/L)					
Tartaric acid	3.15 ± 0.21	3.05 ± 0.04	3.18 ± 0.20	3.2 ± 0.40	ns
Malic acid	2.75 ± 0.11 [a]	2.45 ± 0.10 [b]	2.25 ± 0.04 [c]	2.45 ± 0.02 [b]	*

Note: ** = as tartaric acid; *** = as acetic acid. Data are means ± standard deviations. Data with different superscript letters ([a,b,c]) within each line are significantly different (Duncan test; $p < 0.05$); ns = not significant; * = $p < 0.05$ level.

3.3. Volatile Compositions of Wines

GC/MS analysis of Narince wines produced with control strain, autochthonous 1044, 1088, and 1281 strains, which allowed the identification and quantification of 50 volatile compounds belonging to seven different groups, namely higher alcohols, esters, volatile acids, terpenes, lactones, volatile phenols, and carbonyl compounds (Table 3). Major volatile compounds of ethyl acetate and acetaldehyde concentrations were calculated by GC/FID. The volatile compounds detected in higher amounts in the present study were higher alcohols (isoamyl alcohol, 2-phenyl ethanol), esters (ethy acetate, isoamyl acetate, ethyl hexanoate, ethyl-4-hydroxybutanoate), volatile acids (hexanoic acid, octanoic acid, and decanoic acid), as well as acetaldehyde.

Table 3. Aroma composition of Narince wines produced with different *S. cerevisiae* yeast strains.

				Aroma Compounds (µg/L)				
	Higher alcohols	**RI**	**ID**	**Control**	**1044**	**1088**	**1281**	**F**
1	1-Propanol	1037	RI, MS, Std	682.44 ± 24 [a]	716.23 ± 7 [a]	294.86 ± 17 [c]	558.46 ± 44 [b]	*
2	Isobutyl alcohol	1085	RI, MS, Std	11,036.17 ± 157 [a]	11,628.40 ± 359 [a]	5849.93 ± 394 [c]	9567.90 ± 843 [b]	*
3	1-Butanol	1165	RI, MS, Std	339.02 ± 64 [b]	443.88 ± 32 [a]	28.73 ± 0 [c]	366.42 ± 25 [b]	*
4	Isoamyl alcohol	1210	RI, MS, Std	144,604.80 ± 497 [b]	165,956.55 ± 2805 [a]	111,408.47 ± 1245 [c]	144,826.57 ± 2537 [b]	*
5	2-Hexanol	1226	RI, MS, Std	259.13 ± 16 [a]	188.43 ± 91 [a,b]	24.66 ± 2 [c]	130.09 ± 1 [b]	*
6	4-Methyl-1-pentanol	1301	RI, MS, Std	318.98 ± 12 [a]	185.07 ± 11 [c]	153.23 ± 11 [d]	253.94 ± 20 [b]	*
7	1-Hexanol	1370	RI, MS, Std	1375.73 ± 135 [b]	1518.18 ± 155 [a]	1126.27 ± 73 [b]	1326.59 ± 70 [a,b]	*
8	(Z)-3-Hexen-1-ol	1401	RI, MS, Std	157.93 ± 12 [a]	113.92 ± 15 [b]	15.86 ± 1 [c]	32.65 ± 7 [c]	*
9	2,3-Butanediol	1495	RI, MS, Std	694.55 ± 96 [b]	942.93 ± 12 [a]	712.03 ± 82 [b]	811.68 ± 6 [b]	*
10	Methionol	1737	RI, MS, Std	44.92 ± 6 [c]	34.35 ± 2 [c]	298.61 ± 27 [a]	252.94 ± 22 [b]	*
11	Benzylalcohol	1804	RI, MS, Std	57.49 ± 8 [a]	47.31 ± 0 [a,b]	38.22 ± 5 [b]	55.83 ± 3 [a]	*
12	2-Phenyl ethanol	1916	RI, MS, Std	28,519.52 ± 674 [b]	28,202.12 ± 72 [b]	15,580.23 ± 468 [c]	33,597.08 ± 476 [a]	*
	Sum			**188,090**	**209,977**	**135,526**	**191,780**	
	Esters							
13	Ethyl acetate **	895	RI, MS, Std	27,727.45 ± 753 [a]	26,303 ± 455 [c]	24,254.05 ± 200 [d]	27,127.40 ± 350 [b]	*
14	Ethyl-2-methyl propaonate	960	RI, MS, Std	ND	ND	ND	287.14 ± 4	*
15	Ethyl butyrate	1037	RI, MS, Std	359.31 ± 79 [c]	682.37 ± 29 [a]	173.51 ± 10 [c]	513.92 ± 63 [b]	*
16	Isoamyl acetate	1119	RI, MS, Std	1635.03 ± 85 [c]	1815.57 ± 158 [a,b]	1951.53 ± 50 [b]	2249.17 ± 91 [a]	*
17	Ethyl hexanoate	1241	RI, MS, Std	1534.30 ± 137 [b]	1896.27 ± 47 [a]	441.37 ± 10 [c]	1680.05 ± 131 [b]	*
18	Hexyl acetate	1250	RI, MS, Std	280.00 ± 22 [a]	406.39 ± 31 [a]	40.36 ± 2 [b]	305.22 ± 3 [a]	*
19	Ethyl lactate	1353	RI, MS, Std	465.64 ± 15 [a]	604.12 ± 46 [a]	251.27 ± 21 [b]	465.78 ± 9 [a]	*
20	Ethyl octanoate	1430	RI, MS, Std	669.08 ± 44 [b]	739.52 ± 3 [a,b]	775.27 ± 56 [a]	792.90 ± 56 [a]	*
22	Ethyl decanoate	1635	RI, MS, Std	270.67 ± 20 [a]	276.17 ± 7 [a]	211.0 ± 53 [b]	301.94 ± 16 [a]	*
21	Diethyl succinate	1690	RI, MS, Std	71.99 ± 13 [b]	86.77 ± 4 [a]	31.29 ± 3 [c]	68.76 ± 2 [b]	*
23	Ethyl-9-decenoate	1709	RI, MS	106.17 ± 20 [a]	119.66 ± 3 [a]	42.83 ± 5 [b]	115.46 ± 14 [a]	*
24	2-Phenylethyl acetate	1785	RI, MS, Std	295.57 ± 20 [b]	315.03 ± 2 [b]	236.42 ± 18 [c]	429.71 ± 36 [a]	*
25	Ethyl-4-hydroxybutyrate	1819	RI, MS	3282.75 ± 145 [c]	5632.62 ± 98 [b]	2615.24 ± 234 [b]	6025.05 ± 107 [a]	*
26	Diethyl -DL-malate	2041	RI, MS, Std	124.31 ± 15 [a]	60.68 ± 5 [b]	12.35 ± 1 [c]	71.28 ± 5 [b]	*
27	Ehyl-2-hydroxy-3-phenyl propionate	2246	RI, MS	94.87 ± 10 [a]	62.531 ± 7 [b]	72.59 ± 6 [b]	69.57 ± 1 [b]	*
28	Ethyl hydrogen succinate	2331	RI, MS	845.07 ± 57 [a]	639.56 ± 42 [b]	460.80 ± 82 [c]	930.71 ± 82 [a]	*
	Sum			**37,762**	**39,640**	**31,569**	**41,434**	

Table 3. *Cont.*

				Aroma Compounds (µg/L)				
	Volatile acids	**RI**	**ID**	**Control**	**1044**	**1088**	**1281**	**F**
29	Propanoic acid	1538	RI, MS, Std	70.94 ± 25 [a]	68.11 ± 6 [a]	32.96 ± 1 [b]	48.68 ± 1 [a,b]	*
30	Isobutyric acid	1584	RI, MS, Std	432.14 ± 26 [c]	634.92 ± 51 [a]	202.86 ± 18 [d]	470.32 ± 15 [b]	*
31	Butyric acid	1628	RI, MS, Std	260.79 ± 13 [a]	270.36 ± 11 [a]	149.68 ± 15 [b]	250.86 ± 19 [a]	*
32	Isovaleric acid	1608	RI, MS, Std	488.71 ± 17 [c]	832.67 ± 12 [a]	439.62 ± 38 [d]	668.45 ± 28 [b]	*
33	Hexanoic acid	1840	RI, MS, Std	1822.76 ± 14 [c]	2135.87 ± 22 [b]	857.19 ± 40 [d]	2356.65 ± 193 [a]	*
34	(E)-2-Hexanoic acid	1962	RI, MS	169.01 ± 6 [b]	138.97 ± 6 [c]	57.49 ± 4 [d]	195.55 ± 6 [a]	*
35	Octanoic acid	2060	RI, MS, Std	1638.37 ± 151 [b]	129.83 ± 4 [c]	4180.85 ± 75 [a]	3878.85 ± 282 [a]	*
36	Decanoic acid	2183	RI, MS, Std	993.61 ± 70 [b]	1086.43 ± 58 [b]	1525.73 ± 145 [a]	1104.96 ± 163 [b]	*
37	9-Decenoic acid	2237	RI, MS	253.46 ± 14 [c]	324.50 ± 38 [b]	452.11 ± 46 [a]	284.82 ± 27 [b,c]	*
38	Hexadecanoic acid	2910	RI, MS, Std	465.79 ± 32 [a]	94.52 ± 5 [c]	125.19 ± 16 [c]	181.54 ± 12 [b]	*
	Sum			6595	5716	8023	9440	
	Terpenes							
39	Linalool	1551	RI, MS, Std	1.88 ± 0 [b]	ND	ND	37.60 ± 3 [a]	*
40	cis-Farnesol	1648	RI, MS, Std	208.13 ± 4 [a]	149.20 ± 10 [b]	26.12 ± 1 [c]	145.89 ± 2 [b]	*
	Sum			210	149	26	183	
	Lactones							
41	ɣ-Butyrolactone	1635	RI, MS, Std	1172.93 ± 64 [a]	1244.98 ± 19 [a]	503.63 ± 55 [c]	1000.72 ± 15 [b]	*
42	ɣ-Caprolactone	1694	RI, MS, Std	87.19 ± 4 [a]	23.12 ± 2 [c]	32.85 ± 6 [b]	87.97 ± 2 [a]	*
43	Pantolactone	2414	RI, MS, Std	161.30 ± 6 [a]	79.92 ± 3 [b]	9.57 ± 1 [c]	80.63 ± 3 [b]	*
44	4-Ethoxycarbonyl-ɣ-butyrolactone	2673	RI, MS	202.80 ± 4 [a]	96.37 ± 2 [b]	37.65 ± 2 [d]	90.76 ± 1 [c]	*
	Sum			1624	1444	583	1260	
	Volatile phenols							
45	4-Vinyguaiacol	2091	RI, MS, Std	259.02 ± 11 [a]	138.65 ± 2 [c]	166.66 ± 10 [b]	142.61 ± 8 [c]	*
46	4-Vinylphenol	2415	RI, MS, Std	273.60 ± 9 [a]	156.90 ± 3 [b]	29.84 ± 4 [d]	45.38 ± 5 [c]	*
47	Propiovanillone	2693	RI, MS	91.18 ± 13 [a]	37.38 ± 3 [b]	29.31 ± 2 [b]	33.04 ± 13 [b]	*
48	Acetovanillone	2995	RI, MS, Std	78.52 ± 20 [b]	25.58 ± 2 [d]	57.36 ± 4 [c]	164.12 ± 11 [a]	*
	Sum			702	358	283	385	
	Carbonyl compounds							
49	Acetoin	1291	RI, MS, Std	538.94 ± 17 [c]	825.60 ± 37 [a]	49.36 ± 13 [d]	592.73 ± 18 [b]	*
50	Acetaldehyde **	500	RI, MS, Std	4556.10 ± 150 [d]	22,071.3 ± 85 [a]	12,321.65 ± 200 [b]	10,203.35 ± 130 [c]	*
	Sum			5095	22,896	12,371	10,796	
	TOTAL SUM			240,078	280,180	188,381	255,278	

Note: Control = wine fermented with control strain; 1044 = wine fermented with autochthonous 1044 yeast; 1088 = wine fermented with autochthonous 1088 yeast; 1281 = wine fermented with autochthonous 1281 yeast; RI = retention index calculated on DB-Wax capillary column; ID = identification; MS = mass spectrometry; Std = chemical standard; ± = standard deviation of triplicate analysis of three wines for each strain; ND = not detected; F = significance at which means differ as shown using analysis of variance; * = $p < 0.05$ level. Data with different superscript letters ([a,b,c,d]) within each line are significantly different($p < 0.05$ level); ** = determined by direct injection to GC.

Higher alcohols were found in the quantitatively largest group of volatile compounds in Narince wines. Major higher alcohols, isobutyl alcohol, 1-butanol, isoamyl alcohol, benzyl alcohol, and methionol can be distinguished by their strong and pungent odor and taste. During alcoholic fermentation, the use of different yeast strains significantly contributes to the concentrations and variations of higher alcohol profiles [21]. In this study, the total amount of higher alcohols showed differences between autochthonous and commercial strains used and their concentrations ranged from 135.5 mg/L to 209.9 mg/L in Narince wines. Higher alcohols positively affect the wine aroma when present in concentrations below 300 mg/L, whereas concentrations that exceed 400 mg/L have a detrimental effect [1]. The wines produced during this study show the optimal values of these compounds. Isoamyl alcohol (3-methyl-1-butanol) was the most abundant compound in all of the wines. Strain 1044 produced the highest amount of total higher alcohols and isoamyl alcohol (165.9 mg/L) compared to other autochthonous and control strains. In contrast, strain 1088 produced the lowest amount of total higher alcohols and isoamyl alcohol (111.4 mg/L). Among the alcohols identified, 2- phenyl ethanol, contributing to wine aroma with sweet and flowery notes, was the second most abundant alcohol. While the strain 1281 (33.5 mg/L) produced the highest amount of 2-phenyl ethanol, strain 1088 (15.5 mg/L) produced the lowest amount. However, all three autochthonous strains and commercial strains produced 2-phenyl ethanol, higher than its threshold value of 10 mg/L [1]. Isobutyl alcohol and 1-propanol were also produced by all yeasts. The higher alcohols with six carbon atoms, which provide "vegetal" and "herbaceous" notes to wine, usually have a negative effect on wine quality when their concentration is above their odor threshold values [22,23]. However, these compounds (1-hexanol, Z-3-hexen-1-ol) produced concentrations lower than their threshold value by autochthonous and commercial strains. Methionol is generally described as an off-flavor with cauliflower or baked cabbage odor [24]. In Narince wines, strain 1088 produced the highest amount of methionol, followed by strain 1281. However, it did not exceed its threshold value of 1 mg/L [25] in all Narince wines. Torrens et al. [24] reported that the amount of methionol in Cava sparkling wines was influenced by the yeast strain used.

The majority of esters are produced by yeast during alcoholic fermentation and they have an important effect on the fruity characteristics of the wine. The important contribution of ethyl esters of fatty acids and acetates of higher alcohols to the sensory composition of young wine has been known for some time [1,21,26]. In terms of the number of components quantified, esters and acetates represent the largest group (16 individual compounds) of volatiles in Narince wines. Ethyl acetate was the main ester produced by autochthonous and commercial strains in the production of Narince wines. The highest amount of ethyl acetate was produced by control strain (27.7 mg/L), while the lowest was produced by 1088 (24.2 mg/L). The odor threshold value of ethyl acetate is 7.5 mg/L [1] and all strains used in this study produced this compound in concentrations higher than its odor threshold value. This compound may contribute a pleasant, fruity fragrance to the general wine aroma at concentrations lower than 150 mg/L. Contrary to this, when its concentration is greater than 150–200 mg/L, it may spoil the character of the wine [1]. Autochthonous strains and commercial strain used during this study produced ethyl acetate at optimal values. Other important acetate esters are isoamyl acetate and 2-phenylethyl acetate, which give wine banana and flowery rose aromas, respectively. They were produced by all yeasts, but strain 1281 produced a higher amount of isoamyl acetate (2.2. mg/L) and 2-phenylethyl acetate (0.42 mg/L) than the others. All strains used in this study produced a higher amount of isoamyl acetate than the threshold value of 0.03 mg/L. The 2-Phenyl acetate produced in concentrations higher than its threshold value of 0.25 mg/L by all strains used (except by the strain 1088). Strain 1044 produced the highest amount of ethyl hexanoate and ethyl octanoate. These compounds are ethyl esters of C_6 and C_8 fatty acids and they are responsible for fruity, floral, wine-like aroma [1]. Ethyl hexanoate and ethyl octanoate produced in concentrations higher than their threshold values of 0.05 mg/L and 0.02 mg/L [1] by all yeasts.

Volatile fatty acids are related to negative properties, such as rancid, fatty, cheesy notes, but also they are important for the aromatic equilibrium and complexity of wine [4]. As seen in Table 3, the production of volatile fatty acids in the wine analyzed in the present study was dependant on the yeast strains inoculated. While autochthonous strain 1044 produced the highest concentration of isobutyric, butyric, and isovaleric acids, control strain produced the highest concentration of propionic acid. The strains 1044 and 1281 stand out for their levels of hexanoic acid, while the highest concentrations of octanoic and decanoic acids were produced by the strain 1088. Hexanoic acid was produced in concentrations higher than its threshold value of 420 µg/L by all yeast strains used in this study. All strains (except 1044) produced octanoic acid in concentrations higher than its threshold value of 500 µg/L. In addition, decanoic acid was produced by all strains but it was only produced in concentrations higher than its threshold value (1000 µg/L) by autochthonous strains.

Terpenes are responsible for some of the most characteristic and important aromas in grapes and wines. It has been reported that besides grapes, yeasts are also capable of producing terpenes [24,27]. Two terpene compounds, linalool and cis-farnesol, were produced. Between two terpene compounds, linalool was produced only by control strain and autochthonous 1281. Strain 1281 produced (37.60 µg/L) linalool at a higher concentration than its threshold value of 25 µg/L. Linalool has a rose-like floral aroma and contributes positively to wine aroma. It is generally accepted that linalool, the most powerful odorant in monoterpene compounds, is an important component in the aroma of many white wines [26]. Cis-farnesol was produced by all strains, ranging from 26 µg/L to 208 µg/L

Four lactones were identified in Narince wines. The most abundant lactone was γ-butyrolactone. This compound is associated with fruity, buttery, and rubbery descriptors [28]. However, the concentration of γ-butyrolactone was found to be lower than its threshold value of 35 mg/L [29] in all Narince wines.

Among the volatile phenols in white wines, vinyl phenols play the most important role [30]. In Narince wines, five volatile phenols were identified. Among them, 4-vinylguaicol and 4-vinylphenol were produced in highest concentrations by control strain, and 4-vinylphenol exceeded its threshold value of 180 µg/L in control wine; 4-Vinylguaiacol and 4-vinylphenol produce a pharmaceutical odor, particularly in white wines [1].

Two carbonyl compounds were detected in Narince wines and acetaldehyde was found to be the most abundant carbonyl compound in this study. Yeast strains show differences in their ability to produce acetaldehyde depending on the activity of the enzyme (alcoholic dehydrogenase) involved in the synthesis [24,31]. In the present study, acetaldehyde showed significant differences related to yeast. While strain 1044 produced a higher concentration (22 mg/L) of acetaldehyde, control strain produced a lower amount (4 mg/L). The aroma threshold value of acetaldehyde is 100 mg/L and at low levels, acetaldehyde contributes fruity flavors, while high levels (200 mg/L) cause flatness in wines [1,11]. In Narince wines it did not exceed its threshold value.

The principal component analysis was carried out to separate wines fermented with different yeasts. The first two components, PC1 and PC2, explained 81.79% of the variance (Figure 3). The distribution of samples in the PC1 and PC2 components displayed a clear separation among wines from different yeast strains. Autochthonous 1088 was characterized by the presence of octanoic acid (V35), decanoic acid (V36), 9-decenoic acid (V37), and methionol (V10) (Table 3), and plotted on the negative side of PC1. Wines obtained with autochthonous 1044 and 1281 grouped together in the positive portion of PC1 and negative portion of PC2. Those yeasts characterized by the highest amount of some important volatiles (e.g., isoamyl alcohol (V4), 2-phenylethanol (V12), ethyl hexanoate (V17), 2-phenylethyl acetate (V24) and isovaleric acid (V32), and 2,3-butanediol (V9)). Control strain was plotted on the positive side of both PC1 and PC2 and was characterized by the highest amounts of 2-hexanol (V5) and (Z)-3-hexen-1-ol (V8), which provide "vegetal" and "herbaceous" notes to wine, volatile phenols (V45, V46, V47) which are usually considered as off-flavors, and some volatile acids (such as propionic acid (V29), hexadecanoic acid (V38)), lactones (pantolactone (V43), 4-ethoxycarbonyl-γ-butyrolactone (V44)), and cis-farnesol (V40). Isoamyl acetate (V16) and ethyl octanoate (V20) were negatively correlated with hexadecanoic acid, 4-vinylguaiacol, propiovanillone, and also control strain. The wine fermented with control strain presented lower contents of isoamyl acetate and ethyl octanoate.

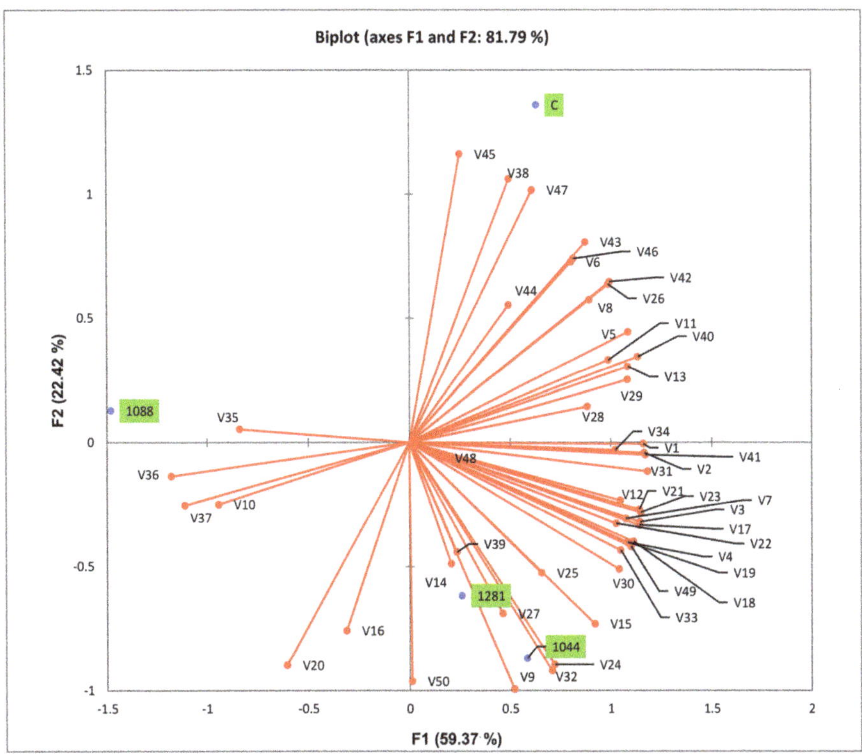

Figure 3. Principle component analysis according to yeast strain, based on volatile compositions of wines. Note: C = wine fermented with control strain; 1044 = wine fermented with autochthonous 1044 yeast; 1088 = wine fermented with autochthonous 1088 yeast; 1281 = wine fermented with autochthonous 1281 yeast; V = variable—the numbers that correspond to each compound are shown in Table 3.

The differentiation of volatile profiles of wines according to yeast strain has been widely reported [4,21,24,32]. The results of this study for Narince wines confirm those findings. Furthermore, autochthonous yeasts 1044 and 1281 produced the highest levels of important sensory volatile compounds, such as 2-phenylethanol, ethyl acetate, isoamyl acetate, 2-phenylethyl acetate, ethyl hexanoate, and ethyl octanoate, compared to control strain. Also, the wines obtained with these yeasts were clearly separated by PCA. In addition, PC1 allowed us to distinguish between wines made with autochthonous strains.

3.4. Sensory Evaluations of Wines

The sensory properties of the four experimental wines considered in this study were performed by a sensory panel using eight attributes: floral, fruity, honey, herbaceous, acidity, persistence, balance, and global impression (Figure 4). Regarding fruity and floral properties, wine 1281 achieved the best score for both attributes, followed by 1044 and control, respectively. The wine fermented with autochthonous 1088 strain achieved the lowest score for those attributes. This result was in agreement with the aroma composition of this wine, because the wine fermented with 1088 contains the lowest amount of acetate and ethyl esters (except ethyl octanoate and isoamyl acetate), which contribute fruity and floral characteristics. Further, 1044 strain achieved the best score for honey attribute, while 1088 achieved the best for the acidity attribute. Wine produced with 1281 also achieved the best score for both balance and global impression. These results were in agreement with the differences found at the chemical level. Wine fermented with autochthonous 1281 and 1044 contain higher amounts of acetates (isoamyl acetate (fruity), hexyl acetate (sweet, perfume), 2-phenylethyl acetate (floral)) and most of the ethyl esters (ethyl butyrate (fruity), ethyl hexanoate (green apple), ethyl lactate (lactic, fruity), ethyl decanoate (floral, soapy)), (Table 3), followed by control strain. However, the chemical composition of the wine and the interaction between the compounds and their effects on the sensory properties are still very complex and not well known [4].

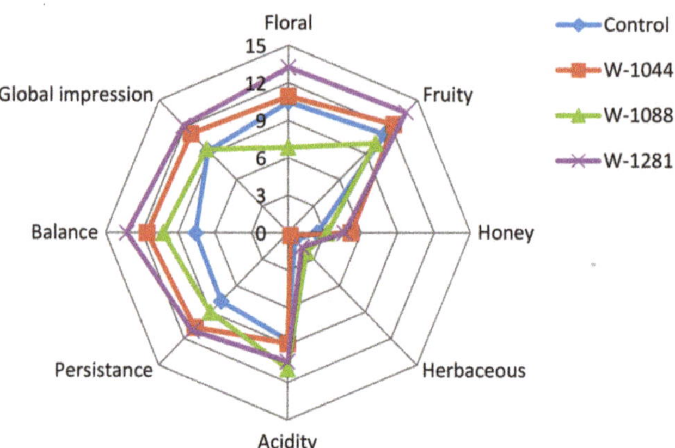

Figure 4. Sensory profile of Narince wines made with autochthonous and commercial *S. cerevisiae* yeasts. Note: C = wine fermented with control strain; 1044 = wine fermented with autochthonous 1044 yeast; 1088 = wine fermented with autochthonous 1088 yeast; 1281 = wine fermented with autochthonous 1281 yeast.

4. Conclusions

Winemaking is a highly industrialized process and different *S. cerevisiae* starter cultures are commercially available for its control. However, several investigations have underlined that using autochthonous yeasts during fermentation is able to give unique organoleptic properties to the produced wines. The present study investigated the effects of three different autochthonous yeast strains on the physicochemical and sensory properties of Narince wines. The autochthonous *S. cerevisiae* strains used in this study presented good fermentative ability. From a chemical point of view, certain trends were detected among strains used, as follows. Aroma compound analyses showed that autochthonous yeast strains 1044 and 1281 were able to produce a higher concentration of ethyl esters and acetates, which are responsible for fresh/fruit attributes. Sensory data were in agreement with chemical compositions. The discrimination analysis allowed the autochthonous strains 1281 and 1044 to be clearly distinguished by their volatile composition. To our knowledge, this is the first study on the effects of autochthonous *S. cerevisiae* yeast strains on the volatile and sensory properties of Narince wines. However, it would be best to confirm these results with industrial large-scale fermentation.

Author Contributions: All authors participated in the design and discussion of the research. Z.D.Ç. carried out the experimental part of work and analysis of the data. Z.D.Ç. wrote the original draft. The work was supervised by T.C. All the authors have read and approved the final manuscript.

Funding: This research was funded by The Scientific and Technical Research Council of Turkey (TUBİTAK), grant number 2140173.

Acknowledgments: The authors would like to thank to Scientific Project Unit of Cukurova University (Project No. FDK-2014-3049) and Kavaklıdere Wine Company for provide the grape samples.

Conflicts of Interest: The authors declare no conflict of interest.

References

1. Swiegers, J.H.; Bartowsky, E.J.; Henschke, P.A.; Pretorius, I.S. Yeast and bacterial modulation of wine aroma and flavour. *Aust. J. Grape Wine Res.* **2005**, *11*, 139–173. [CrossRef]
2. Callejón, R.M.; Clavijo, A.; Ortigueira, P.; Troncoso, A.M.; Paneque, P.; Morales, M.L. Volatile and sensory profile of organic red wines produced by different selected autochthonous and commercial *Saccharomyces cerevisiae* strains. *Anal. Chim. Acta* **2010**, *660*, 68–75. [CrossRef] [PubMed]
3. Lambrechts, M.G.; Pretorius, I.S. Yeasts and its importance to wine aroma. *S. Afr. J. Enol. Vitic.* **2000**, *21*, 97–129. [CrossRef]
4. Blanco, P.; Miras-Avalos, J.M.; Ignacio, O. Modulation of chemical and sensory chracteristics of red wine from Mencía by using indigenous *Saccharomyces cerevisaie* yeast strains. *J. Int. Sci. Vigne Vin* **2014**, *48*, 63–74.
5. Belda, I.; Ruiz, J.; Esteban-Fernández, A.; Navascués, E.; Marquina, D.; Santos, A.; Moreno-Arribas, V. Microbial contribution to wine aroma and its intended use for wine quality improvement. *Molecules* **2017**, *22*, 189. [CrossRef] [PubMed]
6. Abargati, A.; Canonico, L.; Ciani, M.; Comitini, F. Fitness of selected indigenous *Saccharomyces cerevisae* strains for white Piceno DOC wines production. *Fermentation* **2018**, *4*, 37.
7. Capece, A.; Romaniello, R.; Siesto, G.; Romano, P. Diversity of *Saccharomyces cerevisiae* yeasts associated to spontaneously fermenting grapes from an Italian 'heroic vine-growing area'. *Food Microbiol.* **2012**, *31*, 159–166. [CrossRef] [PubMed]
8. Lorca, G.; Uribe, S.; Martinez, C.; Godoy, L.; Ganga, M.A. Screening of native *S. cerevisiae* strains in the production of Pajarete wine: A tradition of Atacama Region, Chile. *J. Wine Res.* **2018**, *29*, 130–142. [CrossRef]
9. Tufariello, M.; Maiorano, G.; Rampino, P.; Spano, G.; Grieco, F.; Perrotta, C.; Capozzi, V.; Grieco, F. Selection of an autochthonous yeast starter culture for indurstrial production of Primitivo 'Gioia del Colle' PDO/DOC in Apulia (Southern Italy). *Food Sci. Technol.* **2019**, *99*, 188–196.
10. Liu, N.; Qin, Y.; Song, Y.Y.; Tao, Y.S.; Sun, Y.; Liu, Y.L. Aroma composition and sensory quality of Cabernet sauvignon wines fermented by indigenous *Saccharomyces cerevisiae* strains in the eastern base of the Helan Mountain, China. *Int. J. Food Prop.* **2016**, *19*, 2417–2431. [CrossRef]
11. Arslan, E.; Çelik, Z.D.; Cabaroğlu, T. Effects of pure and mixed autochthonous *Torulaspora delbrueckii* and *Saccharomyces cerevisiae* fermentation and volatile compounds of Narince wines. *Foods* **2018**, *7*, 147. [CrossRef]
12. Çelik, Z.D.; Erten, H.; Darici, M.; Cabaroğlu, T. Molercular caharacterization and technological properties of wine yeasts isolated during spontaneous fermentation of *Vitis vinifera* L.cv. Narince grape must grown in ancient winemaking area Tokat, Anatolia. *BIO Web Conf.* **2017**, *9*, 1–7. [CrossRef]
13. *Compendium of International Methods of Analysis of Wine and Musts*; International Organisation of Vine and Wine: Paris, France, 2015; Volume 1.
14. Ough, C.S.; Amerine, M.A. *Methods for Analysis of Musts and Wines*; Wiley: New York, NY, USA, 1988.
15. Erten, H. Metabolism of fructose as an electron acceptor by *Leuconostoc mesenteroides*. *Process Biochem.* **1998**, *33*, 735–739. [CrossRef]
16. Esteve-Zarzosa, B.; Gostíncar, A.; Bobet, R.; Uruburu, F.; Querol, A. Selection and molecular characterization of wine yeasts isolated from the' El Penèdes' area Spain. *Food Microbiol.* **2000**, *17*, 553–562. [CrossRef]
17. Selli, S.; Cabaroğlu, T.; Canbaş, A.; Erten, H.; Nurgel, C.; Lepoutre, J.P.; Günata, Z. Volatile composition of red wine from cv. Kalecik karası grown in central Anatolia. *Food Chem.* **2004**, *85*, 207–213. [CrossRef]
18. Schneider, R.; Baumes, R.; Bayanove, C.; Razungles, A. Volatile compounds involved in the aroma of sweet offortified wines (Vins Doux Naturels) from Grenache Noir. *J. Agric. Food Chem.* **1998**, *46*, 3230–3323. [CrossRef]
19. Lawless, H.T.; Heymann, H. *Sensory Evaluation of Food*, 2nd ed.; Springer: New York, NY, USA, 2010; pp. 231–238. ISBN 978-1-4419-6487-8.
20. Cordente, A.G.; Cordero-Bueso, G.; Pretorius, I.S.; Curtin, C.D. Novel wine yeast with mutations in YAP1 that produce less acetic acid during fermentation. *FEMS Yeast Res.* **2013**, *13*, 62–73. [CrossRef]
21. Álvarez-Pérez, J.M.; Campo, E.; San-Juan, F.; Coque, J.J.R.; Ferreira, V.; Hernández-Orte, P. Sensory and chemical characterisation of the aroma of Prieto Picudo rosé wines: The differential role of autochthonous yeast strains on aroma profiles. *Food Chem.* **2012**, *133*, 184–192. [CrossRef] [PubMed]

22. Ferreia, V.; Fernandez, P.; Pena, C.; Escudero, A.; Cacho, J.F. Investigation on the role played by fermentation esters in the aroma younf Spanish wines by multivariate analysis. *J. Sci. Food Agric.* **1995**, *67*, 381–392. [CrossRef]

23. Gómez-Míguez, M.J.; Cacho, J.F.; Ferreira, V.; Vicario, I.M.; Heredia, F.J. Volatile componenets of Zalema white wines. *Food Chem.* **2007**, *100*, 1464–1473. [CrossRef]

24. Torrens, J.; Riu-Aumatell, M.; Vichi, S.; López-Tamames, E.; Buxaderas, S. Different commercial yeast stains affecting the volatile and sensory profile of cava base wine. *Int. J. Food Microbiol.* **2008**, *124*, 48–57. [CrossRef] [PubMed]

25. Clemente-Jiménez, J.M.; Mingorance-Cazorla, L.; Martínez-Rodríguez, S.; Heras-Vázquez, F.J.L.; Rodriguez-Vico, F. Molecular characterization and oenological properties of wine yeasts isolated during spontaneousfermentation of six varieties of grape must. *Food Microbiol.* **2004**, *21*, 149–155. [CrossRef]

26. Ugliano, M.; Henschke, P.A. Yeasts and wine flavour. In *Wine Chemistry and Biochemistry*, 1st ed.; Moreno-Aribas, M.V., Polo, M.C., Eds.; Springer: New York, NY, USA, 2009; pp. 313–392.

27. Mina, M.; Tsaltas, D. Contribution of yeast in wine aroma and flavour. In *Yeast-Industrial Applications*, 1st ed.; Morato, A., Loira, I., Eds.; Intechopen: London, UK, 2017; pp. 117–134.

28. Rocha, S.M.; Countinho, F.; Delgadillo, P.; Coimbra, M.A. Volatile composition of Baga red wine. Assessment of the identification of the impact odourants. *Analytica Chimica Acta* **2004**, *513*, 257–262. [CrossRef]

29. Azzolini, M.; Tosi, E.; Lorenzini, M.; Finato, F.; Zapparoli, G. Contribution to the aroma of white wines by controlled *Torulaspora delbrueckii* cultures in association with *Saccharomyces cerevisiae*. *World J. Microbiol. Biotechnol.* **2015**, *31*, 243–253. [CrossRef] [PubMed]

30. Chatonnet, P.; Dubourdieu, D.; Boidron, J.N.; Lavigne, V. Synthesis of volatile phenols by *Saccharomyces cerevisiae* in wines. *J. Sci. Food Agric.* **1993**, *61*, 191–202. [CrossRef]

31. Pérez-Coello, M.S.; Briones Pérez, A.I.; Ubeda Iranzo, J.F.; Martin Alvarez, P.J. Characteristics of wines fermented with different *Saccharomyces cerevisiae* strains isolated from La Mancha region. *Food Microbiol.* **1999**, *16*, 563–573. [CrossRef]

32. Alves, Z.; Melo, A.; Raquel Figueiredo, R.; Coimbra, M.A.; Gomes, A.C.; Rocha, S.M. Exploring the *Saccharomyces cerevisiae* volatile metabolome: Indigenous versus commercial strains. *PLoS ONE* **2015**, *10*, e0143641. [CrossRef]

 *fermentation*

Article

Preliminary Characterization of Yeasts from Bombino Bianco, a Grape Variety of Apulian Region, and Selection of an Isolate as a Potential Starter

Barbara Speranza, Daniela Campaniello, Leonardo Petruzzi, Milena Sinigaglia, Maria Rosaria Corbo and Antonio Bevilacqua *

Department of the Science of Agriculture, Food and Environment, University of Foggia, Via Napoli 25, 71121 Foggia, Italy; barbara.speranza@unifg.it (B.S.); daniela.campaniello@unifg.it (D.C.); leonardo.petruzzi@unifg.it (L.P.); milena.sinigaglia@unifg.it (M.S.); mariarosaria.corbo@unifg.it (M.R.C.)
* Correspondence: antonio.bevilacqua@unifg.it

Received: 6 November 2019; Accepted: 12 December 2019; Published: 15 December 2019

Abstract: Eighty-seven yeasts were isolated from Bombino bianco, a white grape variety from Apulian Region (Southern Italy). The isolates were characterized for the splitting of arbutin, the hydrolysis of pectins, sulphite production, the resistance to acetic acid, SO_2, and ethanol. An enhanced arbutin splitting (β-glucosidase) and a moderate pectolytic activity were found. Concerning ethanol resistance, the most of yeast population showed a low-to-moderate resistance, but some isolates, identified as *Saccharomyces cerevisiae*, were able to grow in presence of 15% *v/v* of ethanol. Four isolates were selected (coded as 43D, 44D, 45D, and 46D), studied for their ability to decarboxylate amino acids and used in small-scale fermentation trial; for this last experiment a reference strain was used (*S. cerevisiae* EC1118). This experiment suggested the existence of an isolate (*S. cerevisiae* 46D) with interesting traits and performances, which could be potentially proposed as a starter for Bombino bianco.

Keywords: yeasts; Bombino bianco; technological characterization; enzymatic patterns; amino acid decarboxylation

---

1. Introduction

Apulia (Southern Italy) is the second Italian area for wine production. The Apulian wines detain several peculiarities because of pedologic features, climatic conditions, and technologies, all contributing to the definition of a unique "terroir" [1]. Among the innovative trends in the wine sector, the continuous exploration of oenological properties associated with wine microbial resources represents a cornerstone driver of quality improvement [2]. Autochthonous starter cultures have a potential important role on wine quality because of their key-role on organoleptic properties [3].

Yeast strains involved in winemaking influence fermentation speed, nature and quantity of secondary products, and aromatic characters of wine [4]. Yeast microbiota generally comprise oxidative yeasts, which belong to the genera *Rhodotorula*, and *Hanseniaspora*. They comprise up to the 99% of the yeasts isolated in certain grape samples [5]. Other yeasts usually found on grape are *Metschnikowia pulcherrima*, *Candida famata*, *Candida stellata*, *Pichia membranifaciens*, *Pichia fermentans*, *Hansenula anomala*, and, in small proportions, *Saccharomyces* spp. [5].

Many researchers focused on the oenological performances of indigenous yeasts isolated from red Apulian grapes, must, and spontaneous fermentation from Primitivo [6], Negroamaro [7], Uva di Troia [8,9], and Susumaniello, an ancient and recently rediscovered grape cultivar [10]. However, to the best of our knowledge, little is known on yeast microbiota of Bombino bianco. *Vitis vinifera* L. Bombino bianco is a cultivar widespread since ancient times in the Apulia region where is present with a surface of 2000 ha, mainly located in Foggia and Bari counties [11]. This variety can be also found in other regions of Southern Italy, where it is usually referred as Bonvino, Ottenese, Trebbiano d'oro, Trebbiano d'Abruzzo, Uva d'Oro, or Gold Trauben. The main traits are a high yield, a good resistance to bad weather conditions as well as to grape diseases, such as *Plasmopara* or *Botrytis* [12]. Bombino bianco grapes are typically blended with grapes of other varieties for the production of many DOC and IGT wines, but they can be also used alone to produce sparkling wines through the Champenoise method [13].

This paper represents a first approach for the evaluation of yeast diversity and characteristics on Bombino bianco grapes, with a special focus on enzymatic patterns, technological characteristics and safety issues (amino acid decarboxylation), as a prodromal to select and design a wild starter culture for this grape variety.

2. Materials and Methods

2.1. Yeast Isolation

Wine grapes (*Vitis vinifera* L.; Bombino bianco variety) were harvested from a local farm in the Apulian region (Foggia, Italy). Skin and inner part were analyzed. For skins, the sample was treated as follows: 25 g of grapes were detached from different clusters, immersed in 225 mL sterile isotonic solution (0.9% NaCl) and shaken for 30 min at 200 rpm. For the inner part, 25 g of grapes were diluted in 225 mL sterile isotonic solution (0.9% NaCl) and homogenized through a steril-mixer. The serial dilutions of homogenates (skin and inner part) were spread onto appropriate media and incubated at 25 °C for 48 h. The media were the following: Sabouraud Dextrose Agar (Oxoid, Milan, Italy); YPG Agar (bacteriological peptone 20 g/L, yeast extract 10 g/L, glucose 20 g/L, agar 15 g/L; ingredients were purchased from Oxoid); YM Agar (bacteriological peptone 5 g/L, malt extract 3 g/L, yeast extract 3 g/L, dextrose 10 g/L, agar 15 g/L; ingredients were purchased from Oxoid); WL Nutrient Agar (Oxoid). All media were supplemented with 0.1 g/L of chloramphenicol (C. Erba, Milan, Italy).

From each plate and at each time of analysis, 5 colonies were randomly selected, isolated and stored on YPG Agar at 4 °C until the identification; a preliminary characterization between Saccharomyces and non-Saccharomyces yeasts was done by streaking the isolates on WL medium.

2.2. Arbutin Splitting Test

β-D-glucosidase activity was evaluated by using the arbutin splitting test. YNB-Agar (Yeast Nitrogen Base, Oxoid), supplemented with 0.5% arbutin (Sigma-Aldrich, Milan, Italy) and 2% ferric ammonium citrate (J.T. Baker, Milan, Italy) solution, was used [14]. The plates were incubated at 26 °C for 2 to 4 days. The outcome of the experiment was evaluated by means of color change to brown. A qualitative code was used to classify the results as follows: no activity (no halo, –); weak activity (diameter of the halo <17 mm, +); medium/strong activity (diameter of the halo >18 mm, ++) [15].

2.3. Hydrolysis of Pectins

The extracellular pectolytic activity was assessed by measuring the growth zones of yeasts on YM Agar, without glucose and supplemented with 12.5 g/L of apple pectins (Sigma-Aldrich) and adjusted to pH 4.0 with HCl 1.0 N. The plates were incubated at 25 °C for 10 days [16]. The outcome of the experiment was evaluated as a function of the diameter of the growth zone, as reported by Hernández et al. [17]: the activity was classified as weak (+), when the growth zone was <5.5 mm or strong (++), when the growth zone was >5.5 mm.

2.4. Sulphite Production

The test was done on BiGGY Agar medium (Oxoid), as reported by Barbosa et al. [13]. After yeast inoculation, the plates were incubated for 2–4 days at 26 °C. The different intensity of the color of colonies (from white to heavy brown) was used to evaluate the quantitative outcome of the test (white, no production of H_2S; cream, weak production; light and dark brown, medium-to-high production) [15].

2.5. Resistance to Acetic Acid, Ethanol and SO_2

The tests were performed by using YPG Agar, supplemented with 3%, 6%, 9%, 12%, and 15% (*v/v*) ethanol (C. Erba, Milan, Italy), 0.05%, 0.1%, 0.15%, 0.20%, 0.25%, or 0.3% acetic acid, and 50, 100 and 150 mg/L SO_2. After streaking the strains onto the surface of the medium, the plates were incubated at 25 °C for 7 days.

2.6. Preliminary Identification

Some selected isolates were identified as reported by Sinigaglia et al. [18].

2.7. Amino Acid Decarboxylation

Yeasts were streaked onto the surface of a laboratory medium containing amino acids as precursors of amines and a pH indicator (bromocresol purple): an increase of pH due to decarboxylation of amino acids causes a color turning from green to purple. The composition of the medium, modified from a substrate proposed by Bover-Cid and Holzapfel [19] for lactic acid bacteria, was the following: bacteriological peptone, 5 g/L; yeast extract, 5 g/L; glucose, 1 g/L; amino acid, 10 g/L; agar, 12 g/L; bromocresol purple, 0.06 g/L (Sigma-Aldrich); pyridoxal-5-phosphate hydrate (Sigma-Aldrich), 0.05 g/L. The pH of the medium was adjusted to 5.2. Arginine hydrochloride, cysteine, ʟ-histidine, ʟ-tyrosine, ʟ-phenlalanine, serine, gliycine, ʟ-proline, and ʟ-lysine (Sigma-Aldrich) were used; after yeast inoculation, the plates were incubated at 25 °C for 72 h and checked every day. Laboratory media without amino acids but inoculated with yeasts were used as controls.

2.8. Lab-Scale Fermentations and Confirmation of the Technological Performances

The assays were done only for the isolates 43D, 44D, 45D, and 46D; the strain *S. cerevisiae* EC1118 (Lallemand Inc., Castel D'Azzano, VR, Italy) was used as reference. The first experiment was a confirmation of the technological performances in a model medium, as reported by Petruzzi et al. [15]. A synthetic medium was prepared as follows: 100 g/L of glucose; 100 g/L of fructose (Sigma-Aldrich, Milan, Italy); 10 g/L of yeast extract; 1 g/L of 21 ammonium sulfate (J. T. Baker, Milan, Italy); 1 g/L of potassium phosphate (C. Erba); and 1 g/L of magnesium sulfate (J. T. Baker).

Flasks of 150 mL, containing 100 mL of medium, were used for the experiments. After sterilization, citric acid (10 g/L) was added to the medium to decrease the pH to 3.5; then the medium was inoculated with yeasts (ca. 6 log cfu/mL), and the surface was covered with a thin layer of sterilized paraffin oil (10 mL per flask) in order to avoid air contact. The samples were incubated at 25 °C without shaking. Residual sugars, ethanol, glycerol, and volatile acidity were determined by Fourier transform infrared spectroscopy using a WineScan FT120 instrument (software version 2.2.1, FOSS Analytical, Hillerød, Denmark) according to the supplier's instructions.

A second experiment was done in a real must. The fermentation was carried out in duplicate on two independent batches in 250 mL Erlenmeyer flasks containing 150 mL of pasteurized Bombino bianco must (sugar content, 15.75 ± 0.17° Bx; titrable acidity, 4.09 ± 0.53 g of tartaric acid; pH, 3.76 ± 0.08). Yeasts were grown in YPG broth at 25 °C for 72 h, centrifuged at 4 °C for 10 min at $1200 \times g$ and suspended in distilled water. The inocula represented 1% of the total fermentation volume of flasks.

The flasks were stoppered with cotton plugs to allow the CO_2 to escape from the system, and the weight loss of the flasks due to CO_2 production was evaluated every day, until the end of fermentation (constant weight for three consecutive days). The fermentations were carried out under static conditions at 25 °C on two independent batches; weight loss and cell viability (YPG agar plates, incubated at 25 °C for 4 days) were assessed.

2.9. Statistic

The phenotypic tests were performed in triplicate; the results were plotted as frequency histograms. The outcome of the qualitative test was assessed as positive, if at least 2 replicates were positive. For the quantitative assays (pectin hydrolysis), for each isolate the average of the growth zone was evaluated.

The results of the analytical determinations in the synthetic must were analyzed by means of one-way analysis of variance and Tukey's test ($p < 0.05$).

The data of weight loss were modelled as weight loss (mg CO_2 per ml of must) through the lag-exponential model by van Gerwen and Zwietering [20] and by Baty and Delignette-Muller [21], cast in the following form:

$$Y = \begin{cases} 0 & t \leq \alpha \\ y_{max} - \log\{1 + (10^{y_{max}} - 1) \times \exp\left[-d_{max}(t - \alpha)\right]\} & t > \alpha \end{cases}. \tag{1}$$

where: y and t are the dependent and independent variables, respectively (weight loss and time-day); α is the time before the beginning of the fermentation (day); d_{max} is the maximal fermentation rate; y_{max} is the maximum level of weight loss.

When the kinetic did not show the parameter α, the lag-exponential model was used as follows [22]:

$$y = y_{max} - \log\{1 + (10^{y_{max}} - 1) \times \exp(-d_{max}\, t)\} \tag{2}$$

Statistic was performed through the software Statistica for Windows ver. 12.0 (Statsoft, Tulsa, Oklha).

3. Results

As a result of yeast selection, 87 isolates were recovered; 44 showed a green color on WL medium (non-*Saccharomyces*), and 43 were yellow. Figure 1 shows the results for the preliminary phenotyping (arbutin splitting, sulphite production, and pectolytic activity). Focusing on β-glucosidase activity (arbutin splitting), 68 strains could perform the splitting of the arbutin; for 43 isolates this ability was low-to-moderate (weak response) and for 25 isolates strong. Only 19 isolates were negative to the splitting of the arbutin, all of them belonging to non-*Saccharomyces* yeasts.

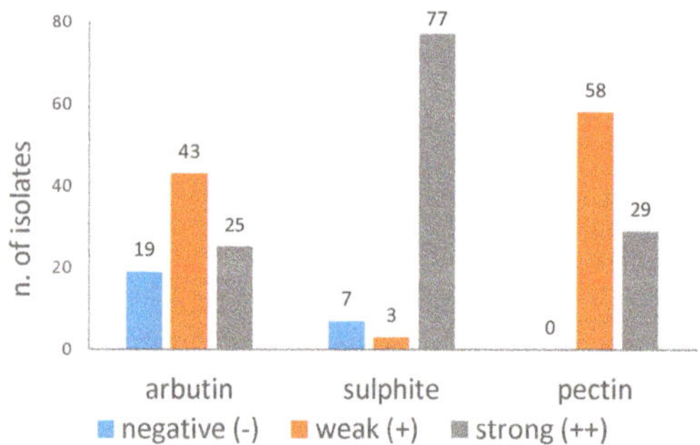

Figure 1. Arbutin splitting, sulphite production, and pectolytic activity. The numbers on the bars represent the number of isolates per each class (for the details on the classification see Materials and Methods).

The production of H_2S is a negative property for its strong impact on the sensorial quality of wine. Yeasts isolated from Bombino Bianco possessed this ability and 77 isolates (ca. 90% of the population) expressed this ability at the highest level (++); otherwise for three isolates this kind of metabolism appeared moderate (yellow colonies on WL), and only seven isolates were negative (non-*Saccharomyces*).

Another trait assessed was the pectolytic activity, recovered in all isolates from a weak response (58 yeasts) to a strong one (29 isolates).

Isolates were also studied in relation to resistance to acetic acid, ethanol, and SO_2. Concerning the resistance to acetic acid, all isolates were able to grow till a maximum concentration of 0.10%. The results for ethanol resistance are reported in Figure 2, as number of yeasts able to grow for each ethanol amount; 76 isolates were able to grow in presence of 3% ethanol, while only 13 were able to grow at 15%. SO_2 exerted a strong impact on yeasts and only four isolates were able to grow in the medium containing 150 ppm of this compound (Figure 3).

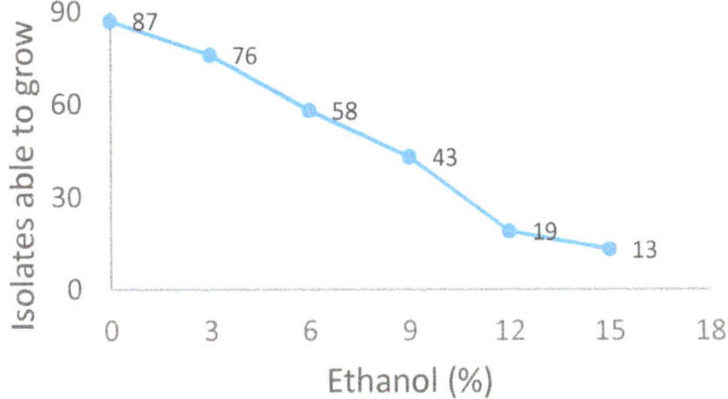

Figure 2. Growth of yeasts on lab media containing ethanol. The results are expressed as number of isolates able (positive) to grow for each ethanol amount.

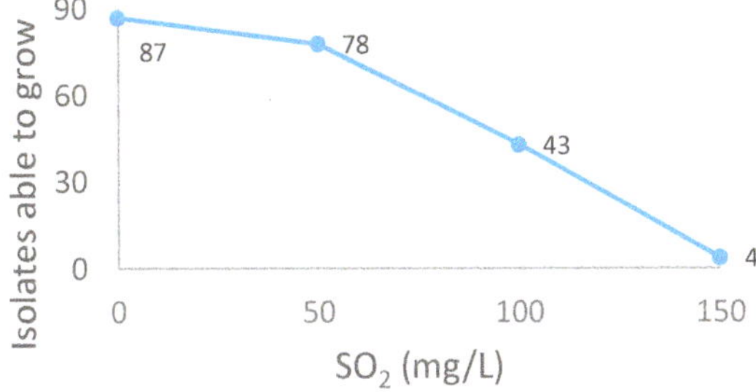

Figure 3. Growth of yeast isolates on lab media SO_2. The results are expressed as number of isolates (positive) able to grow for each amount of SO_2.

By using phenotyping, a selection of isolates for the second step was done; ethanol and SO_2 tolerance were used as primary criteria and only yeasts able to grow in media containing 15% ethanol and 150 ppm of SO_2 were selected. Thus, four isolates were used for the second step (43D, 44D, 45D, and 46D); they all possessed a moderate arbutin splitting, a strong pectolytic activity, while sulphite production was negative for the isolates 43D and 44D and weak for the isolates 45D and 46D (Table 1).

Table 1. Enzymatic activities of selected isolates. Cat, catalase; Arb, arbutin splitting; Sul, sulphite production; Pec, pectolytic activity; SO_2, resistance to SO_2; EtOH, resistance to ethanol; Acetic, resistance to acetic acid. −, negative; +, moderate; ++, strong.

Yeast Isolates	Cat	Arb	Sul	Pec	SO_2	EtOH	Acetic
43D	++	+	−	++	150 ppm	15%	0.10%
44D	++	+	−	++	150 ppm	15%	0.10%
45D	+	+	+	++	150 ppm	15%	0.10%
46D	+	+	+	++	150 ppm	15%	0.10%

These isolates were identified as *S. cerevisiae* and studied for the decarboxylation of amino acids; Table 2 shows the data of decarboxylation. The isolates 43D, 44D, and 45D were negative, whilst the isolate 46 D was positive to the decarboxylation of arginine, lysine, and tyrosine.

Table 2. Results for the test on the decarboxylation activity. A, arginine; B, cysteine; C, phenylalanine; D, glycine; E, histidine; F, lysine; G, proline; H, serine; I, tyrosine. −, negative; +, positive to the assay; +/−, variable.

Yeast Isolates	A	B	C	D	E	F	G	H	I
43D	−	−	−	−	−	−	−	−	−
44D	−	−	−	−	−	−	−	−	−
45D	−	−	−	−	−	−	−	−	−
46D	+	−	−	−	−	+/−	−	−	+

The last experiments focused on the assessment of the technological performances through a small-scale fermentation trials; the strain EC1118 was used as a reference. A first assay was done in a synthetic medium for the evaluation of some target compounds (ethanol, glycerol, and volatile acidity). The isolates showed similar traits, and produced 8.78–11.20 g/L of ethanol, 5.45–6.21 g/L of glycerol and 0.36–0.45 g/L of acetic acid; sugar was always <2 g/L. The differences were not significant (Table 3).

Table 3. Technological performances. Ethanol, glycerol, and acetic acid produced in a synthetic medium.

Yeast Isolates	Ethanol (g/L)	Glycerol (g/L)	Volatile Acidity (Acetic Acid, g/L)
43D	9.76 ± 0.98	5.64 ± 0.89	0.45 ± 0.09
44D	8.78 ± 1.21	5.45 ± 0.63	0.36 ± 0.12
45D	9.89 ± 0.06	6.12 ± 0.65	0.41 ± 0.07
46D	10.25 ± 0.34	6.21 ± 0.09	0.42 ± 0.08
Ec1118	11.20 ± 0.43	6.01 ± 0.11	0.39 ± 0.12

In a second experiment, the fermentation kinetic in a Bombino bianco must was studied; Figure 4 shows the results as standardized weight loss (mg of CO_2 per mL of must).

The kinetic followed a logistic-like model; however, two different trends were recovered—the first one for the isolates 43D and 44D, and the second one for the isolates 45D, 46D, and the reference strains. The isolates 43D and 44D experienced a logistic-like trend with a lag phase (parameter α), where no weight loss occurred; the duration of this period was 4.40 days for the isolate 43D and 2.30 day for the isolate 44D. On the other hand, the isolate 45 and 46D showed the same trend of the reference strain and the fermentation started immediately after the inoculation and no lag phase was found. All isolates experienced a similar maximum weight loss (ca. 55 mg of CO_2 per mL of must) and fermentation rate (18–20 mg of CO_2 per ml of must and per day).

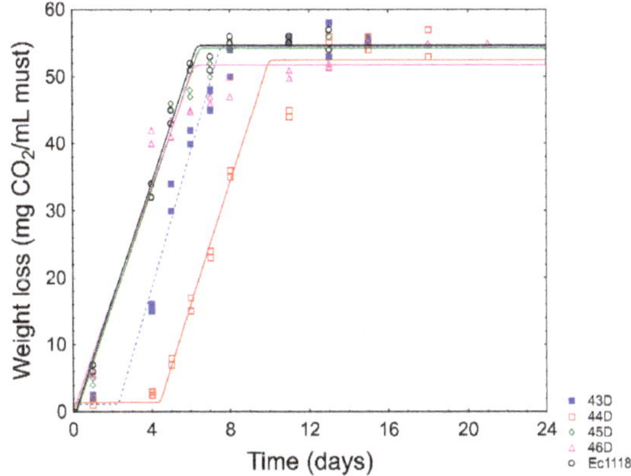

Figure 4. Fermentation kinetics of isolates 43D, 44D, 44D, and 45D compared to *S. cerevisiae* EC1118; the lines represent the best fit through the lag-exponential model.

4. Discussion

The selection of a starter is a complex process, as it involves different steps; however, the interest towards wild strains and the selection of potential starter microorganisms from the natural occurring microflora of many foods increased, as it is well known that autochthonous microorganisms could contribute to quality and safety of final products. Concerning yeasts for oenological use, we focused on a traditional grape variety of Apulian region (Bombino bianco) and performed the first steps of strain selection. Some varieties of Apulian region were studied for the qualitative composition of yeast microbiota [8–10,23]; to the best of our knowledge, little is known on Bombino bianco. After strain isolation, we characterized yeasts for some enzymatic activities (β-glucosidases, hydrolysis of pectins, sulphite production) and technological properties (ethanol tolerance, resistance to acetic acid). Concerning β-glucosidase, the potential applications of this enzyme include the prevention of sediments in the bottles during storage, as well as the production of volatile compounds [24]. Research on β-glucosidase in yeasts has revealed that most *Saccharomyces* isolates do not show activity in a natural substrate and that such activity is more frequent in non-*Saccharomyces* strains [25], from a weak to a strong activity as reported by Fia et al. [26] for some strains belonging to *Brettanomyces*, *Metschnikowia* spp. and *Hanseniaspora* genera. However, some authors found in the past [9,27,28] that *Saccharomyces* strains also possessed this activity. The recovery of some isolates able to perform arbutin splitting is a promising result, for the implication of this metabolism in wine flavour and taste. Concerning sulphite production, this trait is unfavorable, because it is related to the production of off-flavours and off-odours; however, it is well known that this metabolic activity is common amongst wine yeasts [17] and Mendes-Ferreira et al. [29] also reported that production could be strongly affected by the physiological conditions of yeasts. Therefore, strain selection would focus on the choice of low-producer strains.

The impact of pectolytic activity is different in relation to yeast use and destination; therefore, in fermented vegetables, above all in table olives, this activity is negative as it could contribute to olive softening and spoilage [30]. Otherwise, its role in wine is not clear; many authors consider this ability not relevant [17], but in some cases, it is a common idea that it could be useful (for example to produce cider). Pectinolytic enzymes are polysaccharidases that degrade pectins present in middle lamella and primary cell walls of plants; this ability is widely used in winemaking as pectinases can help to improve liquefaction, juice yield, clarification, filterability, and to increase the release of color and flavor compounds entrapped in grape skins. Although this trait was found in some isolates of *Aureobasidium pullulans*, *Hanseniaspora* sp., *Metschnikowia* sp. [31], some experiments done in the past with *S. cerevisiae* strains from a red grape variety (Uva di Troia) suggest that also isolates from this species could possess this trait [9].

We also focused on some technological traits of wine yeasts (tolerance to ethanol and resistance to acetic acid); the results were generally in line with the literature, and only few isolates tolerated up to 15% of ethanol.

After the study of some phenotypic traits, a preliminary selection was done based on two criteria (ability to grow in presence of 15% ethanol and 150 ppm of SO_2). As expected, these criteria were strongly restrictive only four isolates were selected; the genotyping confirmed their attribution to the species *S. cerevisiae*.

The last trait assessed for the technological characterization was the evaluation of the decarboxylation of amino acids, as this this is the major pathway for the synthesis of biogenic amines. The production of biogenic amines is of concern, because EFSA and other Regulatory Agencies require that strains for human use must not possess toxicogenic activity and/or negative effect on health [32]. The results of this assay confirm the possibility of recovering this trait in yeast microbiota, as reported by other authors [23,33]; it poses some safety issues for the selection of starter cultures from the wild microbiota, as biogenic amines could have deleterious effects on the well-being. Moreover, it also stressed that interesting microorganism (like the isolate 46D, with high ethanol and SO_2 tolerance) could exhibit this property.

The last assay was a small-scale fermentation with the selected isolates and a reference strain was added (EC1118); the isolates 45D and 46D showed fermentative performances like the reference strain (high fermentation rate, significant weight loss). The other two isolates (43D and 44D) experienced a variable time (from two to four days) before the beginning of the fermentation. In a primary fermentation this trend could be a problem, as the lack of ethanol in the first days could induce the growth of a competitive microbiota. The technological traits in terms of ethanol, acetic acid, and glycerol produced were in line with the data recovered for some isolates from grape varieties of Sourthern Italy [1,3,15].

Therefore, as a final selection only the isolate 45D could be proposed as a potential starter for Bombino bianco, because of its phenotypic traits (fermentative trend similar to a refence strain, glucosidase activity, high resistance to ethanol), while the other three isolates should be excluded for the potential production of biogenic amines (46D) or for the presence of a lag phase before the beginning of fermentation (isolates 43D and 44D).

5. Conclusions

This paper represents a first approach for the characterization of yeast microbiota of Bombino grape variety, with a focus on some technological properties (ethanol, resistance to SO_2 and acetic acid), safety issues (biogenic amine production) and enzymatic patterns (pectolytic activity, glucosidase, catalase). This preliminary survey highlights the existence of yeasts with glucosidase activity and a medium pectolytic patterns; in addition, the resistance to ethanol is variable, and only few isolates were able to grow in presence of 15% ethanol. These isolates were all identified as *S. cerevisiae*, but in a yeast a decarboxylation activity was found thus suggesting the possibility of recovering amine producing strains from the natural microbiota. A final fermentation trial suggested the existence of an isolate (*S. cerevisiae* 46D) with interesting traits and performances which could be potentially proposed as a starter for Bombino bianco. Further investigations are required for a better characterization of this isolate in terms of secondary compounds produced and fermentative performances in real conditions, as well as typing at strain level and resistance to the common practices of strain storage (e.g., dehydration).

Author Contributions: Conceptualization, A.B., M.S., and M.R.C.; investigation, B.S., D.C., and L.P.; data curation, A.B.; writing—original draft preparation, A.B. and L.P.; writing—review and editing, A.B., M.S., and M.R.C.

Funding: This research received no external funding.

Conflicts of Interest: The authors declare no conflict of interest.

References

1. Tufariello, M.; Maiorano, G.; Rampino, P.; Spano, G.; Grieco, F.; Perrotta, C.; Capozzi, V.; Grieco, F. Selection of an autochthonous yeast starter culture for industrial production of Primitivo "Gioia del Colle" PDO/DOC in Apulia (Southern Italy). *LWT* **2019**, *99*, 188–196. [CrossRef]
2. Petruzzi, L.; Capozzi, V.; Berbegal, C.; Corbo, M.R.; Bevilacqua, A.; Spano, G.; Sinigaglia, M. Microbial resources and enological significance: Opportunities and benefits. *Front. Microbiol.* **2017**, *8*, 995. [CrossRef] [PubMed]
3. Garofalo, C.; Berbegal, C.; Grieco, F.; Tufariello, M.; Spano, G.; Capozzi, V. Selection of indigenous yeast strains for the production of sparkling wines from native Apulian grape varieties. *Int. J. Food Microbiol.* **2018**, *285*, 7–17. [CrossRef] [PubMed]
4. Romano, P.; Capece, A. Wine microbiology. In *Starter Cultures in Food Production*, 1st ed.; Speranza, B., Bevilacqua, A., Corbo, M.R., Sinigaglia, M., Eds.; Wiley Blackwell: Oxford, UK, 2017; pp. 255–282.
5. Ribéreau-Gayon, P.; Denis Dubourdieu, D.; Donèche, B.; Lonvaud, A. Handbook of Enology. In *The Microbiology of Wine and Vinifications*, 2nd ed.; John Wiley & Sons: New York, NY, USA, 2006; Volume 1, p. 497.
6. Grieco, F.; Tristezza, M.; Vetrano, C.; Bleve, G.; Panico, E.; Grieco, F.; Mita, G.; Logrieco, A. Exploitation of autochthonous micro-organism potential to enhance the quality of Apulian wine. *Ann. Microbiol.* **2011**, *61*, 67–73. [CrossRef]
7. Tristezza, M.; Vetrano, C.; Bleve, G.; Grieco, F.; Tufariello, M.; Quarta, A.; Mita, G.; Spano, G.; Grieco, F. Autochthonous fermentation starters for the industrial production of Negroamaro wines. *J. Ind. Microbiol. Biotechnol.* **2012**, *39*, 81–92. [CrossRef] [PubMed]
8. Garofalo, C.; Tristezza, M.; Grieco, F.; Spano, G.; Capozzi, V. From grape berries to wine: Population dynamics of cultivable yeasts associated to "Nero di Troia" autochthonous grape cultivar. *World. J. Microbiol. Biotechnol.* **2016**, *32*, 59. [CrossRef] [PubMed]
9. Petruzzi, L.; Bevilacqua, A.; Corbo, M.R.; Speranza, B.; Capozzi, V.; Sinigaglia, M. A focus on quality and safety traits of *Saccharomyces cerevisiae* isolated from Uva di Troia grape variety. *J. Food Sci.* **2017**, *82*, 124–133. [CrossRef]
10. Tristezza, M.; Fantastico, L.; Vetrano, C.; Bleve, G.; Corallo, D.; Grieco, F.; Mita, G.; Grieco, F. Molecular and technological characterization of *Saccharomyces cerevisiae* strains isolated from natural fermentation of Susumaniello grape must in Apulia, Southern Italy. *Int. J. Microbiol.* **2014**, *897428*. [CrossRef]
11. Antonacci, D.; Perniola, R. la vitivinicoltura pugliese. *Vigneevini* **2005**, *XXXII*, 116–122.
12. Masino, F.; Montevecchi, G.; Arfelli, G.; Antonelli, A. Evaluation of the combined effects of enzymatic treatment and aging on lees on the aroma of wine from Bombino bianco grapes. *J. Agric. Food Chem.* **2008**, *56*, 9495–9501. [CrossRef]
13. Baiano, A.; Varva, G.; De Gianni, A.; Terracone, C.; Viggiani, I.; Del Nobile, M.A. Effects of different vinification technologies on physico-chemical properties and antioxidant activity of 'Falanghina' and 'Bombino bianco' wines. *Eur. Food Res. Technol.* **2013**, *237*, 831–842. [CrossRef]
14. Vincenzini, M.; Romano, P.; Farris, G.A. *Microbiologia del Vino (Wine Microbiology)*; Casa Editrice Ambrosiana: Milan, Italy, 2006; p. 520.
15. Petruzzi, L.; Bevilacqua, A.; Corbo, M.R.; Garofalo, C.; Baiano, A.; Sinigaglia, M. Selection of autochthonous Saccharomyces cerevisiae strains as wine starters using a polyphasic approach and ochratoxin A removal. *J. Food Prot.* **2014**, *77*, 1168–1177. [CrossRef] [PubMed]
16. Barbosa, A.; Lage, P.; Vilela, A.; Mendes-Faia, A.; Mendes-Ferreira, A. Phenotypic and metabolic traits of commercial *Saccharomyces cerevisiae* yeasts. *AMB Express* **2014**, *4*, 39. [CrossRef] [PubMed]
17. Hernández, A.; Martin, A.; Aranda, E.; Pérez-Nevado, F.; Córdoba, M.G. Identification and characterization of yeast isolated from the elaboration of seasoned green table olives. *Food Microbiol.* **2007**, *24*, 346–352. [CrossRef]
18. Sinigaglia, M.; Di Benedetto, N.; Bevilacqua, A.; Corbo, M.R.; Capece, A.; Romano, P. Yeasts isolated from olive mill wastewaters from southern Italy: Technological characterization and potential use for phenol removal. *Appl. Microbiol. Biotechnol.* **2010**, *87*, 2345–2354. [CrossRef]
19. Bover-Cid, S.; Holzapfel, W.H. Improved screening procedure for biogenic amine production by lactic acid bacteria. *Int. J. Food Microbiol.* **1999**, *53*, 33–41. [CrossRef]

20. Van Gerwen, S.J.C.; Zwietering, M.H. Growth and inactivation models to be used in quantitative risk assessments. *J. Food Prot.* **1998**, *6*, 1541–1549. [CrossRef]

21. Baty, F.; Delignette-Muller, M.L. Estimating the bacterial lag time: Which model, which precision? *Int. J. Food Microbiol.* **2004**, *91*, 261–277. [CrossRef]

22. Delignette-Muller, M.L.; Cornu, M.; Pouillot, R.; Denis, J.B. Use of the Bayesian modelling in risk assessment: Application to growth of *Listeria monocytogenes* and food flora in cold-smoked salmon. *Int. J. Food Microbiol.* **2006**, *106*, 195–208. [CrossRef]

23. Tristezza, M.; Vetrano, C.; Bleve, G.; Spano, G.; Capozzi, V.; Logrieco, A.; Mita, G.; Grieco, F. Biodiversity and safety aspects of yeast strains characterized from vineyards and spontaneous fermentations in the Apulia Region, Italy. *Food Microbiol.* **2013**, *36*, 335–342. [CrossRef]

24. Arevalo-Villena, M.; Briones-Perez, A.; Corbo, M.R.; Sinigaglia, M.; Bevilacqua, A. Biotechnological application of yeasts in food science. Starter cultures, probiotic, and enzyme production. *J. Appl. Microbiol.* **2017**, *123*, 1360–1372. [CrossRef] [PubMed]

25. Pando Bedriñana, R.; Querol Simón, A.; Suárez Valles, B. Genetic and phenotypic diversity of autochthonous cider yeasts in a cellar from Asturias. *Food Microbiol.* **2010**, *27*, 503–508. [CrossRef] [PubMed]

26. Fia, G.; Giovani, G.; Rosi, I. Study of β-glucosidase production by wine-related yeasts during alcoholic fermentation. A new rapid fluorimetric method to determine enzymatic activity. *J. Appl. Microbiol.* **2005**, *99*, 509–517. [CrossRef] [PubMed]

27. Mateo, J.J.; Di Stefano, R. Description of the beta-glucosidase activity of wine yeast. *Food Microbiol.* **1997**, *14*, 583–591. [CrossRef]

28. Chaiyasut, C.; Pengkumsri, N.; Sirilun, S.; Peerajan, S.; Khongtan, S.; Sundaram Sivamaruthi, S. Assessment of changes in the content of anthocyanins, phenolic acids, and antioxidant property of *Saccharomyces cerevisiae* mediated fermented black rice bran. *AMB Express* **2017**, *7*, 114. [CrossRef]

29. Mendes-Ferreira, A.; Mendes-Faia, A.; Leão, C. Survey of hydrogen sulphide production by wine yeasts. *J. Food Prot.* **2002**, *65*, 1033–1037. [CrossRef]

30. Bevilacqua, A.; Beneduce, L.; Sinigaglia, M.; Corbo, M.R. Selection of yeasts as starter cultures for table olives. *J. Food Sci.* **2013**, *78*, M742–M751. [CrossRef]

31. Merín, M.G.; Martín, M.C.; Rantsiou, K.; Cocolin, L.; de Ambrosini, V.I. Characterization of pectinase activity for enology from yeasts occurring in Argentine Bonarda grape. *Braz. J. Microbiol.* **2015**, *46*, 815–823. [CrossRef]

32. Laulund, S.; Wind, A.; Derkx, P.M.F.; Zuliani, V. Regulatory and safety requirements for food cultures. *Microorganisms* **2017**, *5*, 28. [CrossRef]

33. Galgano, F.; Caruso, M.; Condelli, N.; Favati, F. Agmatine in fermented foods. *Front. Microbiol.* **2012**, *3*, 199. [CrossRef]

Article

Looking at the Origin: Some Insights into the General and Fermentative Microbiota of Vineyard Soils

Alejandro Alonso [1], Miguel de Celis [1], Javier Ruiz [1], Javier Vicente [1], Eva Navascués [2], Alberto Acedo [3], Rüdiger Ortiz-Álvarez [3], Ignacio Belda [3,4], Antonio Santos [1,*], María Ángeles Gómez-Flechoso [5] and Domingo Marquina [1]

[1] Department of Genetics, Physiology and Microbiology, Unit of Microbiology, Biology Faculty, Complutense University of Madrid, 28040 Madrid, Spain
[2] Pago de Carraovejas, S.L.U., Camino de Carraovejas, s/n, 47300 Peñafiel, Valladolid, Spain
[3] Science Department, Biome Makers Spain, 47011 Valladolid, Spain
[4] Department of Biology, Geology, Physics & Inorganic Chemistry, Unit of Biodiversity and Conservation, Rey Juan Carlos University, 28933 Móstoles, Spain
[5] Departament of Earth Physics & Astrophysics, Physics Faculty, Complutense University of Madrid, 28040 Madrid, Spain
*   Correspondence: ansantos@ucm.es; Tel.: +34-913-944-962

Received: 8 July 2019; Accepted: 27 August 2019; Published: 29 August 2019

Abstract: In winemaking processes, there is a current tendency to develop spontaneous fermentations taking advantage of the metabolic diversity of derived from the great microbial diversity present in grape musts. This enological practice enhances wine complexity, but undesirable consequences or deviations could appear on wine quality. Soil is a reservoir of important microorganisms for different beneficial processes, especially for plant nutrition, but it is also the origin of many of the phytopathogenic microorganisms that affect vines. In this study, a meta-taxonomic analysis of the microbial communities inhabiting vineyard soils was realized. A significant impact of the soil type and climate aspects (seasonal patterns) was observed in terms of alpha and beta bacterial diversity, but fungal populations appeared as more stable communities in vineyard soils, especially in terms of alpha diversity. Focusing on the presence and abundance of wine-related microorganisms present in the studied soils, some seasonal and soil-dependent patterns were observed. The *Lactobacillaceae* family, containing species responsible for the malolactic fermentation, was only present in non-calcareous soils samples and during the summer season. The study of wine-related fungi indicated that the *Debaryomycetaceae* family dominates the winter yeast population, whereas the *Saccharomycetaceae* family, containing the most important fermentative yeast species for winemaking, was detected as dominant in summer.

Keywords: meta-taxonomic analysis; vineyard soil; wine-related bacteria; wine-related fungi

---

1. Introduction

Microorganisms are very successful inhabitants of the soil due to their adaptability and plasticity to cope with adverse conditions [1]. There is a general assumption that, in many ecosystems, a high biodiversity enhances stability and productivity, and it is regulated by climate, soil properties and soil management aspects [1–3]. Since most biodiversity–productivity studies focus on plant diversity, this relationship requires a better understanding within the microbial populations inhabiting soils, as microorganisms play a crucial role in many key ecosystem functions involved in soil fertility [4–6]. Plants are dependent on the growth of soil microbes, which possess the metabolic machinery to access soil nutrients such as N, P, and S that, usually, are minimally bioavailable for them [7]. With hundreds to thousands of taxa per gram of soil, it has been demonstrated that functional redundancy within the soil microbial community is high, indicating that microbial community diversity is dissociated from functioning [8,9]. Such evidence is considered to be highly relevant to infer the impact of climate changes and anthropic practices on soil microbial diversity and, in consequence, biogeochemical cycles in soils [6].

As indicated above, microbial communities are associated with plants, playing a role in soil productivity but also causing phytopathogenic diseases [10]. Numerous studies concerning soilborne microorganisms have been carried out, however, taking into consideration the agricultural, industrial, alimentary and economic implications of soil microorganisms, the development of new tools and approaches for determining their diversity and functions in soils is a continuous task [11].

The interface between roots and soil is probably the most important interaction between plants and their environment [12]. Soil microbes that colonize the plant at the root can move through the plant to colonize the rest of the tissues, promoting plant health or causing different diseases. To help the plant microbiome fight against pathogens, microorganism inoculation has been used in several crops, including vineyards, in an attempt to control plant pathogens using biological agents [11,13,14]. Moreover, the possibility that plant inhabiting microorganisms could influence the flavor and productivity of grapes, impacting the organoleptic characteristics of wine, has been reported [12]. On the other hand, one aspect of the relationship between plants and microorganisms that remains unclear is whether soil microorganisms could be related with postharvest processes, such as fermentative ones, including those related to the production of wine.

With the current tendency to recover past practices in winemaking, the wine industry is now frequently producing wines by spontaneous fermentation. This reformulated enology is emerging and aiming to combine the advantages of spontaneous fermentations with those of monitored fermentations. In such fermentations, the microbiota coming from the vineyard takes the leading role of the fermentation process, being the soil the main reservoir of wine-related microorganisms, inhabiting grape berries and thus the later grape must [15]. In this context, a clear connection has been demonstrated between winery and vineyard fermentative microbiota, with a transference of yeasts from the winery to the surrounding vineyards, influencing the native yeast communities [16–21]. As the number of spontaneous fermentation studies increases, the importance of the autochthonous microbiota of the vineyard studies increases too. Recent studies have indicated that grape and wine microbiome from different grape-growing regions correlate with wine metabolome, suggesting that the grape microbiome may influence regional wine characteristics [22].

To date, few studies have analyzed the relationship between the soil microbiota and its influence on the winemaking processes. Currently, microbiome analyzes start interconnecting multiple "omics" studies, leading to unprecedented opportunities to comprehensively characterize microbial communities and their relationships with their environments or subsequent processes [23]. To understand the crucial roles of microorganisms on the entire winemaking process, we should understand the relationship between vineyard and wine microbiomes, also paying attention to the soil microbiome [24]. The use of soil microbiota as an early predictor of wine *terroir* is unprecedented and poses a potential new challenge for quality control of wine [24].

This study analyzed the microbial, fungal and bacterial communities inhabiting the soils of different blocks of a unique vineyard, in which a relationship between vineyard and wine microbiota has been observed [20]. Thus, we aimed to determine the influence of soil properties in the inhabitant general and fermentative microbiota, and how it changes in a seasonal comparison: summer against winter.

2. Materials and Methods

2.1. Site Description and Weather Data

This research was carried out in a vineyard which belongs to Ribera del Duero Geographical Indication (VCPRD). The entire vineyard covers an area of approximately 1.80 km^2 and has a Mediterranean with Oceanic influence climate, corresponding to Csb on the Köppen–Geiger climate classification. The annual mean temperature in this region is 12.1 °C, and the multi-year average precipitation is about 434 mm (Spanish Meteorological Agency AEMET, 2166Y station). The main landform is of hills formed by calcareous deposition and windy sands, with an altitude ranging from 753 to 900 m. Soil types are comprised of sandy, clayey and calcareous ones. Sandy soils show a sand percentage of 59.8–75.3%. The clayey ones present a clay content ranging 22.5–24.5% and the calcareous soils have a limestone active fraction of 9.8–11.7%. Generally, the soils studied were fairly alkaline, with pH values around 8.58. In some samples, the detected pH values were higher due to the high percentage of limestone.

2.2. Soil Sampling

This study included 36 vineyard soil samples, collected in vineyard plots with sandy, clay and limestone soils. Five samples of sandy soils, two of clay soils and two of calcareous soils were taken. From each sample, a replica was made 30 m away. Soil samples were taken by previously removing the surface layer of leaves that might be on the ground. The 5–25 cm-depth samples were taken to collect the maximum cellular density [25]. This sampling process was done in two seasons: summer and winter.

2.3. DNA Extraction and Sequencing

Soil samples, collected as described, were analyzed following a 16S-ITS metabarcoding strategy for determining bacterial and fungal populations. Samples were stored at −80 °C until DNA extraction was performed using different bead-beating cycles based on DNA extraction kits such as DNeasy® Powerlyzer® Powersoil® Kit (Qiagen, Hilden, Germany). Libraries were prepared following the two-steps PCR Illumina® protocol and these were subsequently sequenced on Illumina® MiSeq instrument (Illumina®, San Diego, CA, USA) using 2 × 301 paired-end reads.

All PCR reactions were prepared using sterilized materials and negative controls were run alongside the samples. In addition, PCR conditions such as number of cycles, annealing temperature, thermocycler and Master-mix composition were done according to the WineSeq® technology procedures. The library was performed using a two-step PCR protocol as described by Feld et al. [26] and Albers et al. [27] and then it was analyzed by amplifying and sequencing the V4 16S rRNA V4 gene region and the ITS1 (ITS) regions using WineSeq® custom primers (patent WO2017096385 [28]).

2.4. Bioinformatic Analysis

The raw fastq sequences (available at https://data.mendeley.com/datasets/yf5mk58kwz/2) were analyzed using DADA2 algorithm [29] implemented in R pipeline [30]. DADA2 implements an error correction model that allows the differentiation of a single nucleotide [31], giving an amplicon sequence variant (ASV) table as a final output. The reads were truncated at their low-quality ends, forward and reverse paired, and chimeras removed. The total good quality reads were 1,636,020 for bacteria and 2,260,792 for fungi. The taxonomic assignment was performed using the naïve Bayesian classifier implemented in DADA2 using as reference Silva (release 132) reference database [32].

2.5. Functional Profiles Prediction

Functional predictions based on representative genomes are a useful tool for the estimation of metabolic potential [33]. Although it has limitations regarding strain-specific functional signatures, environmental distributions, or real magnitude of a process, the functional simulations allow the comparison of communities in terms of their predicted functional potential [34]. For that purpose, we applied an adaptation of the Tax4Fun routine [35] using presence/absence of genes rather than a normalized weighted value per taxa (https://sourceforge.net/projects/Tax4Fun2/). To obtain the proportion of each community containing each specific function, we filtered a total of 25 KEGGs (functional orthologs) within 14 metabolic pathways related to carbon, nitrogen, phosphorus and sulfur cycles pathways (Table A2). We estimated the distribution of each metabolism and their mean proportions in the microbial population of each soil sample.

2.6. Statistical Analysis

Statistical analysis was performed on R (version 3.5.1) using the phyloseq package, version 1.26.1 [36] and vegan, version 2.5.5 [37]. Alpha diversity was calculated as estimated community diversity using Shannon index [38] and ANOVA test was used to calculate significant differences among sample groups (Figure 1). Beta diversity (differences between samples) was calculated using Bray–Curtis distance matrix on proportion transformed data [39,40] and permutational multi-variate analysis of variance (PERMANOVA). Non-Metric Multi-Dimensional Scaling (NMDS) was computed from the resulting distance matrices to compress dimensionality into two-dimensional plots (Figure 2). For heat map plots, pheatmap package version 1.0.12 R was used (Figure 3).

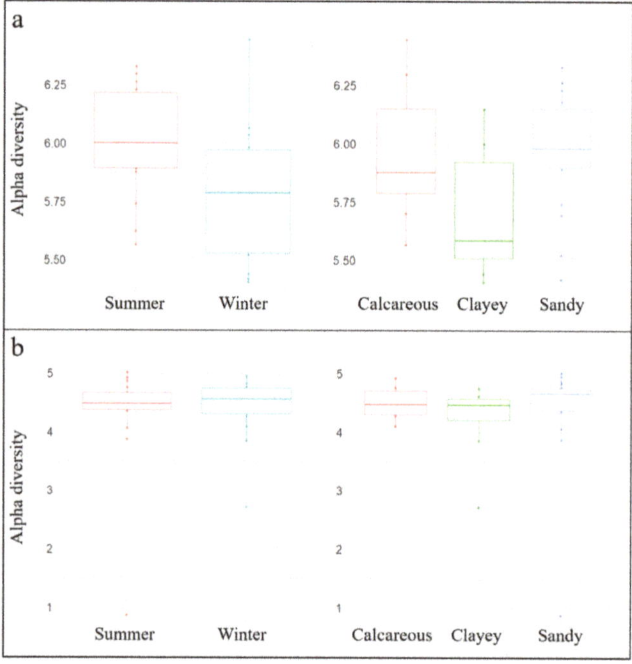

Figure 1. Alpha diversity, measured as Shannon index, calculated on bacterial subset of the dataset (**a**) showed significant differences (*p*-value = 0.012) among seasons, but not among soil types (*p*-value = 0.056). When alpha diversity was calculated on the fungal subset of the dataset (**b**), no statistically significant differences were found for seasons (*p*-value = 0.716) and for soil type (*p*-value = 0.771).

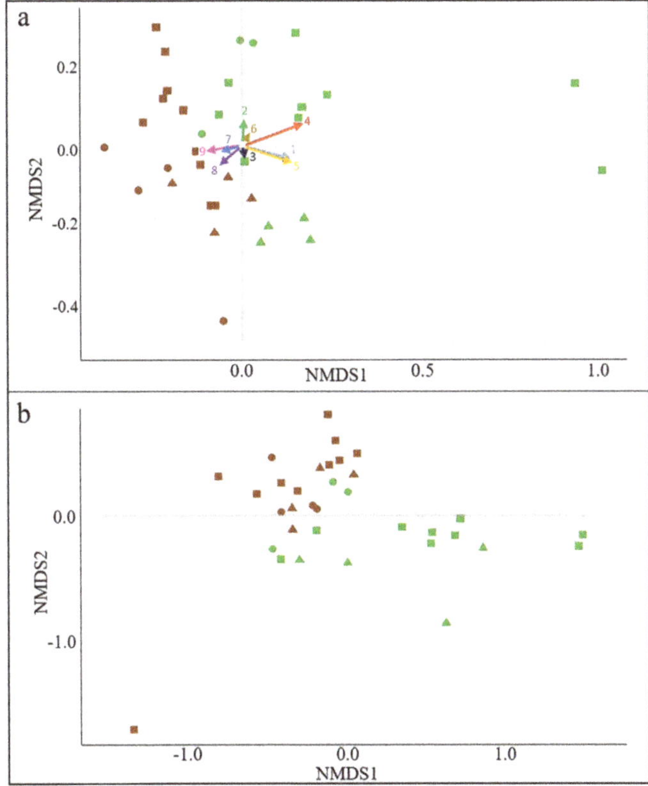

Figure 2. Beta diversity calculated on bacterial (**a**) and fungal (**b**) datasets shown in a non-parametric multi-dimensional scale (NMDS) using Bray-Curtis distance. The stress parameter used in the bacterial analysis to define the ordination quality was 0.160. Significant differences were found among different season samples (p-value = 0.001) and for samples of different soil types (p-value = 0.058). Stress parameter used in the fungal analysis to define the ordination quality was 0.145. Significant differences were found among different season samples (p-value = 0.001) and for samples of different soil types (p-value = 0.052). Seasons (summer (brown) and winter (green)), oil types (calcareous ($\bigcirc$), clayey ($\triangle$) and sandy ($\square$)). Figure 2a includes information on the contribution of some bacterial-derived soil-related metabolic functions, inferred from the taxonomical bacterial diversity using the Tax4Fun routine. The nine vectors showed were calculated from the relative abundance of metabolic enzymes (KEGG) corresponding to: (1) carbon organic formation; (2) carbon organic use; (3) nitrogen organic formation; (4) nitrogen organic use; (5) other; (6) phosphorus inorganic transport; (7) phosphorus organic transport; (8) sulfur organic formation; and (9) sulfur organic use. A detailed list of the metabolic enzymes (KEGG) included on each group is reported in Table A2.

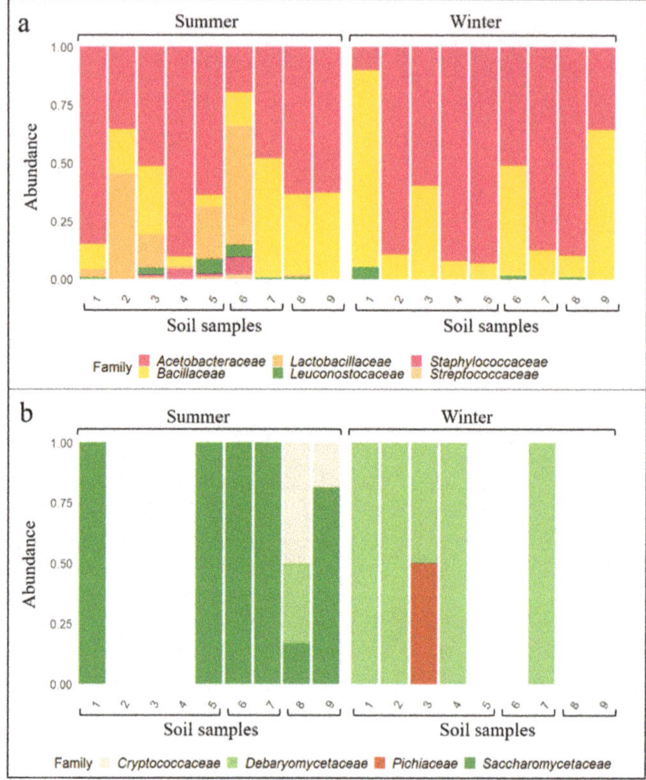

Figure 3. Abundance of wine-related bacterial (**a**) and fungal (**b**) families in soil samples and classified according to the soil characteristics (Sandy (1–5), clayey (6–7) and calcareous (8–9)) and the season (summer and winter).

Physical-chemical data of the soil were analyzed by Infostat© statistical program [41], using the analysis of main components for the classification of different soils (clay, sandy and limestone) and the subsequent representation of the variables in the two-dimensional principal subspace (Figure A1).

3. Results and Discussion

3.1. General Biodiversity

There is a general assumption which indicates that the larger the environmental heterogeneity is the larger the diversity of complex organisms is, indicating that more heterogeneous habitats usually present higher species alpha diversity [42,43]. However, despite the recognized central role of bacteria in the soils' fertility, less knowledge has been reported concerning the link between the environmental heterogeneity and bacterial diversity [44,45]. Several investigations have reported that locations highly different in their environmental and physicochemical parameters usually tend to be very different in their bacterial community composition too [46,47], suggesting that soil heterogeneity increases bacterial beta diversity. Even though the relationship between soil environments and fungal diversity is less known [48], some meta-analysis studies have indicated that, in addition to bacterial alpha diversity, fungal alpha diversity is higher in fields with crop rotations [49] or in temperate deciduous forests [50].

In this study, we analyzed the general microbial diversity (bacterial and fungal population), with the aim of determining the potential connections between soil and wine-related microbiota from different vineyards. The microbial diversity as alpha diversity of the vineyard soils was measured using the Shannon index (measure of the species richness and abundance), comparing the effect of two seasons (summer and winter), with very different conditions of temperature and humidity in the sampled region. The climate conditions in the center of Spain are characterized by cold and humid winters, while the summers are hot with very little and occasional rainfall. In addition, the impact of the textural characteristics of the soils (Figure A1) were determined, studying their impact over microbial biodiversity [51] (Figure 1).

The seasonal effect on bacterial alpha diversity changes significantly between seasons. The complexity of bacterial communities has been generally described to be lower in winter than summer [52,53], and our results are in agreement with this observation (Figure 1a). As a result, we hypothesize about the possibility whether the soil bacterial community can be used as a new biological parameter to be considered in vineyard soils zoning strategies in viticulture soils or not. In contrast to bacterial diversity, fungal diversity did not change with the seasons (Figure 1b) and showed a lower Shannon index and greater dispersion than the bacterial subset. The fungal community inhabiting soils was more homogeneous during seasons, maintaining the regional homogeneity of the studied soils. Bacterial populations showed a microscale effect due to their heterogeneity in summer and winter. Although fungi populations could become an indicator of regional character in vineyard blocks.

In this study, we analyzed bacterial diversity as a function of the textural characteristics: calcareous, clayey and sandy soils (Table A1). A lower bacteria diversity was observed in the different types of soil (Figure 1a). Although the differences in the Shannon index among clayey, calcareous and sandy soils using ANOVA test were slightly statistically significant (p-value = 0.056), it was observed that the Shannon index is different between sandy and clayey soils. No differences were observed in soil types in the fungal subset analyses (Figure 1b).

Beta diversity was calculated as dissimilarity between soil samples, according to the ASVs extracted from the raw data curation process. In the non-parametric multi-dimensional scale ordering (NMDS), the ASVs of the group of bacteria (Figure 2a) and the fungi subset (Figure 2b) show the distances of each soil sample.

The bacterial population separates into two groups defined in the NMDS1 component (p-value = 0.001). The bacterial subset present in soils in winter was observed for NMDS1 > 0, approximately. Furthermore, the samples whose bacterial population defined the summer season were located for NMDS1 < 0 (Figure 2a). The textural characteristics of the soil were also statistically significant (p-value = 0.001), having NMDS2 > −0.1 for sandy soils, approximately, and NMDS2 < −0.1 for the other two soils (calcareous and clayey).

Sorting based on stress index for the fungi group allowed separation in the NMDS2 component (p-value = 0.001). The population of fungi linked to winter were found in NMDS2 < 0. In samples collected in summer, the fungal population was observed in NMDS2 > 0. Statistical analyses of the textural characteristics were also slightly significant (p-value = 0.052) since this separation was not too clear to define as a function of the NMDS values.

Based on the genomic sequencing of the V4 16S rRNA gene region, it was possible to estimate the functional genes that the bacterial population could express in the soil. The estimated metabolic functions include enzymes involved, among others, in the biogeochemical cycles of carbon, nitrogen, phosphorus and sulfur (vectors at Figure 2a). Based on that, the metabolic routes involved in organic carbon formation, organic nitrogen use and others (see Table A2 for a detailed list of the metabolic routes included) appeared more represented in winter samples. We can hypothesize that this could be because winter samples were collected in January, and a greater concentration of organic matter is accumulated in the soil (coming from fall autumn leaves). On the other hand, summer samples cluster matched the direction of the contribution of metabolic routes involved in sulfur metabolism (organic formation and use). This can be explained as summer samples were collected in early June and some routine sulfur-based treatments were applied in April and May for guaranteeing a healthy grape ripening.

Nevertheless, contrary to what was observed at taxonomic (alpha and beta diversity) level, there is not a clear pattern clustering the soils samples coming from different soil types or collected at different seasons (Figure A2). This can indicate that the taxonomic differences found between vineyard blocks are buffered at a functional level due to the high functional redundancy commonly found within soil microbial communities [8,9].

3.2. Wine-Related Microbial Diversity

Since the soil has been reported as the main reservoir of microorganisms in the vineyard, and a notable co-occurrence of microorganisms exists among vineyard soils, grapes and musts [15,24], it is of interest to study the presence, diversity and abundance of wine-related bacterial and fungal species in the studied soils. Soil microbiota has been described as important, not only for the chemical and nutritional properties of soils, but also for health, yield, and quality of the grapevine. Apart from being the origin of the fermentative microbiota that will reach the winery as part of the microbial consortia established in the grapes—which would be responsible for positive flavor compounds production or in the production of undesirable molecules (off-flavors, biogenic amines, etc.)—the soil microbiome has been directly co-related with some flavor characteristics of wines (via plant-microbiome interactions), such as the rotundone concentration found in Shiraz grapes from Australian Cool Climate areas [54]. Thus, in response to the current trend of elaborating "single-vineyard" wines as a way to enhance the *terroir* characteristics of each vineyard block, understanding the microbial signature of soils should be considered in future vineyard zoning works, when trying to define their fermentative potential. The raw data from the sequencing process were filtered, obtaining the abundances of the microorganisms previously described to be isolated from wine-related samples (Table A3).

The WRB found in the meta-taxonomic studies of soils were filtered at the taxonomic level of family due to the limitations showed by the NGS-technique used in this work [55]. The soil samples collected in winter and summer differ in the presence of the family *Lactobacillaceae*, being of greater presence in summer and absent in winter, while *Leuconostocaeae* appears in more plots in summer samples. Some examples of species from these families are *Oenococcus*, *Leuconostoc*, *Weissella* (*Leuconostocaceae*), *Lactobacillus* and *Pediococcus* (*Lactobacillaceae*), mainly responsible for malolactic fermentation [56]. In addition, various species of *Lactobacillus*, *Pediococcus* and *Leuconostoc* can cause spoilage of wine during bulk storage in the cellar and after bottling [57]. No differences were observed by soil type, although in summer calcareous Samples 8 and 9 showed a similar abundance pattern. However, it is possible that in Plots 7–9 the absence of the *Lactobacillaceae* family was due to an active limestone concentration of more than 5.1% (Table A1). The pronounced prevalence of the *Acetobacteraceae* family observed in winter stood out. The ability of acetic acid bacteria to convert ethanol in acetic acid is one of the main sources of wine spoilage. Both grapes and wine are subject to spoilage by this bacteria at different stages of the grape ripening and the winemaking processes [58].

The wine-related fungi (WRF) present in the soils were the *Cryptococcaceae*, *Debaryomycetaceae*, *Pichiaceae* and *Saccharomycetaceae* families (Figure 4b). However, within some samples, no representatives of these families were found. In summer soils samples from Plots 2–4 and winter samples from Plots 5, 6, 8 and 9, no WRF families were detected. In the summer season, a clear prevalence of the family *Saccharomycetaceae* was observed in Plots 1, 5, 6 and 9. Plot 8 did not present fungi of the family *Cryptococcaceae* and *Debaryomycetaceae*. Is important to highlight that the calcareous soils of Plots 8 and 9 showing the presence of the WRF family *Cryptococcaceae* were the only ones that presented this family during summer. In winter, a high frequency of the family *Debaryomycetaceae* was observed in Plots 1, 2, 4 and 7. The soil of Plot 3 showed the *Pichiaceae* and *Debaryomycetaceae* families, which were equally represented. Due to the succession of families *Saccharomycetaceae* and *Cryptococcaceae* between summer and winter, WRF seems to be a better indicator for differentiating the seasonal fermentative potential among plots.

The beta diversity analyzed in the WRB families shows a clear distinction between winter and summer (Figure 4). The component NMDS2 allowed good separation between the variations in the subset of bacteria. The winter samples were mainly arranged in NMDS2 < 0, while variations in summer samples qwew disposed in NMDS2 > 0.

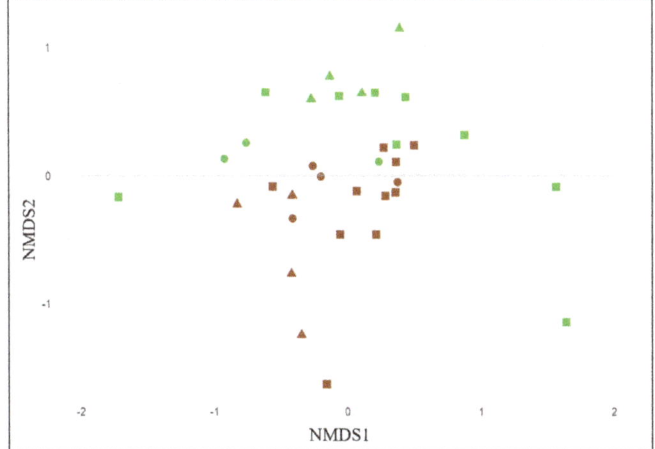

Figure 4. Beta diversity calculated on bacterial datasets shown in a non-parametric multi-dimensional scale (NMDS) using Bray–Curtis distance. Wine related family bacteria exhibited significant seasonal (*p*-value = 0.001) variations in soil samples, but this variation was not evident between soil types (*p*-value = 0.058). Seasons (summer (brown) and winter (green)). Soil types (calcareous (○), clayey (△) and sandy (□)).

4. Conclusions

The microbial alpha diversity of the vineyard soils determined in this study varied between seasons (Figure 1), being bacteria a better indicator than fungi in the vineyard zoning and allowing to differentiate the sandy soils from the clayey ones. Furthermore, beta diversity allowed us to separate populations between seasons (summer vs. winter) from bacteria and fungi (Figure 2). The microbial *terroir*, at a single vineyard scale, could be a tendency in each season, although the bacterial function remained constant (Figure 3). Wine-related bacteria (WRB) remained constant between seasons, except for the family *Lactobacillaceae*. This family, moreover, was not found in soils with a percentage of limestone higher than 5.1% (Figure 4a). Wine-related fungi (WRF) described a summer population dominated by the *Saccharomycetaceae* family and another winter population represented mainly by the *Debaryomycetaceae* family (Figure 4b).

Future Perspectives

Although the relationship between the soil microbiome and in fruit-associated microbial composition is difficult to establish [59], it is possible to indicate that the microorganisms present in the vineyard influence through many routes the vine development and the quality and fermentative potential of grapes [60]. Analyzing the microbial particularities of different blocks of a single vineyard, we here highlight the concept of microbial *terroir*. Thus, our results can be used as a starting point for future scientific studies and in-field works considering the microbial aspects of soils in vineyards zoning works trying to define homogeneous *terroir* units. Apart from the direct importance of the microorganisms in soil health and vine yield, in the present work, we tried to establish a parallelism between the microorganisms that can be detected in the vineyard soils (as the main microbial reservoir in agricultural environments) and the microorganisms reported in different studies that could be found during wine fermentation. The great inter-blocks variability found here highlights the important of including the biological aspects of *terroir* for a complete understanding of the enological potential of vineyards. This study represents an advance in the knowledge of how the microorganisms detected in the vineyard environment, mainly present in the microbial reservoir of the soil, that could affect vine and grape development, and, through this, positively or negatively influence the resulting wine. In addition, taking into account that many of these microorganisms are not only detected but could play a role during fermentation, in this study, we identified the abundance of these microorganisms in the soil microbial reservoir. Future studies in this area will go through the analysis of how the presence of a certain microorganism or a particular microbial consortium present in the soil can influence the quality of a wine in a certain way and, how, through the precise use of appropriate viticulture techniques, we can favor or counteract the presence of these microorganisms. Additionally, this type of studies can contribute to the discovery of undetected microorganisms with optimal fermentation properties and, therefore, could be used as new microorganisms in oenology. Furthermore, they could also help to detect microorganisms for the biological control of pests or phytopathogenic fungi that affect the wood of the vine.

Author Contributions: A.A. (Alejandro Alonso), I.B., E.N., A.S. and D.M. were involved in the sampling and experimental design; A.A. (Alejandro Alonso), M.d.C. and M.A.G.-F. performed the formal data analysis; A.A. (Alejandro Alonso), I.B. and A.S. wrote the article; M.A.G.-F., E.N., A.S. and D.M. supervised the project; and E.N., J.R., J.V., R.O.-Á and A.A. (Alberto Acedo) contributed to data acquisition. All authors discussed the results and contributed to the final manuscript.

Funding: Funding for the research in this paper was provided by Pago de Carraovejas Estate Winery (GLOBALVITI, IDI-20160750) and Biome Makers Spain S.L. (IDI-20180120) under the framework of the MINECO CDTI projects (Spanish Ministry of Economy, Industry and Competitiveness, Centre for Industrial Technological Development).

Acknowledgments: Ignacio Belda acknowledges for his postdoctoral "Torres Quevedo" grant (PTQ-16-08253) provided by the Spanish Ministry of Economy, Industry and Competitiveness.

Conflicts of Interest: Eva Navascués is an employee of Pago de Carraovejas Estate Winery. Alberto Acedo is an employee of Biome Makers, and Rüdiger Ortiz-Álvarez and Ignacio Belda developed part of this work employed by Biome Makers

Abbreviations

The following abbreviations are used in this manuscript:

ASV	amplicon sequence variant
ITS	internal transcribed spacer
NMDS	non-parametric multi-dimensional scale
PCA	principal component analysis
VCPRD	a quality wine psr
WRB	wine-related bacteria
WRF	wine-related fungi

Appendix A. Soil Compositional Characteristics

For this analysis, the database of physical-chemical parameters of the vineyard soils was used (http://dx.doi.org/10.17632/yf5mk58kwz.2#file-0a4b4597-abb7-4df9-96f3-3ea7f44e5cd5). The eigenvectors that implied a greater explanation in each component were analyzed and then the most relevant variables were taken. The statistical on PCA test (Figure A1) allowed classifying by means of different percentages of sand, clay and limestone in the soil (Table A1).

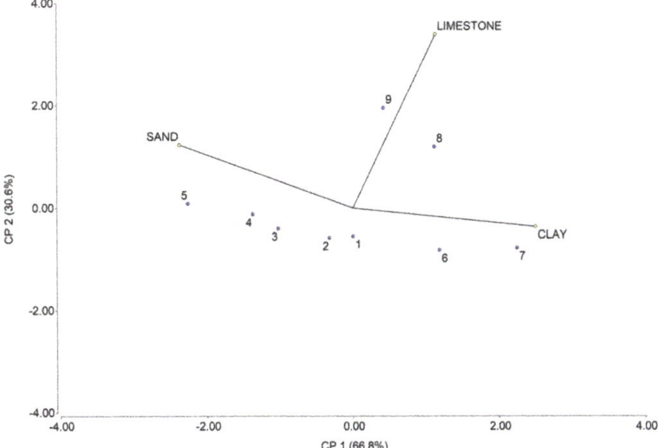

Figure A1. The two-dimensional principal subspace for the different textural soil percentages (correlation matrix PCA).

Table A1. Different compositional soils.

Soil Sample	Sand (%)	Clay (%)	Limestone (%)
1	59.8	20.4	3.1
2	62.5	19.8	2.6
3	63.7	16.4	2.8
4	65.9	15.0	3.4
5	75.3	13.7	2.9
6	48.3	22.5	3.7
7	39.5	24.5	5.1
8	58.0	22.4	9.8
9	62.5	19.3	11.7

Appendix B. KEGGs and Metabolism Pathways

From the raw data, a functional estimation of the bacterial population was carried out using Tax4Fun (https://sourceforge.net/projects/Tax4Fun2/). This process allowed us to estimate the functional status of the plots studied with respect to winter and summer.

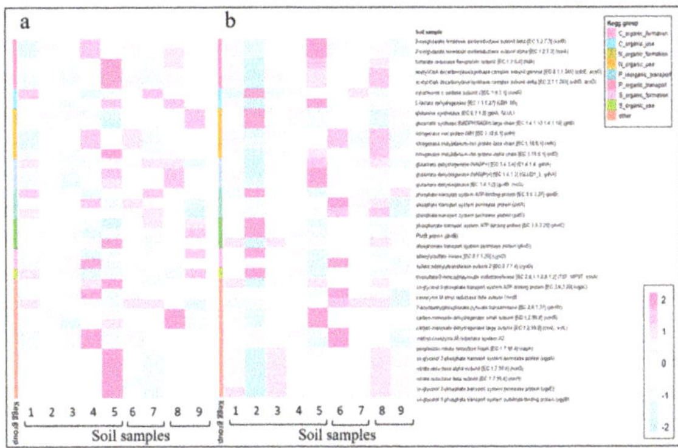

Figure A2. Functional prediction of bacterial populations in different season. Summer (**a**) and winter (**b**).

Table A2. KEGG's table.

KEGG	Functional Description: Name [EC] (gen)	Metabolism
K02274	cytochrome c oxidase subunit I [EC:1.9.3.1] (coxA)	Aerobic Respiration
K00174	2-oxoglutarate ferredoxin oxidoreductase subunit alpha [EC:1.2.7.3] (korA)	Arnon Carbon Fixation
K00175	2-oxoglutarate ferredoxin oxidoreductase subunit beta [EC:1.2.7.3] (korB)	Arnon Carbon Fixation
K00244	fumarate reductase flavoprotein subunit [EC:1.3.5.4] (frdA)	Arnon Carbon Fixation
K00860	adenylylsulfate kinase [EC:2.7.1.25] (cysC)	Assimilatory Sulfate Reduction
K00957	sulfate adenylyltransferase subunit 2 [EC:2.7.7.4] (cysD)	Assimilatory Sulfate Reduction
K00016	L-lactate dehydrogenase [EC:1.1.1.27] (LDH, ldh)	Fermentation
K05816	sn-glycerol 3-phosphate transport system ATP-binding protein [EC:3.6.3.20] (ugpC)	G3P Transporter
K00400	coenzyme Methyl reductase beta subunit (mrcB)	Methanogenesis
K00401	methyl coenzyme M reductase system A2	Methanogenesis
K00265	glutamate synthase (NADPH/NADH) large chain [EC:1.4.1.13, 1.4.1.14] (gltB)	Nitrogen Assimilation
K01915	glutamine synthetase [EC:6.3.1.2] (glnA, GLUL)	Nitrogen Assimilation
K02588	nitrogenase iron protein NifH [EC:1.18.6.1] (nifH)	Nitrogen Fixation
K02591	nitrogenase molybdenum-iron protein beta chain [EC:1.18.6.1] (nifK)	Nitrogen Fixation
K00261	glutamate dehydrogenase (NAD(P)+) [EC:1.4.1.3] (GLUD1 2, gdhA)	Nitrogen Mineralization
K00262	glutamate dehydrogenase (NADP+) [EC:1.4.1.4] (gdhA)	Nitrogen Mineralization
K00260	glutamate dehydrogenase [EC:1.4.1.2] (gudB, rocG)	Nitrogen Mineralization
K02567	periplasmic nitrate reductase NapA [EC:1.7.99.4] (napA)	Nitrogen Reduction
K02036	phosphate transport system ATP-binding protein [EC:3.6.3.27] (pstB)	Phosphate Transport High
K02038	phosphate transport system permease protein (pstA)	Phosphate Transport High
K02037	phosphate transport system permease protein (pstC)	Phosphate Transport High
K03430	2-aminoethylphosphonate-pyruvate transaminase [EC:2.6.1.37] (phnW)	Phosphonate Metabolism
K04750	PhnB protein (phnB)	Phosphonate Transport
K02041	phosphonate transport system ATP-binding protein [EC:3.6.3.28] (phnC)	Phosphonate Transport
K01011	thiosulfate/3-mercaptopyruvate sulfurtransferase [EC:2.8.1.1, 2.8.1.2] (TST, MPST, sseA)	Sulfur Mineralitation

Appendix C. Wine-Related Microorganism

The following families of microorganisms have been used to define WRBs and WRFs. The presented data were elaborated from the description of these microorganisms in the bibliography.

Table A3. Wine-related microorganisms taxonomy.

Kingdom	Phylum	Class	Order	Family
Bacteria	*Firmicutes*	*Bacilli*	*Bacillales*	*Bacillaceae*
Bacteria	*Firmicutes*	*Bacilli*	*Bacillales*	*Staphylococcaceae*
Bacteria	*Firmicutes*	*Bacilli*	*Lactobacillales*	*Enterococcaceae*
Bacteria	*Firmicutes*	*Bacilli*	*Lactobacillales*	*Lactobacillaceae*
Bacteria	*Firmicutes*	*Bacilli*	*Lactobacillales*	*Leuconostocaceae*
Bacteria	*Firmicutes*	*Bacilli*	*Lactobacillales*	*Streptococcaceae*
Bacteria	*Proteobacteria*	*Alphaproteobacteria*	*Rhodospirillales*	*Acetobacteraceae*
Fungi	*Ascomycota*	*Saccharomycetes*	*Saccharomycetales*	*Debaryomycetaceae*
Fungi	*Ascomycota*	*Saccharomycetes*	*Saccharomycetales*	*Metschnikowiaceae*
Fungi	*Ascomycota*	*Saccharomycetes*	*Saccharomycetales*	*Pichiaceae*
Fungi	*Ascomycota*	*Saccharomycetes*	*Saccharomycetales*	*Saccharomycetaceae*
Fungi	*Basidiomycota*	*Tremellomycetes*	*Tremellales*	*Cryptococcaceae*

* The information on the origin, occurrence and potential role of the wine related microorganisms (bacteria and yeasts) considered in this table can be found in the following references: Fleet (1993) [61]; König et al. (2009) [62]; Capozzi et al. (2011) [63]; and Benavent-Gil et al. (2016) [64].

References

1. Frąc, M.; Hannula, S.E.; Bełka, M.; Jędryczka, M. Fungal biodiversity and their role in soil health. *Front. Microbiol.* **2018**, *9*, 707. [CrossRef] [PubMed]

2. Isbell, F.I.; Polley, H.W.; Wilsey, B.J. Biodiversity, productivity and the temporal stability of productivity: Patterns and processes. *Ecol. Lett.* **2009**, *12*, 443–451. [CrossRef] [PubMed]

3. Jactel, H.; Gritti, E.; Drössler, L.; Forrester, D.; Mason, W.; Morin, X.; Pretzsch, H.; Castagneyrol, B. Positive biodiversity–productivity relationships in forests: Climate matters. *Biol. Lett.* **2018**, *14*, 20170747. [CrossRef] [PubMed]

4. Balvanera, P.; Pfisterer, A.B.; Buchmann, N.; He, J.S.; Nakashizuka, T.; Raffaelli, D.; Schmid, B. Quantifying the evidence for biodiversity effects on ecosystem functioning and services. *Ecol. Lett.* **2006**, *9*, 1146–1156. [CrossRef] [PubMed]

5. Reed, H.E.; Martiny, J.B. Testing the functional significance of microbial composition in natural communities. *FEMS Microbiol. Ecol.* **2007**, *62*, 161–170. [CrossRef] [PubMed]

6. Maron, P.A.; Sarr, A.; Kaisermann, A.; Lévêque, J.; Mathieu, O.; Guigue, J.; Karimi, B.; Bernard, L.; Dequiedt, S.; Terrat, S.; et al. High microbial diversity promotes soil ecosystem functioning. *Appl. Environ. Microbiol.* **2018**, *84*, e02738-17. [CrossRef] [PubMed]

7. Van Der Heijden, M.G.; Bardgett, R.D.; Van Straalen, N.M. The unseen majority: Soil microbes as drivers of plant diversity and productivity in terrestrial ecosystems. *Ecol. Lett.* **2008**, *11*, 296–310. [CrossRef]

8. Allison, S.D.; Martiny, J.B. Resistance, resilience, and redundancy in microbial communities. *Proc. Natl. Acad. Sci. USA* **2008**, *105*, 11512–11519. [CrossRef]

9. Bardgett, R.D.; Van Der Putten, W.H. Belowground biodiversity and ecosystem functioning. *Nature* **2014**, *515*, 505. [CrossRef]

10. Maron, J.L.; Marler, M.; Klironomos, J.N.; Cleveland, C.C. Soil fungal pathogens and the relationship between plant diversity and productivity. *Ecol. Lett.* **2011**, *14*, 36–41. [CrossRef]

11. Katan, J. Diseases caused by soilborne pathogens: Biology, management and challenges. *J. Plant Pathol.* **2017**, *99*, 305–315.

12. Gilbert, J.A.; van der Lelie, D.; Zarraonaindia, I. Microbial terroir for wine grapes. *Proc. Natl. Acad. Sci. USA* **2014**, *111*, 5–6. [CrossRef] [PubMed]

13. Santos, A.; Marquina, D. Killer toxin of Pichia membranifaciens and its possible use as a biocontrol agent against grey mould disease of grapevine. *Microbiology* **2004**, *150*, 2527–2534. [CrossRef] [PubMed]

14. Karuppiah, V.; Li, T.; Vallikkannu, M.; Chen, J. Co-cultivation of Trichoderma asperellum GDFS1009 and Bacillus amyloliquefaciens 1841 causes differential gene expression and improvement in the wheat growth and biocontrol activity. *Front. Microbiol.* **2019**, *10*, 1068. [CrossRef] [PubMed]

15. Zarraonaindia, I.; Owens, S.M.; Weisenhorn, P.; West, K.; Hampton-Marcell, J.; Lax, S.; Bokulich, N.A.; Mills, D.A.; Martin, G.; Taghavi, S.; et al. The soil microbiome influences grapevine-associated microbiota. *MBio* **2015**, *6*, e02527–14. [CrossRef] [PubMed]

16. Cordero-Bueso, G.; Arroyo, T.; Serrano, A.; Valero, E. Remanence and survival of commercial yeast in different ecological niches of the vineyard. *FEMS Microbiol. Ecol.* **2011**, *77*, 429–437. [CrossRef] [PubMed]

17. Tello, J.; Cordero-Bueso, G.; Aporta, I.; Cabellos, J.; Arroyo, T. Genetic diversity in commercial wineries: Effects of the farming system and vinification management on wine yeasts. *J. Appl. Microbiol.* **2012**, *112*, 302–315. [CrossRef]

18. Capozzi, V.; Garofalo, C.; Chiriatti, M.A.; Grieco, F.; Spano, G. Microbial terroir and food innovation: The case of yeast biodiversity in wine. *Microbiol. Res.* **2015**, *181*, 75–83. [CrossRef]

19. Garofalo, C.; El Khoury, M.; Lucas, P.; Bely, M.; Russo, P.; Spano, G.; Capozzi, V. Autochthonous starter cultures and indigenous grape variety for regional wine production. *J. Appl. Microbiol.* **2015**, *118*, 1395–1408. [CrossRef]

20. De Celis, M.; Ruiz, J.; Martín-Santamaría, M.; Alonso, A.; Marquina, D.; Navascués, E.; Gómez-Flechoso, M.Á.; Belda, I.; Santos, A. Diversity of Saccharomyces cerevisiae yeasts associated to spontaneous and inoculated fermenting grapes from Spanish vineyards. *Lett. Appl. Microbiol.* **2019**, *68*, 580–588. [CrossRef]

21. Viel, A.; Legras, J.L.; Nadai, C.; Carlot, M.; Lombardi, A.; Crespan, M.; Migliaro, D.; Giacomini, A.; Corich, V. The geographic distribution of Saccharomyces cerevisiae isolates within three Italian neighboring winemaking regions reveals strong differences in yeast abundance, genetic diversity and industrial strain dissemination. *Front. Microbiol.* **2017**, *8*, 1595. [CrossRef] [PubMed]

22. Bokulich, N.A.; Collins, T.S.; Masarweh, C.; Allen, G.; Heymann, H.; Ebeler, S.E.; Mills, D.A. Associations among wine grape microbiome, metabolome, and fermentation behavior suggest microbial contribution to regional wine characteristics. *MBio* **2016**, *7*, e00631-16. [CrossRef] [PubMed]

23. Sirén, K.; Mak, S.S.T.; Fischer, U.; Hansen, L.H.; Gilbert, M.T.P. Multi-omics and potential applications in wine production. *Curr. Opin. Biotechnol.* **2019**, *56*, 172–178. [CrossRef] [PubMed]

24. Belda, I.; Zarraonaindia, I.; Perisin, M.; Palacios, A.; Acedo, A. From vineyard soil to wine fermentation: Microbiome approximations to explain the "terroir" concept. *Front. Microbiol.* **2017**, *8*, 821. [CrossRef] [PubMed]

25. Raynaud, X.; Nunan, N. Spatial ecology of bacteria at the microscale in soil. *PLoS ONE* **2014**, *9*, e87217. [CrossRef] [PubMed]

26. Feld, L.; Nielsen, T.K.; Hansen, L.H.; Aamand, J.; Albers, C.N. Establishment of bacterial herbicide degraders in a rapid sand filter for bioremediation of phenoxypropionate-polluted groundwater. *Appl. Environ. Microbiol.* **2016**, *82*, 878–887. [CrossRef] [PubMed]

27. Albers, C.N.; Ellegaard-Jensen, L.; Hansen, L.H.; Sørensen, S.R. Bioaugmentation of rapid sand filters by microbiome priming with a nitrifying consortium will optimize production of drinking water from groundwater. *Water Res.* **2018**, *129*, 1–10. [CrossRef]

28. Becares, A.A.; Fernandez, A.F. Microbiome Based Identification, Monitoring and Enhancement Of Fermentation Processes and Products. US Patent Application 15/779,531, 20 December 2018.

29. Callahan, B.J.; McMurdie, P.J.; Rosen, M.J.; Han, A.W.; Johnson, A.J.A.; Holmes, S.P. DADA2: High-resolution sample inference from Illumina amplicon data. *Nat. Methods* **2016**, *13*, 581–583. [CrossRef]

30. Callahan, B.J.; Sankaran, K.; Fukuyama, J.A.; McMurdie, P.J.; Holmes, S.P. Bioconductor workflow for microbiome data analysis: From raw reads to community analyses. *F1000Research* **2016**, *5*, 1492. [CrossRef]

31. Callahan, B.J.; McMurdie, P.J.; Holmes, S.P. Exact sequence variants should replace operational taxonomic units in marker-gene data analysis. *ISME J.* **2017**, *11*, 2639. [CrossRef]

32. Quast, C.; Pruesse, E.; Yilmaz, P.; Gerken, J.; Schweer, T.; Yarza, P.; Peplies, J.; Glöckner, F.O. The SILVA ribosomal RNA gene database project: Improved data processing and web-based tools. *Nucleic Acids Res.* **2012**, *41*, D590–D596. [CrossRef] [PubMed]

33. Langille, M.G.; Zaneveld, J.; Caporaso, J.G.; McDonald, D.; Knights, D.; Reyes, J.A.; Clemente, J.C.; Burkepile, D.E.; Thurber, R.L.V.; Knight, R.; et al. Predictive functional profiling of microbial communities using 16S rRNA marker gene sequences. *Nat. Biotechnol.* **2013**, *31*, 814–821. [CrossRef] [PubMed]

34. Ortiz-Álvarez, R.; Fierer, N.; de los Ríos, A.; Casamayor, E.O.; Barberán, A. Consistent changes in the taxonomic structure and functional attributes of bacterial communities during primary succession. *ISME J.* **2018**, *12*, 1658. [CrossRef] [PubMed]

35. Aßhauer, K.P.; Wemheuer, B.; Daniel, R.; Meinicke, P. Tax4Fun: Predicting functional profiles from metagenomic 16S rRNA data. *Bioinformatics* **2015**, *31*, 2882–2884. [CrossRef] [PubMed]

36. McMurdie, P.J.; Holmes, S. phyloseq: An R package for reproducible interactive analysis and graphics of microbiome census data. *PLoS ONE* **2013**, *8*, e61217. [CrossRef] [PubMed]

37. Oksanen, J.; Blanchet, F.; Kindt, R.; Legendre, P.; Minchin, P.; O'Hara, R.; Simpson, G.; Solymos, P.; Stevens, M.; Wagner, H. Package "vegan." *Community Ecol. Package* **2013**, *12*, 1–295.

38. Shannon-Wiener, C.; Weaver, W.; Weater, W. *The Mathematical Theory of Communication*; EUA, University of Illinois Press: Champaign, IL, USA, 1949.

39. Bray, J.R.; Curtis, J.T. An ordination of the upland forest communities of southern Wisconsin. *Ecol. Monogr.* **1957**, *27*, 325–349. [CrossRef]

40. McMurdie, P.J.; Holmes, S. Waste not, want not: Why rarefying microbiome data is inadmissible. *PLoS Comput. Biol.* **2014**, *10*, e1003531. [CrossRef]

41. Di Rienzo, J.; Casanoves, F.; Balzarina, M.; Gonzalez, L.; Tablada, M.; Robledo, C. *Infostat Versión 2018. Centro de Transferencia Infostat, FCA*; Universidad Nacional de Córdoba: Córdoba, Argentina, 2018.

42. Stein, A.; Gerstner, K.; Kreft, H. Environmental heterogeneity as a universal driver of species richness across taxa, biomes and spatial scales. *Ecol. Lett.* **2014**, *17*, 866–880. [CrossRef]

43. Heino, J.; Melo, A.S.; Bini, L.M. Reconceptualising the beta diversity-environmental heterogeneity relationship in running water systems. *Freshw. Biol.* **2015**, *60*, 223–235. [CrossRef]

44. Dickens, S.; Allen, E.; Santiago, L.; Crowley, D. Exotic annuals reduce soil heterogeneity in coastal sage scrub soil chemical and biological characteristics. *Soil Biol. Biochem.* **2013**, *58*, 70–81. [CrossRef]

45. Weber, C.F.; King, G.M.; Aho, K. Relative abundance of and composition within fungal orders differ between cheatgrass (Bromus tectorum) and sagebrush (Artemisia tridentata)-associated soils. *PloS ONE* **2015**, *10*, e0117026. [CrossRef] [PubMed]

46. Horner-Devine, M.C.; Lage, M.; Hughes, J.B.; Bohannan, B.J. A taxa–area relationship for bacteria. *Nature* **2004**, *432*, 750–753. [CrossRef] [PubMed]

47. Ranjard, L.; Dequiedt, S.; Prévost-Bouré, N.C.; Thioulouse, J.; Saby, N.; Lelievre, M.; Maron, P.; Morin, F.; Bispo, A.; Jolivet, C.; et al. Turnover of soil bacterial diversity driven by wide-scale environmental heterogeneity. *Nat. Commun.* **2013**, *4*, 1434. [CrossRef] [PubMed]

48. Peay, K.G.; Baraloto, C.; Fine, P.V. Strong coupling of plant and fungal community structure across western Amazonian rainforests. *ISME J.* **2013**, *7*, 1852. [CrossRef] [PubMed]

49. Venter, Z.S.; Jacobs, K.; Hawkins, H.J. The impact of crop rotation on soil microbial diversity: A meta-analysis. *Pedobiologia* **2016**, *59*, 215–223. [CrossRef]

50. He, J.; Tedersoo, L.; Hu, A.; Han, C.; He, D.; Wei, H.; Jiao, M.; Anslan, S.; Nie, Y.; Jia, Y.; et al. Greater diversity of soil fungal communities and distinguishable seasonal variation in temperate deciduous forests compared with subtropical evergreen forests of eastern China. *FEMS Microbiol. Ecol.* **2017**, *93*. [CrossRef]

51. Zhou, J.; Xia, B.; Huang, H.; Palumbo, A.V.; Tiedje, J.M. Microbial diversity and heterogeneity in sandy subsurface soils. *Appl. Environ. Microbiol.* **2004**, *70*, 1723–1734. [CrossRef]

52. Koranda, M.; Kaiser, C.; Fuchslueger, L.; Kitzler, B.; Sessitsch, A.; Zechmeister-Boltenstern, S.; Richter, A. Seasonal variation in functional properties of microbial communities in beech forest soil. *Soil Biol. Biochem.* **2013**, *60*, 95–104. [CrossRef]

53. Kaiser, C.; Fuchslueger, L.; Koranda, M.; Gorfer, M.; Stange, C.F.; Kitzler, B.; Rasche, F.; Strauss, J.; Sessitsch, A.; Zechmeister-Boltenstern, S.; et al. Plants control the seasonal dynamics of microbial N cycling in a beech forest soil by belowground C allocation. *Ecology* **2011**, *92*, 1036–1051. [CrossRef]

54. Gupta, V.V.S.R.; Bramley, R.G.V.; Greenfield, P.; Yu, J.; Herderich, M.J. Vineyard Soil Microbiome Composition Related to Rotundone Concentration in Australian Cool Climate 'Peppery' Shiraz Grapes. *Front. Microbiol.* **2019**, *10*, 1607. [CrossRef] [PubMed]

55. Kircher, M.; Kelso, J. High-throughput DNA sequencing–concepts and limitations. *Bioessays* **2010**, *32*, 524–536. [CrossRef] [PubMed]

56. Cappello, M.S.; Zapparoli, G.; Logrieco, A.; Bartowsky, E.J. Linking wine lactic acid bacteria diversity with wine aroma and flavour. *Int. J. Food Microbiol.* **2017**, *243*, 16–27. [CrossRef] [PubMed]

57. Sponholz, W.R. Wine spoilage by microorganisms. In *Wine Microbiology and Biotechnology*; Taylor & Francis: Abingdon, UK, 1993; pp. 395–420.

58. Bartowsky, E.J.; Henschke, P.A. Acetic acid bacteria spoilage of bottled red wine—A review. *Int. J. Food Microbiol.* **2008**, *125*, 60–70. [CrossRef] [PubMed]

59. Chou, M.; Vanden, H.J.; Bell, T.; Panke-Buisse, K.; J, K.K. Vineyard under-vine floor management alters soil microbial composition, while the fruit microbiome shows no corresponding shifts. *Sci. Rep.* **2018**, *8*, e00631-16. [CrossRef] [PubMed]

60. Jackson, D.I.; Lombard, P.B. Environmental and Management Practices Affecting Grape Composition and Wine Quality—A Review. *Am. J. Enol. Viticult.* **1993**, *44*, 409–430.

61. Fleet, G.H. *Wine Microbiology and Biotechnology*; CRC Press: Boca Raton, FL, USA, 1993.

62. König, H.; Unden, G.; Fröhlich, J. *Biology of Microorganisms on Grapes, in Must and in Wine*; Springer: Berlin, Germany, 2009.

63. Capozzi, V.; Ladero, V.; Beneduce, L.; Fernández, M.; Alvarez, M.A.; Benoit, B.; Laurent, B.; Grieco, F.; Spano, G. Isolation and characterization of tyramine-producing Enterococcus faecium strains from red wine. *Food Microbiol.* **2011**, *28*, 434–439. [CrossRef] [PubMed]

64. Benavent-Gil, Y.; Berbegal, C.; Lucio, O.; Pardo, I.; Ferrer, S. A new fear in wine: Isolation of Staphylococcus epidermidis histamine producer. *Food Control* **2016**, *62*, 142–149. [CrossRef]

are factors of great importance [9,10,14,15]. In traditional winemaking, only 40% of the anthocyanins of the grapes are transferred to the wine [4,16]. The limited extraction of anthocyanins is mainly due to the lack of permeability of cell walls and cytoplasmic membranes [17,18], because these compounds are in the skin, in the upper cellular layers of the hypodermis. The composition of cell walls is genetically determined and modifies the changes in the hardness of skin and seed tissues along with ripening. The extraction of anthocyanins and proanthocyanidins during winemaking depends on the grape variety [19,20]. The simultaneous development of maceration and alcoholic fermentation influence the extraction of polyphenols, because the ethanol content determines the disintegration of the vacuolar membranes and the walls of the skin cell [15]. Anthocyanins are compounds easily soluble in water and therefore are dissolved from the beginning of the maceration, independent of the ethanol concentrations [21].

However, wine color not only depends on the anthocyanin concentration [4,22]. Anthocyanins undergo structural transformations depending on the pH of the medium. They present a red color in an acid medium, acquire a violet color when approaching a neutral pH, and decrease the intensity of the color as the pH increases. Under very high pH conditions, anthocyanins are irreversibly destroyed. Further, during the making, conservation, and aging of wine, the formation of new compounds and their polymerization modify the red wine color and determine its stability [23]. These molecules are partially degraded due to hydrolysis or oxidation reactions [24,25], while other molecules participate in cyclo-addition reactions with metabolites produced by yeasts [26]. Other anthocyanins are condensed with catechins [27,28]. A significant fraction of the anthocyanins extracted from grape skins will be adsorbed by yeasts and will precipitate in the lees [29], whereas there is also a fixation of these compounds in the solid parts of the grapes [21].

In the last few decades, several alternative techniques of maceration have been proposed that allow a differentiated extraction of the phenolic and aromatic compounds of the grape to the wine to improve quality and aging potential [11,13,30]. Most of these techniques have had a substantial impact on the color of red wines [13,31]. More recently, some research groups have evaluated different winemaking techniques to regulate the ethanol content and pH of wines in response to the effect of global warming on the composition of grapes [32–34]. The results obtained with the application of these procedures have allowed the reduction of the ethanol content and pH of the wines, but the effects on the sensory characteristics, particularly on the color, have not been conclusive [32,33,35].

In Uruguay, Tannat is the most relevant red cultivar due to its adaptation to the country's eco-physiological conditions. The polyphenolic and anthocyanin richness of Tannat wines is related to the enological potential of their grapes. The grapes have a low extraction capacity of anthocyanins and lower proportions of malvidin and acetylated glycosides compared with other red cultivars, such as Cabernet Sauvignon and Merlot [30]. Consequently, the color stability of Tannat wines is lower than wines of other varieties [3,8], although they maintain the characteristic anthocyanin profile of the grape of origin for a specified period. Additionally, high interannual climate variability has been recorded during the ripening period, which strongly affects the composition of the grape. In particular, high temperatures during the ripening period cause a high accumulation of sugars and degradation of acidity [36] due to the consumption of malic acid [37] and alter the synthesis of polyphenols [9,38]. Thermal stress during the maturation period causes the degradation and inhibition of the accumulation of anthocyanins, compounds responsible for the color of grapes and red wines [9]. Currently, there is a growing concern of winemakers regarding having tools that allow regulation of the contents of ethanol and pH and the concentrations of phenolic compounds without causing detriment to the color of Tannat red wines. The intensity and hue of the color of Tannat red wines determine the target market and commercial value.

This research aims to study the impact of must replacement and hot pre-fermentative maceration in the color of Uruguayan Tannat red wines produced in three consecutive vintages. Both techniques have been proposed to obtain red wines with lower alcohol content and pH and higher phenolic compound concentration [35]. Hot pre-fermentative maceration consists of the degradation of cellular structures, mainly of the grape skins, through the heating of the must before alcoholic fermentation at a temperature and a period variable [39]. These techniques increase the extraction of phenolic compounds. Moreover, must replacement consists of the substitution of a percentage of grape juice of very ripe grapes with the grape juice of unripe grapes before alcoholic fermentation to reduce the alcohol content and the pH of the wines [35].

2. Materials and Methods

2.1. Chemicals and Equipment

Methanol, acetonitrile, formic acid, and acetic acid were of HPLC grade (>99%) and purchased from Panreac (Barcelona, Spain). Acetaldehyde (>99.5%), ascorbic acid (>99%), and sodium acetate (>99%) were purchased from Sigma-Aldrich (Madrid, Spain). Absolute ethanol and hydrochloric acid (37%) were purchased from Panreac. Malvidin-3-O-glucoside chloride ($\geq$95%), was purchased from Extrasynthese (Genay, France). A Winescan TM Autosampler 79,000 infrared analyzer (Foss, USA) and Foss Integrator software version 154 (Foss, Denmark) were used to determine the alcohol content, total acidity, and pH of the wines. The HPLC analyses were performed using an Agilent 1200 series liquid chromatograph equipped with a G1315D diode array detector (DAD), a G1311A quaternary pump, a G1316A column oven, and a G1329A autosampler (Agilent Technologies, Santa Clara, CA, USA). All the spectrophotometric measurements were performed using a Helios Alpha UV–Vis spectrophotometer (Thermo Fisher Scientific Inc., Waltman, MA, USA).

2.2. Grapes and Wines

This research was carried out with grapes of Tannat *Vitis vinifera L.*, Vitis International Variety Catalogue (VIVC) number 12,257 [40], in 2016, 2017, and 2018 vintages. The grapes were manually harvested from a commercial vineyard located in Canelones in the south of Uruguay.

At the beginning of veraison, 100 kg of Tannat grapes were harvested to obtain a must with high acidity and low sugar concentration. The grapes were crushed (Alfa 60 R crusher, Italcom, Piazzola Sul Brenta, Italy) and lightly pressed in a manual press to obtain 50 L of an unripe grape must. The grape must was immediately sulphited with 100 mg/L of $K_2S_2O_2$, settled overnight, packaged in a 50-L recipient, and conserved at 4 °C until use. When the grapes reached technological maturity, 120 kg of grapes were collected and randomly distributed into 12 lots of 10 kg. The grapes were destemmed and crushed (Alfa 60 R crusher, Italcom, Piazzola Sul Brenta, Italy), and the must was sulphited with 100 mg/L of $K_2S_2O_2$ and distributed in 12 polyethylene containers (each of 10-L capacity). The must containers were randomly divided into two groups of six containers each. Six containers were considered to be controls (original must—OM), whereas in the other six containers (must replacement—MR), 3 L of the original grape must were replaced with 3 L of unripe grapes must with the aim of decreasing sugar content and pH.

Next, three containers of each experimental group (OM and MR) were traditionally macerated (TM), whereas the other three were subjected to hot pre-fermentative maceration (HM) for 1 h at a temperature between 60 and 70 °C. The heating was carried out by transferring the pomace to 11-L stainless-steel tanks that were submerged in a hot water bath (at 80–90 °C). During warming, the pomace was homogenized manually. At the end of the heat treatment, the stainless-steel tanks were submerged in a cold water bath in order to refrigerate them to ambient temperature (around 26 °C). After that, the must was transferred to the original 10-L polyethylene containers. Thus, four experimental groups for each cultivar were obtained: control wine with traditional maceration (OM-TM), must replacement with reduced alcohol and pH in the wine obtained by traditional maceration (MR-TM), control wine with hot pre-fermentative maceration (OM-HM), and must replacement and hot pre-fermentative maceration (MR-HM) (Figure 1).

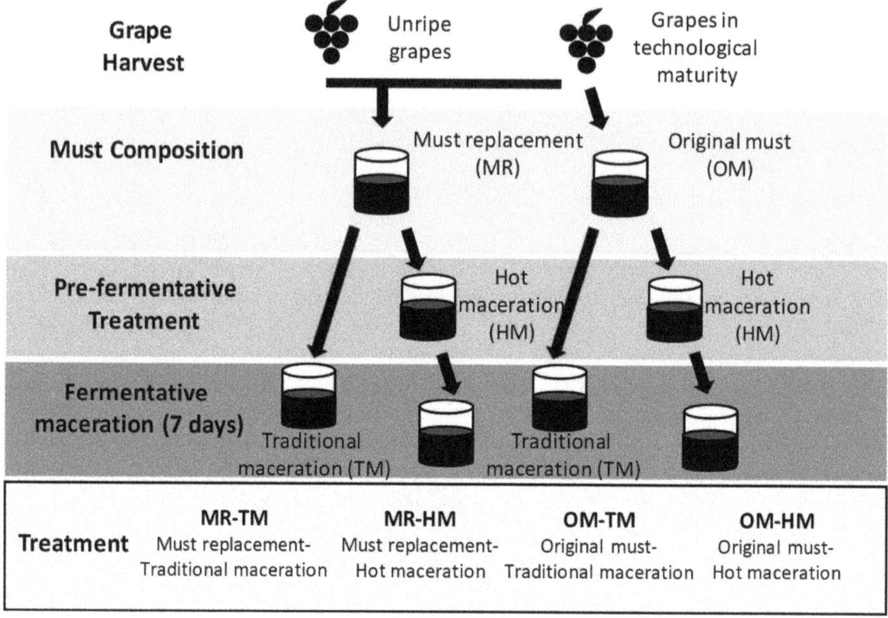

Figure 1. Diagram of the experimental design.

All the containers were inoculated with 200 mg/L of active dry yeast (*Saccharomyces cerevisiae* ex bayanus Natuferm 804; Oenobiotech, Paris, France) and were fermented in contact with the skins and seeds. During maceration, all the containers were manually pumped over once daily, followed by a manual punching down of the cap to favor polyphenol extraction. The fermentation temperature ranged between 26 and 29 °C in the 2016 vintage, between 22 and 27 °C in the 2017 vintage, and between 25 and 29 °C in the 2018 vintage. After 7 days of maceration, the free-run wine was extracted by gravity, and the resting pomace was lightly pressed in a manual press. The free-run wine and the lightly pressed wine of each tank were blended and maintained in 5-L vessels at room temperature (18 ± 2 °C). The alcoholic fermentation was completed when the daily measurements of the must density were less than 998 g/L for three consecutive days. The wines were preserved in polyethylene containers of 5 L of capacity at laboratory room temperature (18 ± 2 °C), and once spontaneous malolactic fermentation was finished (around 35 days later), all the wines were stabilized with 100 mg/L of $K_2S_2O_2$ and 300 mg/L of lysozyme (Delvo®Zyme, Delft, the Netherlands). Finally, the wines were bottled and stored in a dark cellar at laboratory ambient temperature until analysis. The analyses started 2 months after bottling and ended 3 weeks later.

2.3. Standard Grape Juice and Wine Analysis

Analytical methods recommended by the International Organization of Vine and Wine [41] were used to determine the sugar concentration, pH, and titratable acidity of the grape juices. During the fermentation, the temperature and density of the must were monitored daily. The ethanol content, titratable acidity, pH, residual sugars, and volatile acidity of the wines were analyzed using an infrared analyzer Winescan TM Autosampler 79,000 (Foss, USA) and Foss Integrator software version 154 (Foss, Denmark).

2.4. Color Parameters

The color parameters were determined directly on wine samples placed in a 1-mm pathlength cuvette. Color intensity (CI) was estimated using the method proposed by Glories [42]. The CIELAB coordinates, lightness (L*), chroma (C*), hue (h*), red-greenness (a*), and yellow-blueness (b*), were determined according to the method described by Ayala et al. [43]. Thus, data processing was performed with MSCV software [44].

2.5. Spectrophotometric Analysis of Anthocyanins and Related Parameters

The total anthocyanin content of the grapes, their extractability, and their total phenolic index were determined, according to the procedure outlined by González-Neves et al. [45].

The polyphenolic composition was evaluated using classical spectrophotometric indices. The total polyphenols were determined using the Folin–Ciocalteu method, according to Singleton and Rossi [46], and their contents in the wines are expressed in mg of gallic acid per liter. The total pigment and anthocyanin content were analyzed using the technique described by Ribéreau-Gayon and Stonestreet [47], and they are expressed as mg of malvidin-3-glucoside equivalent (EMG) per liter. Catechins were quantified using the method proposed by Swain and Hillis [48], and their concentrations are expressed in mg of D-catechin per liter. Proanthocyanidins were determined according to Ribéreau-Gayon and Stonestreet [49], and their contents are expressed in mg of cyanidin chloride per liter of wine. The ionization index (which indicates the proportion of red-colored anthocyanins at wine pH) and the PVPP index (which indicates the proportion of anthocyanins combined with proanthocyanidins) were determined in line with the method described by Glories [42]. The copigmentation index was measured in accordance with the procedure outlined by Boulton [4].

2.6. HPLC Anthocyanidin Analysis

Reversed-phase HPLC analyses of the anthocyanidins were carried out by injecting 40 µL of wine into an Agilent 1200 series liquid chromatographer (HPLC-DAD) and using an Agilent Zorbax Eclipse XDBC18, 4.6 × 250 mm, 5-µm column (Agilent Technologies). The solvents used were 10% aqueous formic acid (solvent A) and a mixture of 45% methanol, 45% water, and 10% formic acid (solvent B), following the method described by Valls [50]. Chromatograms were recorded at 530 nm, and anthocyanin standard curves were made using malvidin-3-*O*-glucoside chloride. Compounds were identified considering the relative retention times between the compounds and by recording their UV spectra with a diode array detector and comparing these with the UV spectra reported by Valls [50]. The five anthocyanidin-3-monoglucosides of wine (delphinidin, cyanidin, peonidin, petunidin, and malvidin) and their respective acetylated and p-coumarylated anthocyanins were quantified.

2.7. Statistical Analysis

All the data are expressed as the arithmetic average ± standard deviation of three replicates. Multifactorial analysis of variance (MANOVA) was carried out with INFOSTAT [51] (version 2017, Grupo InfoStat, FCA, Universidad Nacional de Córdoba, Argentina), and multiple comparisons between samples were performed by using the Hotelling test.

3. Results and Discussion

3.1. Fermentation Kinetics

Figure 2 shows the fermentation kinetics of the treatments evaluated according to the year of vintage. In the treatments with must replacement (MR-TM and MR-HM), the density was lower due to lower concentrations of sugars. Therefore, these musts finished alcoholic fermentation before the must without substitution and traditional maceration (OM-TM). These results were expected, because the sugar concentrations of the musts were low, and the level of alcohol generated did not affect the development of the yeasts, achieving a complete fermentation of the musts. Moreover, the musts produced by hot pre-fermentative maceration finished alcoholic fermentation before the traditional maceration musts. When a must is subjected to temperatures above 40 °C, the populations of lactic and acetic bacteria, as well as yeasts, disappear [52]. Additionally, the extraction of growth factors during warming favors the subsequent development of inoculated yeasts [53], which explains the results obtained for this treatment. These results are more clearly observed for the wines produced from the 2016 and 2018 vintages, as the climatic conditions allowed the grape to reach a higher degree of maturity. On the contrary, in the vintage of 2017, the ripening stopped, so the harvested grapes were immature.

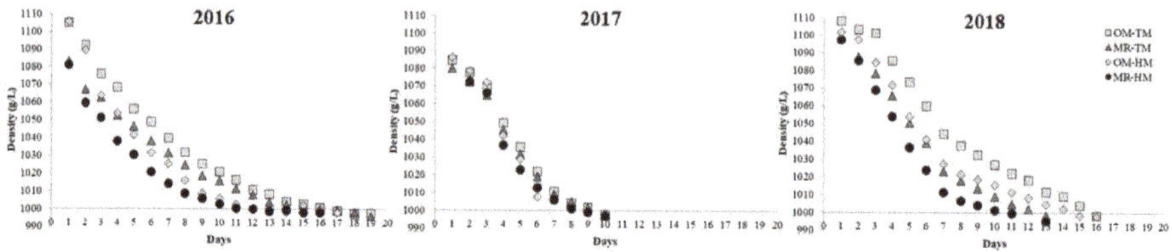

Figure 2. Fermentation kinetics of the treatments by the year of vintage. Average of three wines. OM-TM: original must and traditional maceration; MR-TM: must replacement and traditional maceration; OM-HM: original must and hot pre-fermentative maceration; MR-HM: must replacement and hot pre-fermentative maceration.

3.2. General Composition of Wines

Table 1 shows the effects of the year of vintage, must composition, and winemaking technique factors on the contents of ethanol, titratable acidity, pH, residual sugars, and volatile acidity of wines.

Table 1. General composition of the wines.

Factor Analyzed		Ethanol (% *v/v*)	Titratable Acidity (gH$_2$SO$_4$/L)	pH	Residual Sugars (g/L)	Volatile Acidity (gH$_2$SO$_4$/L)
Year of vintage (*)	2016	14.0 ± 0.1 [b]	4.30 ± 0.27 [a]	3.92 ± 0.16 [a]	1.47 ± 0.41 [c]	0.36 ± 0.07 [b]
	2017	11.2 ± 0.2 [c]	2.93 ± 0.05 [c]	3.86 ± 0.04 [c]	1.85 ± 0.21 [b]	0.43 ± 0.09 [a]
	2018	15.4 ± 0.2 [a]	3.85 ± 0.03 [b]	3.89 ± 0.09 [b]	2.44 ± 0.44 [a]	0.44 ± 0.07 [a]
Must composition (**)	OM	14.0 ± 0.1 [a]	3.51 ± 0.17 [b]	3.95 ± 0.09 [a]	2.07 ± 0.59 [a]	0.43 ± 0.09 [a]
	MR	13.0 ± 0.1 [b]	3.88 ± 0.06 [a]	3.83 ± 0.09 [b]	1.83 ± 0.39 [a]	0.39 ± 0.08 [b]
Maceration technique (***)	TM	13.3 ± 0.2 [b]	3.74 ± 0.19 [a]	3.87 ± 0.09 [a]	2.01 ± 0.55 [a]	0.47 ± 0.06 [a]
	HM	13.7 ± 0.1 [a]	3.64 ± 0.04 [a]	3.92 ± 0.09 [a]	1.89 ± 0.46 [a]	0.35 ± 0.05 [b]
Must composition - Maceration techinque (****)	OM-TM	14.0 ± 0.2 [a]	3.61 ± 0.30 [b]	3.92 ± 0.09 [b]	2.30 ± 0.56 [a]	0.50 ± 0.06 [a]
	MR-TM	12.6 ± 0.2 [c]	3.87 ± 0.09 [a]	3.81 ± 0.12 [d]	1.72 ± 0.37 [c]	0.45 ± 0.06 [b]
	OM-HM	14.0 ± 0.1 [a]	3.40 ± 0.03 [c]	3.98 ± 0.08 [a]	1.84 ± 0.53 [bc]	0.36 ± 0.05 [c]
	MR-HM	13.4 ± 0.1 [b]	3.88 ± 0.04 [a]	3.85 ± 0.09 [c]	1.95 ± 0.39 [b]	0.33 ± 0.05 [c]

(*) Average of 12 wines ± standard deviation regardless of the grape juice composition and the winemaking technique. (**) Average of the 18 wines ± standard deviation regardless of the year of vintage and the winemaking technique. (***) Average of 18 wines ± standard deviation regardless of the year of vintage and the grape juice composition. (****) Average of nine wines ± standard deviation regardless of the year of vintage. Different letters indicate statistical differences ($p < 0.05$). OM: original must; MR: must replacement; TM: traditional maceration; HM: hot pre-fermentative maceration.

The vintage factor expresses the average content of ethanol, titratable acidity, pH, residual sugars, and volatile acidity of all the wines produced in the same vintage, regardless of the must composition and winemaking procedure. Wines produced from the 2018 vintage had the highest ethanol content, and those of the 2017 vintage had the lowest. The highest values of titratable acidity and pH were recorded in the wines produced in 2016 and the lowest in 2017. During the ripening of the grapes, the sugar concentration and the pH increased, whereas titratable acidity decreased. However, climatic conditions during the ripening determine the composition of the grape [14,40]. The ripeness conditions were different between vintages. The grapes harvested in 2016 and 2018 had better maturation conditions, with high concentrations of sugar and an optimum pH. In contrast, in 2017, grape maturation halted, resulting in lower concentrations of sugars and pH. The wines produced from the 2016 and 2017 vintages presented residual sugar concentrations lower than 2 g/L [52], whereas the 2018 wines presented a slightly higher value. These results may be related to a higher concentration of non-fermentable sugars in the 2018 vintage, because the grapes showed a high concentration of sugars. Another possibility may be that the high levels of alcohol generated during alcoholic fermentation affected the development of yeasts in the final stages of alcoholic fermentation [52]. The volatile acidity of the wines elaborated in the different vintages were expected according to the winemaking system used.

The must composition factor expresses the average contents of ethanol, titratable acidity, pH, residual sugars, and volatile acidity of all the wines produced with original must (OM) or must replacement (MR), independent of the vintage or the maceration technique. The MR wines had lower ethanol content and pH and higher titratable acidity than the OM wines. These results were expected, because the must replacement of the well-ripened grapes with the must of unripe grapes implicated a decrease in sugar content and pH and an increase of titratable acidity. These data agree with those obtained by Kontoudakis et al. [32] and Role et al. [33], who proposed a similar but different procedure. Kontoudakis et al. [38] proposed the simultaneous reduction of the ethanol content and the pH of the wine by mixing wines, one of them obtained with green grapes and the other with ripe grapes [32]. Moreover, Role et al. [33] proposed three alternative procedures to achieve alcohol reduction: (i) pre-fermentation addition of liquid derived from grape must (reverse osmosis byproduct); (ii) mixed fermentations with strains of *Starmerella bacillaris* and *Saccharomyces cerevisiae*; and (iii) dealcoholization of wine post-fermentation with a polypropylene membrane. In our research, the partial replacement of grape juice had a low impact on the chemical composition of the wines. The concentration of residual sugars in the wine was not affected by the must replacement, whereas the volatile acidity was slightly lower.

The maceration technique factor expresses the average contents of ethanol, titratable acidity, pH, residual sugars, and volatile acidity of all the wines produced by traditional maceration or hot pre-fermentative maceration, without considering the initial must composition and the vintage. The HM wines presented higher ethanol content than the TM wines, without significant differences in the total acidity or pH. The highest levels of ethanol were observed in the HM wines. These results agree with those obtained by other authors [53,54] and could be explained by two factors, the first of which is due to how the hot pre-fermentation maceration was carried out. Weak evaporation of water could have occurred during the pre-fermentative stage, which may have contributed to the small concentration of all the compounds of the must, particularly the sugars. Second, a higher level of amino acids has been reported in thermovinified musts [53]. This increase in amino acids could contribute to improving ethanol yields [55]. However, the residual sugar concentrations of the wines were not affected by the winemaking technique, whereas the volatile acidity was slightly lower.

The must composition x maceration technique factor expresses the average contents of ethanol, titratable acidity, pH, residual sugars, and volatile acidity of all the wines produced with the original must and traditional maceration (OM-TM), must replacement and traditional maceration (MR-TM), original must and pre-fermentative hot pre-fermentative maceration (OM-HM), or must replacement and hot pre-fermentative maceration (MR-HM), regardless of the vintage. The ethanol content of the OM-TM and OM-HM wines was significantly higher than that of the MR-TM and MR-HM wines, which evidenced significant differences due to the maceration techniques used. In contrast, the ethanol content of the MR-HM wine was significantly higher than that of the MR-TM wine, probably because of the maceration technique described previously. As expected, the MR-TM and MR-HM wines presented the highest values of titratable acidity and the lowest pH values in comparison with the OM-TM and OM-HM wines. When analyzing the combination of both winemaking techniques, changes in pH were observed, associated with the initial composition of the must and the maceration technique. In this sense, it has been reported that wines developed via hot pre-fermentative maceration have shown higher pH values, because, during the pre-fermentative heating, the extraction of cations increases, which results in a rise in the pH mainly given by the salification of tartaric acid [56]. Additionally, the wines produced with must replacement and/or hot pre-fermentative maceration showed the lowest concentrations of residual sugars and lower values of volatile acidity.

3.3. Spectrophotometrical Phenolic Composition and Related Parameters

The phenolic composition of the wines was different according to the vintage (Table 2). Wines produced in 2016 were characterized by the highest concentrations of total polyphenols, anthocyanins, and proanthocyanidins, whereas the wines produced in 2017 presented the lowest values. The concentrations of catechins in the wines produced in 2018 were significantly higher than those in the wines produced in other vintages. The concentration of anthocyanins did not significantly differ from the wines produced in 2016, whereas the concentrations of total polyphenols and proanthocyanidins were intermediate (Table 2). These results indicate that the ripening stage of the grapes strongly determined the wine composition. Fourment et al. [57] reported that for the conditions of Uruguay, the interannual climate variability strongly modifies the composition of the grape, especially in the concentration of secondary metabolites.

Table 2. Polyphenolic composition of the wines.

Factor Analyzed		Total Polyphenol (mg/L)	Anthocyanins (mg/L)	Catechins (mg/L)	Proanthocyanidins (mg/L)
Year of vintage (*)	2016	2479 ± 252 [a]	1052 ± 156 [a]	1769 ± 455 [b]	4172 ± 714 [a]
	2017	1624 ± 68 [c]	614 ± 68 [b]	1420 ± 58 [c]	2690 ± 60 [c]
	2018	2140 ± 43 [b]	1165 ± 43 [a]	1883 ± 86 [a]	3260 ± 80 [b]
Must composition (**)	OM	2045 ± 140 [a]	960 ± 67 [a]	1667 ± 239 [a]	3397 ± 372 [a]
	MR	2117 ± 102 [a]	994 ± 73 [a]	1714 ± 160 [a]	3352 ± 197 [a]
Maceration technique (***)	TM	1784 ± 112 [b]	838 ± 69 b	1281 ± 215 [b]	2764 ± 261 [b]
	HM	2379 ± 129 [a]	1117 ± 71 [a]	2100 ± 184 [a]	3985 ± 308 [a]
Must composition-Maceration techinque (****)	OM-TM	1821 ± 131 [c]	832 ± 69 [c]	1273 ± 268 [b]	2792 ± 352 [b]
	MR-TM	1747 ± 94 [d]	843 ± 69 [c]	1289 ± 161 [b]	2735 ± 170 [b]
	OM-HM	2345 ± 149 [b]	1088 ± 66 [b]	2061 ± 209 [a]	4001 ± 390 [a]
	MR-HM	2413 ± 109 [a]	1146 ± 77 [a]	2141 ± 159 [a]	3968 ± 225 [a]

(*) Average of 12 wines ± standard deviation regardless of the grape juice composition and the winemaking technique. (**) Average of the 18 wines ± standard deviation regardless of the year of vintage and the winemaking technique. (***) Average of 18 wines ± standard deviation regardless of the year of vintage and the grape juice composition. (****) Average of nine wines ± standard deviation regardless of the year of vintage. Different letters indicate statistical differences ($p < 0.05$). OM: original must; MR: must replacement; TM: traditional maceration; HM: hot maceration.

Total polyphenols, anthocyanins, catechins, and proanthocyanidins of the MR wines did not differ significantly from those of the OM wines. The techniques proposed by Role et al. [33] to reduce the alcohol content of the wines reduced the concentration of highly polymerized flavonols without substantially modifying the concentration of anthocyanins. According to these authors, the lower ethanol concentration could be the extraction of high polymerized flavanols from the grapes during fermentation. Moreover, they suggest that although lower concentration of anthocyanins would be expected, because a portion of must was eliminated, this does not necessarily imply anthocyanin losses, because the replacement was done before maceration. With ripe berries, however, these red pigments are more easily extracted from the skins during the crushing process and the short time of skin contact, and therefore, the fraction removed could contain a considerable amount of anthocyanin [58]. This was not observed in our results. Meanwhile, Kontoudakis et al. [32] found that anthocyanins remained almost unchanged when the ethanol concentration was reduced by 3.0% *v/v* by replacing a part of the total volume of the grape juice with the same volume of a low-ethanol wine. These authors reported that proanthocyanidin was less abundant in the reduced alcohol wines than in the control wines.

In contrast, total polyphenols, anthocyanins, catechins, and proanthocyanidins of the HM wines were significantly higher than those of the TM wines (Table 2). These results agree with previous studies [30,39,56,59] and confirm that this technique is useful to improve polyphenol extraction, because pre-fermentative heating contributes to degrading the tissues of the skins, releasing these compounds into the must.

When we analyzed the joint effect of the grape juice composition and the maceration technique, it was observed that the wines produced by hot pre-fermentative maceration presented the highest concentrations of the different phenolic families evaluated. In particular, the HM-OM wines presented lower contents of total polyphenols and anthocyanins than the HM-MR wines, whereas no significant differences were observed in the concentrations of catechins and proanthocyanidins given by the initial composition of the must. Similar results were observed between the OM and MR wines made by traditional maceration. These results indicate that the combination of must replacement and hot pre-fermentative maceration increased the concentration of anthocyanins in wines, whereas the concentration of catechins and proanthocyanidins was affected only by this winemaking technique, as was discussed previously.

Table 3 shows the effects of the vintage, must composition, maceration technique, and the combination of must composition–maceration technique on the ionization, copigmentation, and PVPP indices. The ionization index represents the percentage of anthocyanins colored given the standard pH and free SO_2 concentration of the wine [4], the copigmentation index represents the percentage of color due to the copigmentation process [4], and the PVPP index measures the percentage of anthocyanins combined with proanthocyanidins [48]. These indices were different according to the vintage. These results could be explained by the effects of ripening conditions on the concentration and the relationship between the phenolic compounds that subsequently interact in the wine. Thus, the highest indices of ionization and PVPP were recorded in the 2016 vintage together with the highest concentrations of total polyphenols, anthocyanins, and proanthocyanidins, whereas the lowest values of these indices were recorded in the 2017 harvest. In the 2018 harvest, the highest value of the copigmentation index was probably associated with a higher concentration of catechins, whereas in the 2016 harvest, it was the lowest value.

Table 3. Color fractions of the wines.

Factor Analyzed		Ionization Index (%)	Copigmentation Index (%)	PVPP Index (%)
	2016	33.9 ± 2.3 [a]	16.5 ± 3.7 [c]	45.2 ± 0.8 [a]
Year of vintage (*)	2017	15.7 ± 2.4 [c]	17.8 ± 4.2 [b]	35.9 ± 0.8 [c]
	2018	17.7 ± 0.6 [b]	31.7 ± 3.1 [a]	40.0 ± 1.2 [b]
Must composition (**)	OM	20.1 ± 1.8 [b]	20.9 ± 3.2 [b]	38.2 ± 0.9 [b]
	MR	24.8 ± 1.7 [a]	23.1 ± 4.0 [a]	42.4 ± 0.9 [a]
Maceration technique (***)	TM	18.0 ± 2.1 [b]	18.4 ± 3.4 [b]	35.9 ± 0.9 [b]
	HM	26.9 ± 1.4 [a]	26.6 ± 3.9 [a]	44.7 ± 0.9 [a]
	OM-TM	16.0 ± 2.6 [d]	15.7 ± 3.0 [c]	35.2 ± 0.9 [c]
Must composition -	MR-TM	20.0 ± 1.7 [c]	21.0 ± 3.9 [b]	36.6 ± 1.0 [c]
Maceration techinque (****)	OM-HM	24.2 ± 1.1 [b]	26.0 ± 3.4 [a]	41.3 ± 0.9 [b]
	MR-HM	29.6 ± 1.7 [a]	25.3 ± 4.3 [a]	48.2 ± 0.9 [a]

(*) Average of 12 wines ± standard deviation regardless of the grape juice composition and the winemaking technique. (**) Average of the 18 wines ± standard deviation regardless of the year of vintage and the winemaking technique. (***) Average of 18 wines ± standard deviation regardless of the year of vintage and the grape juice composition. (****) Average of nine wines ± standard deviation regardless of the year of vintage. Different letters indicate statistical differences ($p < 0.05$). OM: original must; MR: must replacement; TM: traditional maceration; HM: hot maceration.

Nevertheless, an effect of the must replacement treatments on the different indices was found. The MR wines presented higher ionization, copigmentation, and PVPP indices. The color of red wine is the result of the concentration of ionized free anthocyanins and the interactions between these and other components of the wine that produce new pigments [22]. During the winemaking, the new pigment produced when anthocyanins combine with tannins is much less sensitive to bleaching by pH and SO_2, so the percentage of coloring increases [12,27].

This effect and the result obtained in the pH (Table 1) of the wines could explain the differences registered in both indices. Further, the HM wines presented higher values of all these indices than the TM wines (Table 4). This effect could be determined by the increase in the concentrations of anthocyanins, catechins, and proanthocyanidins registered in the wines produced with hot pre-fermentative maceration, which could promote their interaction in the wine by increasing copigmentation and condensation between anthocyanins and tannins [24].

Table 4. Color of the wines.

Factor Analyzed		Color Intensity	Lightness (L*)	Chroma (C*)	Hue (h_{ab})
	2016	32.5 ± 1.4 [a]	31.5 ± 1.2 [b]	45.0 ± 1.0 [b]	348.1 ± 1.6 [a]
Year of vintage (*)	2017	16.0 ± 0.5 [c]	60.5 ± 1.5 [a]	28.1 ± 1.5 [c]	10.6 ± 1.3 [c]
	2018	24.2 ± 0.5 [b]	25.5 ± 0.9 [c]	53.1 ± 0.8 [a]	11.8 ± 0.5 [b]
Must composition (**)	OM	23.2 ± 0.9 [b]	40.2 ± 1.3 [a]	41.0 ± 1.2 [b]	3.27 ± 1.0 [a]
	MR	25.1 ± 0.8 [a]	37.9 ± 1.2 [b]	43.1 ± 1.4 [a]	3.74 ± 1.2 [a]
Maceration technique (***)	TM	20.4 ± 0.7 [b]	44.8 ± 1.2 [a]	41.4 ± 1.3 [b]	4.66 ± 0.8 [a]
	HM	27.9 ± 0.9 [a]	33.3 ± 1.3 [b]	42.7 ± 1.3 [a]	2.35 ± 1.4 [a]
	OM-TM	19.6 ± 1.0 [d]	45.9 ± 1.4 [a]	40.4 ± 1.0 [c]	5.11 ± 0.6 [a]
Must composition -	MR-TM	21.2 ± 0.4 [c]	43.8 ± 0.9 [b]	42.6 ± 1.7 [b]	4.21 ± 0.9 [a]
Maceration techinque (****)	OM-HM	26.8 ± 0.8 [b]	34.6 ± 1.2 [c]	41.6 ± 1.5 [bc]	1.43 ± 1.4 [c]
	MR-HM	29.0 ± 1.1 [a]	32.0 ± 1.4 [d]	43.7 ± 1.0 [a]	3.27 ± 1.5 [b]

(*) Average of 12 wines ± standard deviation regardless of the grape juice composition and the winemaking technique. (**) Average of the 18 wines ± standard deviation regardless of the year of vintage and the winemaking technique. (***) Average of 18 wines ± standard deviation regardless of the year of vintage and the grape juice composition. (****) Average of nine wines ± standard deviation regardless of the year of vintage. Different letters indicate statistical differences ($p < 0.05$). OM: original must; MR: must replacement; TM: traditional maceration; HM: hot maceration.

The wines presented significant differences in the evaluated indices given by the initial composition of the must and the winemaking technique with which they were developed. The OM-HM and RM-HM wines presented higher ionization, copigmentation, and PVPP indices than the OM-TM and RM-TM wines, but the highest values recorded were in the wines where pre-fermentative treatment was carried out on the must replacement. The anthocyanin, catechins, and proanthocyanidin contents of the HM wines were higher than those of the TM wines (Table 3). These results suggest that hot pre-fermentation maceration favors the reactions between anthocyanins and tannins, which suggests greater color stability over time, according to [60]. Moreover, when hot pre-fermentative maceration was carried out on the replaced grape juice, the values registered in the indices were substantially higher, suggesting that the combination of both techniques improves the stability of the wine color.

3.4. Wine Anthocyanin Composition

Figure 3a,b shows the average of the levels and profiles of the anthocyanin composition of the wines elaborated in the 2016, 2017, and 2018 vintages, according to treatment. As observed, total anthocyanin concentrations determined by HPLC-DAD were lower than the total anthocyanin concentrations measured by spectrophotometry. It should be considered that spectrophotometric analysis includes contributions from other pigments in the measurement and therefore overestimates the total anthocyanin concentration, whereas the HPLC-DAD analysis only detects free anthocyanins. In general, Tannat wines had a high non-acylated glucosides, delphinidin, and petunidin proportions and low acylated anthocyanin (acetylated and coumarylated) proportions, as has been previously reported [1,8].

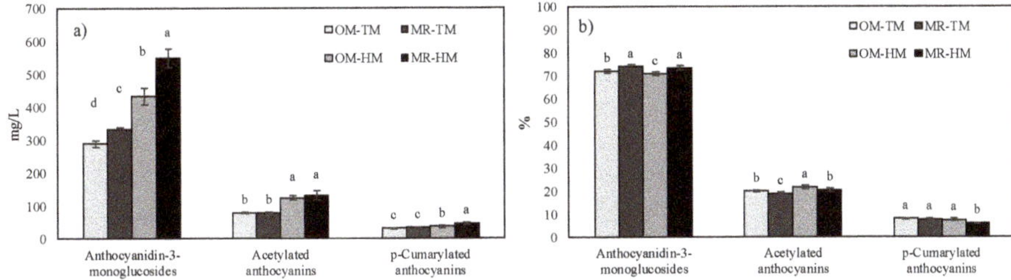

Figure 3. Concentration (**a**) and proportion (**b**) of anthocyanidin-3-monoglucosides, acetylated anthocyanins, and p-coumarylated anthocyanins. Average of nine wines ± standard deviation. Different letters indicate statistical differences ($p < 0.05$). OM-TM: original must and traditional maceration; MR-TM: must replacement and traditional maceration; OM-HM: original must and hot pre-fermentative maceration; MR-HM: must replacement and hot pre-fermentative maceration.

Figure 3a shows the effect of the treatments evaluated on the concentration of monoglucosylated, acetylated, and coumarylated anthocyanins. Both must replacement and the hot pre-fermentative maceration contributed to increase the concentrations of monoglucosylated and p-coumarylated anthocyanins compared with those of the wine produced by original must followed by a traditional maceration. Instead, the concentration of acetylated anthocyanins was differentiated between wines only by the maceration technique used. These results confirm those obtained through spectrophotometric analysis. The must replacement seemed to increase the concentration of monoglucosides, probably because these wines had a lower pH, whereas the hot pre-fermentation maceration seemed to generate an increase in the monoglucosylated, acetylated, and p-coumarylated anthocyanins concentration. However, when analyzing the proportion of different anthocyanins, we observed that the differences between treatments were attenuated (Figure 3b).

In general, it was observed that in the wines produced from must replacement the percentage of monoglucocylated anthocyanins was lower, and the percentage of acetylated anthocyanins was higher compared with the wines produced from the original grape must. In this sense, it could be said that there was a modification in the proportion of the different anthocyanin forms that was more affected by the must replacement than by the hot pre-fermentative maceration. In a previous investigation where must replacement and hot pre-fermentative maceration were evaluated on the composition of Pinot Noir and Tannat wines produced from the 2016 vintage, a differential behavior was observed according to the cultivar [35]. The monoglucosylated anthocyanin concentration of Pinot Noir wines with must replacement was significantly lower in relation to that of the control wines, especially when they were subjected to hot pre-fermentation maceration. This behavior was explained because the lower pH caused by the substitution of must could favor the formation of other pigments at high temperatures. However, in the Tannat wines, the changes in monoglucosylated, acetylated, and p-coumarylated anthocyanin concentrations caused by the must replacement and the hot pre-fermentation maceration were different. In general, no significant effect of the must substitution was observed on the concentration of these anthocyanins, but its concentration was increased when hot pre-fermentative maceration was carried out. The results obtained in this research help to clarify the effect of both winemaking techniques, where must replacement and hot pre-fermentation maceration increase the concentrations of monoglucosylated, acetylated, and p-coumarylated anthocyanins in Tannat wines without modifying their proportions.

The average concentration of the different anthocyanin forms and the anthocyanin profile of wines produced in the 2016, 2017, and 2018 vintages are shown in Figure 4a,b, respectively. As can be observed, the concentrations of the different anthocyanin forms of the wines were increased by the must replacement and the hot pre-fermentative maceration with the sole exception of petunidin-3-glucoside, whose concentration in the MR-TM wines did not differ from that in the OM-TM wines.

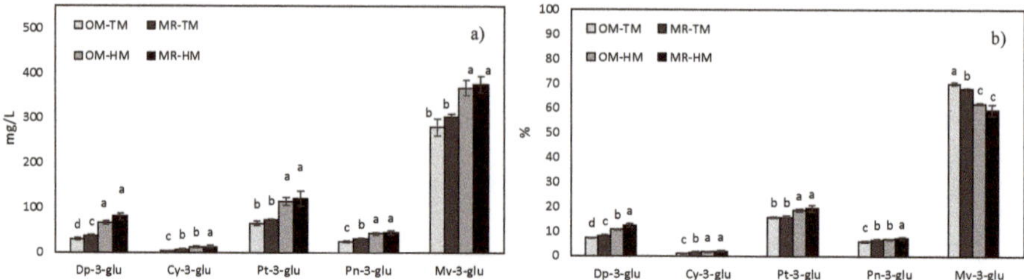

Figure 4. Concentration (**a**) and proportion (**b**) of different anthocyanidin forms. Average of nine wines ± standard deviation. Different letters indicate statistical differences ($p < 0.05$). OM-TM: original must and traditional maceration; MR-TM: must replacement and traditional maceration; OM-HM: original must and hot pre-fermentative maceration; MR-HM: must replacement and hot pre-fermentative maceration.

Wines produced by the combination of both techniques presented the highest concentrations of all anthocyanin forms independent of the composition of the must. It is known that pH and the ethanol content of the medium are factors that contribute to the extraction of the phenolic compounds during the fermentative maceration [24]. As seen in Figure 4b, the anthocyanin profile of the wines was modified by the winemaking techniques used. In general, must replacement and hot pre-fermentative maceration increased the percentages of delphinidin-3-glucoside, cyanidin-3-glucoside, petunidin-3-glucoside, and peonidin-3-glucoside. In particular, the winemaking in which hot pre-fermentation maceration was carried out presented the highest values. In contrast, the percentage of malvidin-3-glucoside was lower in the OM-HM and MR-HM wines than in the MR-TM and OM-TM wines. As previously discussed, hot pre-fermentative maceration allows greater extraction of the anthocyanins by degrading the cellular structures of the skins [34]. The effect of hot pre-fermentation maceration was also observed in the anthocyanin profile of the wines where, in the three vintages, the HM wines had higher percentages of delphinidin, petunidin, and peonidin and a significantly lower percentage of malvidin than the TM wines. At this point, the results obtained in our research are contradictory, because it was shown that wines produced by hot pre-fermentation maceration had a higher percentage of less stable anthocyanidins and a lower percentage of the more stable anthocyanidins. As is known, malvidin is more resistant to thermal degradation than other anthocyanin forms [39], so the idea that hot pre-fermentation maceration affects malvidin more than the other anthocyanidins does not seem to be the correct explanation. On the other hand, it has been shown that pre-fermentative heating above 60 °C degrades polyphenoloxidases enzymes, which are responsible for the oxidation of phenolic compounds in the early stages of winemaking [24,59]. Because the adjacent hydroxyl groups of o-diphenols are sensitive to oxidation, the malvidin-3-*O*-glucoside and peonidin-3-*O*-glucoside that do not possess ortho-positioned hydroxyl groups are comparatively more resistant to oxidation than cyanidin-3-*O*-glucoside [22]. Therefore, it could be thought that the increase in the proportions of petunidin, delphinidin, and cyanidin occurred, because these forms were preserved from enzymatic oxidation during winemaking by hot pre-fermentation maceration.

3.5. Wine Color

Table 4 shows the chromatic parameters of the wines produced. The wines produced from the 2016 vintage were characterized by having the highest coloring intensity and the greatest hue, whereas those produced from the 2017 harvest presented the highest lightness and the lowest speed of coloring intensity, chroma, and hue. The wines produced during the 2018 vintage presented the highest chroma value with intermediate values of coloring intensity and hue. In general, the MR wines had a deeper red color, because the color intensity, chroma, and hue were significant higher and the lightness was significant lower than that of the OM wines, while the HM wines also had a deeper color than the TM wines due to the fact that the color intensity and the chroma were significantly higher and the lightness was significantly lower in the HM wines. No significant differences were observed due to the hot pre-fermentative maceration.

When analyzing the effect of the combination of the initial must composition and the maceration technique, it was observed that the MR-HM wines presented the highest intensity of color and chroma and the lowest lightness, whereas the OM-TM wines presented the lowest values. Meanwhile, the OM-HM wines presented a lower value of hue, which suggests that these wines are more bluish. For the other chromatic parameters, the RM-HM and OM-TM wines presented intermediate values.

The differences in the chromatic parameters of the wines were associated with the differences in the concentrations of phenolic compounds found, in particular, those of the anthocyanins; the pH of the wine and the percentage of ionized, copigmented, and polymerized anthocyanins were also different among the wines produced in different vintages and from different treatments, as was previously discussed. Furthermore, hot pre-fermentative maceration increasing the extraction of anthocyanins explains the differences in the color parameters. Other authors have previously described similar results [31]. Moreover, in this sense, the increase in the extraction of anthocyanins from the first stages of the maceration and the increase in the extraction of tannins allowed a greater association of these molecules, which has been reported as a determining factor to improve the color stabilization [12]. The results obtained in this investigation in the ionization, copigmentation, and PVPP indices support this theory.

While it is true that in a sensory evaluation, the chromatic characteristics of these wines can be challenging to differentiate, even for a panel of experts, it must be considered that the wines were evaluated two months after bottling. As is known, the color of the wine evolves during conservation, decreasing its coloring intensity and increasing its angle. The results obtained in this research suggest that wines made by both winemaking techniques could have a more stable color over time and, consequently, a greater potential for aging.

3.6. Multifactorial Analysis of Variance

Multifactorial analysis of the variance shows the effect of each factor and its interaction on the different components of the wines (Table 5). In general, it was verified that the year of vintage (Y), the composition of the grape must (M), the maceration techniques (T), and their interactions (YxM, YxT, MxT, YxMxT) influenced differently the color and the concentration of the phenolic composition of the wine.

The results obtained in the ethanol content, pH, and titratable acidity of the wines seem logical, because the initial composition of the grape must (concentration of sugars, pH, and titratable acidity) at harvest was very different in the vintages due to the climatic conditions of maturation. In this sense, in the treatments where a must replacement for immature grape must was produced, the initial composition of the must, and therefore the wine, was also affected. Moreover, the maceration technique strongly influenced the ethanol content and the pH of the wines. The results obtained regarding the concentration of residual sugars and the volatile acid content of the wines corresponded to the initial composition of the grape and the conditions in which the alcoholic fermentation took place. The vintage and the maceration technique strongly influenced all the phenolic compounds and the ionization, copigmentation, and PVPP indices. Several authors have shown that the phenolic composition of a grape and a wine is determined by the maturation conditions of each year in particular [15]. Moreover, hot pre-fermentative maceration strongly degrades the cellular structures of the skins, extracting their content toward the grape juice and favoring the interaction between them, as mentioned above. The composition of the grape must influences significantly the concentrations of total polyphenols and anthocyanins and the ionization, copigmentation, and PVPP indices.

Table 5. Multifactorial analysis of variance.

	Year of Vintage (Y)	Must Composition (M)	Vinification Technique (V)	Y × M	Y × V	M × V	Y × M × V
Ethanol	5152.9 ***	939.6 ***	137.5 ***	61.8 ***	52.5 ***	131.1 ***	120.2 ***
Titratable acidity	185.5 ***	38.8 ***	2.93 *	25.1 ***	6.9 ***	3.3 *	3.9 **
pH	10.6 ***	101.0 ***	18.3 ***	41.9 ***	10.8 ***	0.4	21.9 ***
Reducing sugars	80.9 ***	9.1 **	2.2	8.0 **	4.6 **	43.4 ***	9.4 ***
Volatile acidity	21.5 ***	11.7 ***	193.1 ***	7.2 **	10.1 ***	0.2	17.8 ***
Total polyphenols	574.8 ***	11.7 ***	824.7 ***	11.9 ***	25.8 ***	0.1	2.9
Anthocyanins	1232.6 ***	10.8 ***	728.2 ***	14.4 ***	89.5 ***	5.11 **	10.1 ***
Catechins	92.4 ***	2.7	800.3 ***	9.5 ***	12.0 ***	1.2	3.0 *
Proanthocyanidins	193.6 ***	0.5	387.0 ***	2.2	0.2	0.1	0.4
Ionization index	248.7 ***	41.6 ***	149.9 ***	4.1 **	28.2 ***	0.9	3.6 **
Copigmentation index	690.4 ***	36.8 ***	385.1 ***	3.8 **	12.9 ***	66.1 ***	3.6 **
PVPP index	15.4 ***	9.33 ***	41.4 ***	1.6	28.8 ***	4.0 *	6.6 ***
Color intensity	1526.4 ***	60.9 ***	966.5 ***	7.6 *	29.5 ***	2.6	2.7
Lightness (L*)	5272.7 ***	65.0 ***	1519.0 ***	10.8 ***	9.6 ***	0.8	5.5 ***
Chroma (C*)	1180.2 ***	25.3 ***	8.0 ***	6.4 ***	94.7 ***	0.1	5.8 ***
Hue (h_{ab})	2160.5 ***	2.0	48.3 ***	5.8 ***	37.0 ***	17.0 ***	7.4 ***
Anthocyanidin-3-monoglucosides	566.1 ***	25.3 ***	364.3 ***	15.5 ***	53.6 ***	1.4	4.6 **
Acetylated anthocyanins	138.1 ***	1.2	231.5 ***	1.2	25.1 ***	1.3	0.1
p-Coumarylated anthocyanins	439.0 ***	16.3 ***	91.2 ***	22.7 ***	1.7	41.9 ***	3.8 **
Delphinidin-3-glucoside	219.3 ***	31.1 ***	531.8 ***	4.9 *	85.2 ***	3.6 *	11.3 ***
Cyanidin-3-glucoside	58.1 ***	2.4	71.2 ***	1.5	20.0 ***	3.7 *	2.6 *
Petunidin-3-glucoside	117.1 ***	3.3 *	158.0 ***	4.9 **	16.6 ***	0.1	0.6
Peonidin-3-glucoside	208.5 ***	12.9 ***	209.2 ***	0.5	12.3 ***	3.2 *	1.6
Malvidin-3-glucoside	623.8 ***	1.6	207.5 ***	8.4 ***	33.6 ***	7.7 ***	1.6

F values and statistical significance ($p < 0.001$ = ***; $p < 0.01$ = **; $p < 0.1$ = *). OM: original must; MR: must replacement; TM: traditional maceration; HM: hot maceration.

As discussed above, the ethanol content and pH are factors that contribute to the extraction during fermentative maceration, but this effect was only observed in the concentrations of total polyphenols and anthocyanins. A strong interaction between YxT was detected for the phenolic compounds and the indices analyzed, except for the concentration of proanthocyanidins, which was not significant. The YxM interaction was not significant for the concentration of proanthocyanidins or for the PVPP index, whereas the MxT interaction was highly significant only for the anthocyanin concentration and the copigmentation index. The year of harvest and the technique of maceration strongly influenced the concentrations of the different anthocyanin forms, while the initial composition of the grape must only affected the concentrations of monoglucosylated anthocyanins, p-coumarylated, delphinidin-3-glucoside, and petunidin-3-glucoside. Again, a strong interaction was detected in the anthocyanin composition of the wines between the harvest year and the maceration technique (YxT), while the other interactions were significant in the concentrations of some anthocyanin forms.

As discussed earlier, the color of red wine results from the concentration of anthocyanins, their interactions with other phenolic compounds or metabolites of alcoholic fermentation, and the physical–chemical conditions of the medium in which these pigments are found. Therefore, any modification of these factors determines a change in the wine color. The year of vintage, the composition of the grape must, and the maceration technique had a strong impact on all the color parameters, with the only exception being the effect of the composition of the grape must on the hue (h_{ab}), which showed a lower significance. All the interactions were significant with respect to the chromatic parameters, except for the MxT interaction, which was only significant for the hue of the wine.

4. Conclusions

The must replacement of mature grape juice for immature grape juice and hot pre-fermentative maceration are technological alternatives to improve the color of Tannat red wines.

The effect of MR on the color and the general composition of wines is highly dependent on the composition of the grape. In contrast, HM improved the intensity and quality of the wine color by increasing the extraction of phenolic compounds and promoting condensation between anthocyanins and tannins, suggesting greater color stability. The results obtained in our research are relevant, because this winemaking technique allows us to mitigate the limitations in the extractability of anthocyanins presented by the Tannat cultivar. Moreover, this winemaking technique modified the anthocyanin profile of the wines in which a relative increase of the most oxidizable forms was obtained. Further studies should be focused on determining the effect of pre-fermentation heating on the degradation of oxidation enzymes and how that influences the phenolic profile of wines.

Author Contributions: Investigation, D.P., Conceptualization, D.P., J.M.C., F.Z. and G.G.-N., Methodology, D.P., G.F., O.P., F.Z. and G.G.-N., Supervision, F.Z. and G.G.-N., Writing—original draft, D.P. and G.G.-N., Writing—review & editing, D.P., G.F., O.P., J.M.C., F.Z. and G.G.-N.

Funding: This research was funded by CAP (Comisión Académica de Posgrado of the Universidad de la República), ANII (Agencia Nacional de Investigación e Innovación of Uruguay, scholarship MOV_CA_2015_1_107599), and CSIC (Comisión Sectorial de Investigación Científica of the Universidad de la República, scholarships Movilidad 2017 and 2018).

Acknowledgments: All authors are grated to INAVI (Instituto Nacional de Vitivinicultura of Uruguay) for the technical support and to Establecimiento Juanicó and Bodega Olga Silva for the grapes used in the experiments.

Conflicts of Interest: The authors declare no conflicts of interest.

References

1. González-Neves, G.; Favre, G.; Gil, G. Effect of fining on the colour and pigment composition of young red wines. *Food. Chem.* **2014**, *157*, 385–392. [CrossRef] [PubMed]
2. Österbauer, R.A.; Matthews, P.M.; Jenkinson, M.; Beckmann, C.F.; Hansen, P.C.; Calvert, G.A. Color of scents: Chromatic stimuli modulate odor responses in the human brain. *J. Neurophysiol.* **2005**, *93*, 3434–3441. [CrossRef] [PubMed]
3. Alcalde-Eon, C.; Escribano-Bailón, M.; Santos-Buelga, C.; Rivas-Gonzalo, J. Changes in the detailed pigment composition of red wine during maturity and ageing. A comprehensive study. *Anal. Chim. Acta.* **2006**, *563*, 238–254. [CrossRef]
4. Boulton, R. The copigmentation of anthocyanins and its role in the color of red wine: A critical review. *Am. J. Enol. Vitic.* **2001**, *52*, 67–87.
5. Llaudy, M.C.; Canals, R.; Canals, J.M.; Zamora, F. Influence of ripening stage and maceration length on the contribution of grape skins, seeds and stems to phenolic composition and astringency in wine-simulated macerations. *Eur. Food Res. Technol.* **2008**, *226*, 337–344. [CrossRef]
6. Ortega-Regules, A.; Romero-Cascales, I.; Ros-García, J.M.; López-Roca, J.M.; Gómez-Plaza, E. A first approach towards the relationship between grape skin cell-wall composition and anthocyanin extractability. *Anal. Chim. Acta* **2006**, *563*, 26–32. [CrossRef]
7. Mattivi, F.; Vrhovsek, U.; Masuero, D.; Trainotti, D. Differences in the amount and structure of extractable skin and seed tannins amongst red grape varieties. *Aust. J. Grape Wine Res.* **2009**, *15*, 27–35. [CrossRef]
8. González-Neves, G.; Franco, J.; Barreiro, L.; Gil, G.; Moutounet, M.; Carbonneau, A. Varietal differentiation of Tannat, Cabernet-Sauvignon and Merlot grapes and wines according to their anthocyanin composition. *Eur. Food Res. Technol.* **2007**, *225*, 111–117. [CrossRef]
9. Mori, K.; Goto-Yamamoto, N.; Kitayama, M.; Hashizume, K. Loss of anthocyanins in red-wine grape under high temperature. *J. Exp. Botany* **2007**, *58*, 1935–1945. [CrossRef]
10. Rolle, L.; Gerbi, V.; Schneider, A.; Spanna, F.; Río Segade, S. Varietal Relationship between Instrumental Skin Hardness and Climate for Grapevines (*Vitis vinifera* L.). *J. Agric. Food Chem.* **2011**, *59*, 10624–10634. [CrossRef]
11. Sacchi, K.; Bisson, L.; Adams, D. A review of the effect of winemaking techniques on phenolic extraction in red wines. *Am. J. Enol. Vitic.* **2005**, *56*, 197–206.
12. Cheynier, V.; Dueñas-Paton, M.; Souquet, J.M.; Sarni-Manchado, P.; Fulcrand, H. Structure and properties of wine pigments and tannins. *Am. J. Enol. Vitic.* **2006**, *57*, 298–305.
13. González-Neves, G.; Favre, G.; Piccardo, D.; Gil, G. Anthocyanin profile of young red wines of Tannat, Syrah and Merlot made using maceration enzymes and cold soak. *Int. J. Food Sci. Technol.* **2016**, *51*, 260–267. [CrossRef]

14. Castellarin, S.; Matthews, M.; Di Gaspero, G.; Gambetta, G. Water deficits accelerate ripening and induce changes in gene expression regulating flavonoid biosynthesis in grape berries. *Planta* **2007**, *227*, 101–112. [CrossRef] [PubMed]

15. González-Neves, G.; Gil, G.; Favre, G.; Ferrer, M. Influence of grape composition and winemaking on the anthocyanin composition of red wines of Tannat. *Int. J. Food Sci. Technol.* **2012**, *47*, 900–909. [CrossRef]

16. Cerpa-Calderón, F.; Kennedy, J. Effect of berry crushing on skin and seed tannin extraction during fermentation and maceration. *Am. J. Enol. Vitic.* **2008**, *59*, 350.

17. Monagas, M.; Bartolomé, B.; Gómez-Cordovés, C. Updated knowledge about the presence of phenolic compounds in wine. *Crit. Rev. Food Sci.* **2005**, *45*, 85–118. [CrossRef] [PubMed]

18. Pinelo, M.; Arnous, A.; Meyer, A.S. Upgrading of grape skins: Significance of plant cell-wall structural components and extraction techniques for phenol release. *Trends Food Sci. Technol.* **2006**, *17*, 579–590. [CrossRef]

19. Rolle, L.; Torchio, F.; Lorrain-Lorette, B.; Giacosa, S.; Río-Segade, S.; Cagnasso, E.; Gerbi, V.; Teissedre, P.L. Rapid methods for the evaluation of total phenol content and extractability in intact grape seeds of cabernet sauvignon: Instrumental mechanical properties and FT-NIR spectrum. *J. Int. Sci. Vigne Vin.* **2012**, *46*, 29–40. [CrossRef]

20. Bautista-Ortín, A.; Busse-Valverde, N.; Fernández-Fernández, J.; Gómez-Plaza, E.; Gil-Muñoz, R. The extraction kinetics of anthocyanins and proanthocyanidins from grape to wine in three different varieties. *J. Int. Sci. Vigne Vin.* **2016**, *50*, 91–100. [CrossRef]

21. Amrani-Joutei, K.; Glories, Y. Tanins et anthocyanes: Localization dans la baie de raisin et mode d'extraction. *Rev. Franç. Oenol.* **1995**, *153*, 28–31.

22. He, F.; Liang, N.N.; Mu, L.; Pan, Q.H.; Wang, J.; Reeves, M.J.; Duan, C.Q. Anthocyanins and Their Variation in Red Wines II. Anthocyanin derived pigments and their color evolution. *Molecules* **2012**, *17*, 1483–1519. [CrossRef]

23. Fulcrand, H.; Dueñas, M.; Salas, E.; Cheynier, V. Phenolic reactions during winemaking and aging. *Am. J. Enol. Vitic.* **2006**, *57*, 289–297.

24. Cheynier, V.; Souquet, J.; Kontek, A.; Moutounet, M. Anthocyanin degradation in oxidizing grape musts. *J. Sci. Food Agric.* **1994**, *66*, 283–288. [CrossRef]

25. Dallas, C.; Ricardo-Da-Silva, J.; Laureano, O. Degradation of oligomeric procyanidins and anthocyanins in a Tinta Roriz red wine during maturation. *Vitis* **1995**, *34*, 51–56.

26. Fulcrand, H.; Benabdeljalil, C.; Rigaud, J.; Cheynier, V.; Moutounet, M. A new class of wine pigments generated by reaction between pyruvic acid and grape anthocyanins. *Phytochemistry* **1998**, *47*, 1401–1407. [CrossRef]

27. McRae, J.; Kennedy, J. Wine and Grape Tannin Interactions with Salivary Proteins and Their Impact on Astringency: A Review of Current Research. *Molecules* **2011**, *16*, 2348–2364. [CrossRef] [PubMed]

28. Es-Safi, N.; Fulcrand, H.; Cheynier, V.; Moutounet, M. Studies on the acetaldehyde-induced condensation of (-)-epicatechin and malvidin 3-*O*-glucoside in a model solution system. *J. Agric. Food Chem.* **1999**, *47*, 2096–2102. [CrossRef] [PubMed]

29. Morata, A.; Gómez-Cordovés, M.C.; Suberviola, J.; Bartolomé, B.; Colombo, B.; Suárez-Lepe, J. Adsorption of anthocyanins by yeast cell walls during the fermentation of red wines. *J. Agric. Food Chem.* **2003**, *51*, 4084–4088. [CrossRef]

30. Geffroy, O.; Lopez, R.; Feilhes, C.; Violleau, F.; Kleiber, D.; Favarel, J.; Ferreira, V. Modulating analytical characteristics of thermovinified Carignan musts and the volatile composition of the resulting wines through the heating temperature. *Food Chem.* **2018**, *257*, 7–14. [CrossRef] [PubMed]

31. El Darra, N.; Grimi, N.; Maroun, R.; Louka, N.; Vorobiev, E. Pulsed electric field, ultrasound, and thermal pretreatments for better phenolic extraction during red fermentation. *Eur. Food Res. Technol.* **2013**, *236*, 47–56. [CrossRef]

32. Kontoudakis, N.; Esteruelas, M.; Fort, F.; Canals, J.; Zamora, F. Use of unripe grapes harvested during cluster thinning as a method for reducing alcohol content and pH of wine. *Aust. J. Grape Wine Res.* **2011**, *17*, 203–208. [CrossRef]

33. Rolle, L.; Englezos, V.; Torchio, F.; Cravero, F.; Río Segade, S.; Rantsiou, K.; Giacosa, S.; Gambuti, A.; Gerbi, V.; Cocolin, L. Alcohol reduction in red wines by technological and microbiological approaches: A comparative study. *Aust. J. Grape Wine Res.* **2017**, *24*, 1–13. [CrossRef]

34. Piccardo, D.; Gombau, J.; Pascual, O.; Vignault, A.; Pons, P.; Canals, J.M.; González-Neves, G.; Zamora, F. Influence of two prefermentative treatments to reduce the ethanol content and pH of red wines obtained from overripe grapes. *Vitis* **2019**. In Press.

35. Piccardo, D.; Favre, G.; Pascual, O.; Canals, J.M.; Zamora, F.; González-Neves, G. Influence of the use of unripe grapes to reduce ethanol content and pH on the color, polyphenol and polysaccharide composition of conventional and hot macerated Pinot Noir and Tannat wines. *Eur. Food Res. Technol.* **2019**, *245*, 1321–1335. [CrossRef]

36. Sadras, V.; Petrie, P.; Moran, M. Effects of elevated temperature in grapevine. II Juice pH, titratable acidity and wine sensory attributes. *Aust. J. Grape Wine Res.* **2013**, *19*, 107–115. [CrossRef]

37. Jackson, D.I.; Lombard, P.B. Environmental and management practices affecting grape composition and wine quality: A review. *Am. J. Enol. Vitic.* **1993**, *44*, 409–430.

38. Nicholas, K.A.; Matthews, M.A.; Lobell, D.B.; Willits, N.H.; Field, C.B. Effect of vineyard-scale climate variability on Pinot noir phenolic composition. *Agric. For. Meteorol.* **2011**, *151*, 1556–1567. [CrossRef]

39. Atanackovic, M.; Petrovic, A.; Jovic, S.; Gojkovic-Bukarica, L. Influence of winemaking techniques on the resveratrol content, total phenolic content and antioxidant potential of red wines. *Food Chem.* **2012**, *131*, 513–518. [CrossRef]

40. Vitis International Variety Catalogue. 2018. Available online: http://www.vivc.de/OIV (accessed on 1 September 2019).

41. Organización Internationale de la Vigne et du Vin. *International Oenological Codex*; OIV: Paris, France, 2018; 772p.

42. Glories, Y. La couleur des vins rouges. 2e. Partie: Mesure, origine et interpretation. *Conn. Vigne Vin.* **1984**, *18*, 253–271. [CrossRef]

43. Ayala, F.; Echávarri, J.F.; Negueruela, A.I. A new simplified method for measuring the color of wines. I. Red and Rosé Wines. *Am. J. Enol. Vitic.* **1997**, *48*, 357–363.

44. Ayala, F.; Echávarri, J.F.; Negueruela, A.I. MSCVes.zip. 2001. Available online: http://www.unizar.es/negueruela/MSCV.es (accessed on 1 September 2019).

45. González-Neves, G.; Charamelo, D.; Balado, J.; Barreiro, L.; Bochicchio, R.; Gatto, G.; Gil, G.; Tessore, A.; Carbonneau, A.; Moutounet, M. Phenolic potential of Tannat, Cabernet-Sauvignon and Merlot grapes and their correspondence with wine composition. *Anal. Chim. Acta.* **2004**, *513*, 191–196. [CrossRef]

46. Singleton, V.; Rossi, J. Colorimetry of total phenolics with phosphomolybdic and phosphotungstic acid reagents. *Am. J. Enol. Vitic.* **1965**, *16*, 144–158.

47. Ribéreau-Gayon, P.; Stonestreet, E. Le dosage des anthocyanes dans les vins rouges. *Bull. Soc. Chim.* **1965**, *9*, 2649–2653.

48. Swain, T.; Hillis, W. The phenolic constituents of *Prunus domestica*: I. The quantitative analysis of phenolic constituents. *J. Sci. Food Agric.* **1959**, *10*, 63–68. [CrossRef]

49. Ribéreau-Gayon, P.; Stonestreet, E. Dosage des tanins dans du vin rouge et détermination de leur structure. *Chim. Anal.* **1966**, *48*, 188–196.

50. Valls, J. Composició Fenòlica en Varietats Negres de *Vitis vinifera*. Influència de Diferents Factors. Ph.D. Thesis, Universitat Rovira i Virgili, Tarragona, Spain, 2004.

51. Di Rienzo, J.A.; Casanoves, F.; Balzarini, M.G.; Gonzalez, L.; Tablada, M.; Robledo, C.W. InfoStat Versión 2015. Grupo InfoStat, FCA, Universidad Nacional de Córdoba, Argentina. Available online: http://www.infostat.com.ar (accessed on 1 September 2019).

52. Ribéreau-Gayon, P.; Glories, Y.; Maujean, A.; Dubourdieu, D. *Tratado de Enología: Química del Vino Estabilización y Tratamientos*, 1st ed.; Hemisferio Sur: Buenos Aires, Argentina, 2003; p. 537.

53. Geffroy, O.; Lopez, R.; Serrano, E.; Dufourcq, T.; Gracia-Moreno, E.; Cacho, J.; Ferreira, V. Changes in analytical and volatile compositions of red wines induced by pre-fermentation heat treatment of grapes. *Food Chem.* **2015**, *187*, 243–253. [CrossRef]

54. Lukic', I.; Budic'-Leto, I.; Bubola, M.; Damijanic', K.; Staver, M. Pre-fermentative cold maceration, saignée, and various thermal treatments as options for modulating volatile aroma and phenol profiles of red wine. *Food Chem.* **2017**, *224*, 251–261. [CrossRef]

55. Albers, E.; Larsson, C.; Lidén, G.; Niklasson, C.; Gustafsson, L. Influence of the nitrogen source on *Saccharomyces cerevisiae* anaerobic growth and product formation. *Appl. Environ. Microbiol.* **1996**, *62*, 3187–3195.

56. Baiano, A.; Terracone, G.; Gambacorta, G.; La Notte, E. Phenolic content and antioxidant activity of Primitivo wine: Comparison among winemaking technologies. *J. Food Sci.* **2009**, *74*, 258–267. [CrossRef]

57. Fourment, M.; Ferrer, M.; González-Neves, G.; Barbeau, G.; Bonnardot, V.; Quénol, H. Tannat grape composition responses to spatial variability of temperature in a Uruguay's coastal wine region. *Int. J. Biomet.* **2017**, *61*, 1617–1628. [CrossRef] [PubMed]

58. Canals, R.; Llaudy, M.C.; Valls, J.; Canals, J.M.; Zamora, F. Influence of ethanol concentration on the extraction of color and phenolic compounds from the skin and seeds of Tempranillo grapes at different stages of ripening. *J. Agric. Food Chem.* **2005**, *53*, 4019–4025. [CrossRef] [PubMed]

59. Andrade Neves, N.; Pantoja, L.; Soares dos Santos, A. Thermovinification of grapes from the Cabernet sauvignon and Pinot noir varieties using immobilized yeasts. *Eur. Food Res. Technol.* **2014**, *238*, 79–84. [CrossRef]

60. Casassa, L.F.; Beaver, C.W.; Mireles, M.S.; Harbertson, J.F. Effect of extended maceration and ethanol concentration on the extraction and evolution of phenolics, color components and sensory attributes of Merlot wines. *Aust. J. Grape Wine Res.* **2013**, *19*, 25–39. [CrossRef]

 *fermentation*

Article

Optimized pH and Its Control Strategy Lead to Enhanced Itaconic Acid Fermentation by *Aspergillus terreus* on Glucose Substrate

Péter Komáromy, Péter Bakonyi *, Adrienn Kucska, Gábor Tóth, László Gubicza, Katalin Bélafi-Bakó and Nándor Nemestóthy

Research Institute on Bioengineering, Membrane Technology and Energetics, University of Pannonia, Egyetem ut 10, 8200 Veszprém, Hungary; komaromy.peter@mk.uni-pannon.hu (P.K.); kucskaadrienn@yahoo.com (A.K.); tothgabor2@gmail.com (G.T.); gubiczal@almos.uni-pannon.hu (L.G.); bako@almos.uni-pannon.hu (K.B.-B.); nemesn@almos.uni-pannon.hu (N.N.)
* Correspondence: bakonyip@almos.uni-pannon.hu or bakonyipeter85@gmail.com; Tel.: +36-88-624385

Received: 11 March 2019; Accepted: 7 April 2019; Published: 8 April 2019

Abstract: Biological itaconic acid production can by catalyzed by *Aspergillus terreus* (a filamentous fungi) where the fermentation medium pH is of prominent importance. Therefore, in this work, we investigated what benefits the different pH regulation options might offer in enhancing the process. The batch itaconic acid fermentation data underwent a kinetic analysis and the pH control alternatives were ranked subsequently. It would appear that the pH-shift strategy (initial adjustment of pH to 3 and its maintenance at 2.5 after 48 h) resulted in the most attractive fermentation pattern and could hence be recommended to achieve itaconic acid production with an improved performance using *A. terreus* from carbohydrate, such as glucose. Under this condition, the itaconic acid titer potential, the maximal itaconic acid (titer) production rate, the length of lag-phase and itaconic acid yield were 87.32 g/L, 0.22 g/L/h, 56.04 h and 0.35 g/g glucose, respectively.

Keywords: itaconic acid; *A. terreus*; pH control; glucose; kinetic analysis; Gompertz-model

1. Introduction

Microbial fermentation has been demonstrated as an efficient technology to produce a variety of organic acids such as malic acid, succinic acid, propionic acid, itaconic acid, etc. [1–4]. The latter, itaconic acid, is taken into account as an important compound since it can serve as a platform molecule for the synthesis of industrially-relevant chemicals, such as plastics, etc. [5,6]. Nowadays, itaconic acid is mainly generated through biological pathways by the assistance of filamentous fungi, particularly *Aspergillus terreus* [7]. The fermentation of itaconic acid can be carried out on numerous feedstocks, including complex agro-industrial wastes such as lignocelluloses as well as simple (monomeric) sugars, e.g., glucose [8]. Certainly, the properties of the actual starting material will influence the achievable itaconic acid formation efficiency [9–11], and besides that, process control via the maintenance of adequate environmental conditions will play a key role. As a matter of fact, ensuring suitable aeration, broth composition, mixing, temperature and pH are crucial criteria for the improved formation of itaconic acid by *A. terreus* [12–15].

For the recovery of itaconic acid from the fermentation liquor, membrane electrodialysis (MED) has been proven to be a plausible solution [16–18]. Furthermore, MED (depending on the trait of the membrane) was found to be an approach that provides additional process benefits, such as in the case of citric acid downstreaming [19]. It was concluded by scientists such as Tongwen and Waihua [20], as well as Pinacci and Radaelli [21], studying the separation of fermentatively-generated citric acid, that MED equipped with a bipolar membrane enabled the production of caustic soda. This chemical,

NaOH, can be recycled to the bioreactor unit in order to adjust and keep the pH at the desired level during the fermentation [12].

From the viewpoint of itaconic acid biosynthesis catalyzed by *A. terreus* strains, the pH is usually set to the acidic range. However, the appropriate pH adjustment strategy leading to a better fermentation efficiency could be worthy for investigation, since in the literature there is no clear justification of whether pH should be controlled or not and what pH value is the most appropriate. Actually, various studies have come to different conclusions regarding this aspect [8,12,22], implying the need for further (case-specific) examination. Therefore, in this work, we aimed to study how different acidic pH values (2.5–4) and the pH regulation strategy (only initial pH setting vs. continuous pH control; one-step vs. stepwise pH setting) might make any difference in governing the itaconic acid fermentation towards a higher efficacy.

In this respect, analogously to the case of citric acid [20,21], the NaOH—obtainable by a bipolar MED process [23]—may be employed for the regulation of pH during itaconic acid production. To evaluate and rank the various pH regulation scenarios, the progress curves of batch itaconic acid fermentations by *A. terreus* on a glucose substrate were kinetically analyzed to deliver process performance indicators (lag-phase time, itaconic acid production rate and titer potential).

The importance of this work can be explained by the inconsistencies regarding the effect of the pH on itaconic acid production by *A. terreus* and, hence, the findings could demonstrate an added-value to this segment of the literature.

2. Materials and Methods

2.1. Microbial Catalyst

A. terreus NRRL 1960 strain [13] was employed for itaconic acid fermentation in this study. The fungus was sustained on Petri-dishes at 37 °C using a solid medium comprising of (g/L): glucose–10; NaCl–20, potato dextrose agar–40; pH = 5. For the inoculation of the fermenter (Section 2.2.), liquid cultures of *A. terreus* (grown under pH = 3 and a 150 rpm agitation rate on (g/L): glucose–10, KH_2PO_4–0.1, NH_4NO_3–3, $MgSO_4 \times 7\,H_2O$–1, $CaCl_2 \times 2\,H_2O$–5, $FeCl_3 \times 6\,H_2O$–1.67×10^{-3}, $ZnSO_4 \times 7\,H_2O$–8×10^{-3} and $CuSO_4 \times 7\,H_2O$–15×10^{-3}) were harvested after 72 h.

2.2. Bioreactor System for Itaconic Acid Production

To aerobically produce itaconic acid under batch conditions, a Lambda Minifor bioreactor system (available online: https://www.lambda-instruments.com/fileadmin/user_upload/PDF/MINIFOR/ Operation_manual_of_LAMBDA_MINIFOR_laboratory_Fermentor_and_Bioreactor.pdf; accessed on 07.01.2019) was applied. The bioreactor with a 1.8 L working volume was filled with a medium (similar to the one used for preparing the inoculum) and autoclaved before commencing the actual experiment. The fermentation conditions tested in this investigation can be seen in Table 1. The concentration of the glucose substrate was fixed at 120 g/L thoroughly, and the temperature was controlled at 37 °C. The inoculation rate was 5% in all cases. The pH (Table 1) was adjusted using NaOH and HCl solutions. The term "STP" in Table 1 refer to the standard temperature and pressure conditions.

Table 1. The pH setting strategies tested in this study.

Experimental Setting	pH	Aeration (L (STP)/min)	Agitation (Hz)	Substrate
A	Initial pH set to 3 and left uncontrolled	1.5	2	glucose
B	Initial pH set to 3 and maintained	1.5	2	glucose
C	Initial pH set to 2.5 and maintained	1.5	2	glucose
D	Initial pH set to 4 and maintained	1.5	2	glucose
E	Initial pH set to 3 and, after 48 h, maintained at 2.5	1.5	2	glucose

2.3. Analytical Procedure

In this study, the itaconic acid production was monitored by the High Performance Liquid Chromatography (HPLC) technique on a Young Lin Instrument Co., Ltd. (YL9100-type) device. The unit contained a Hamilton × 300 HPLC column (length: 15 cm, inner diameter: 4.6 mm, particle size: 5 µm) as well as a UV/VIS detector. The analytical method employed a gradient elution (2 mL/min flow rate) where the moving phase was comprised of A (0.01 M H_2SO_4) and B (methanol) solutions (2 min–100% A; 5 min–50% A, 50% B; 8 min–20% A, 80% B). The samples taken at various spots of the fermentation were treated by membrane filtration (0.22 µm PVDF) and thereafter diluted 1000× using 0.01 M sulfuric acid. The itaconic acid yield (as seen in the Results and Discussion section) was estimated on the grounds of the substrate that was added initially. Fermentation metabolites that were possibly competitive to itaconic acid (e.g., itatartaric acid, gluconic acid, oxalic acid, etc.) were not assessed.

3. Results and Discussion

The pH is one of the most crucial among the fermentation variables, therefore requiring special attention for submerged fungal cultures producing organic acids, such as itaconic acid, with a sufficient performance. Basically, the impact of the pH is associated with the (i) activity of enzymes taking part in the biosynthesis of itaconic acid, and additionally with (ii) the subsequent transfer mechanism to the extracellular space/out of the cell. As could be deduced from the literature, itaconic acid generation by filamentous fungi such as *A. terreus* favors lower pH conditions, mostly around pH = 2–3 [8]. It has been argued that besides enabling the appropriate growth of *A. terreus* [24], such a fermentation environment can be useful for suppressing the formation of by-products that would lower the final itaconic acid yield and productivity [25]. Typical by-products of fungal itaconic acid fermentation can be itatartaric acid, gluconic acid and oxalic acid, depending on the pH conditions, due to mechanisms reviewed by Mondala [10]. The advantages of a low pH can also originate from the (i) limited threat of microbiological contamination, (ii) the avoidance of an extreme mycelial network expansion facilitating itaconic acid conversion because of an improved carbon flux as well as (iii) the proper morphology of the strain, (iv) the increased transfer of oxygen gas and (v) aided downstream [10,25]. Although it seems to be established that an adequate pH adjustment is a key-step, the results obtained by various studies that apply the same strain of *A. terreus* are still frequently divergent [12,22]. In fact, although the optimum pH is basically a strain-specific feature, various studies suggest that the pH should be optimized by taking into account other process parameters characterizing the particular bioreactor unit. For instance, Riscaldati et al. [26] demonstrated that the pH and stirring rate together govern itaconic acid fermentation, while Vassilev et al. [27] found by a response surface methodology that itaconic acid production was notably influenced by the complex relationship of the pH, substrate concentration and nitrogen source (e.g., ammonium nitrate). These examples and observations imply the need for investigating the impact of the pH under the actual circumstances of a particular study.

Accordingly, as can be seen in Table 1, the effect of various pH setting strategies was sought. The experiments were planned on the grounds of relevant concepts reported in the literature: one common practice considers only the setting of the initial pH, where afterwards it is allowed to decrease automatically [13,27,28] (Table 1A), while others propose a well-controlled pH throughout the fermentation to prevent the depression of the itaconic acid production efficiency [29] (Table 1B–D). Apart from that, researchers such as Hevekerl et al. [12] found potential in the pH-shift approach, where the pH is initially set, runs freely for a certain period of time and is controlled only from a given point of the biological conversion, e.g., when itaconic acid production begins (Table 1E). In accordance with the previous argument, the pH was varied between 2.5 and 4 (Table 1). To characterize the batch itaconic acid fermentation kinetics, under the conditions listed in Table 1, the modified Gompertz-model (Equation (1)) was adopted [30]:

$$IA(t) = P \exp\left\{-\exp\left[\frac{R_m e}{P}(\lambda - t) + 1\right]\right\} \tag{1}$$

This approach enables the user to determine important process parameters (Table 2) from the evaluation of fermentation time profiles (Figure 1), where $IA(t)$ is the actual itaconic acid titer (g/L) at time t (h); P is the itaconic acid titer potential (g/L); R_m denotes the maximal itaconic acid (titer) production rate (g/L/h), λ is the length of the lag-phase time (h); and e is 2.718. To obtain the best fitting of the model and experimental curves, this work relied on the least-squares regression method using the Solver tool in MS Excel. The basic statistical assessment of the results in Table 2 is shown in Table 3.

Table 2. Experimental itaconic acid production data.

	Experimental Setting				
Time (h)	**A**	**B**	**C**	**D**	**E**
	Itaconic Acid Titer (g/L)				
0	0	0	0	0	0
24	0	1.63	0	0.80	0
48	2.88	3.99		0.94	0
72	5.63	5.07			17.17
96	8.32	6.07	9.20		
120			12.18	0.90	
144			18.48	1.79	17.84
168	18.69	8.94	20.37	1.98	24.40
192	21.01	12.48	19.81	3.39	29.47
216	24.24	14.12		3.43	
240	26.28	13.87			41.4
264		13.99	19.98		
288				5.10	

Table 3. Descriptive statistics for the experimental data presented in Table 2.

Statistical Data	Experimental Setting				
	A	**B**	**C**	**D**	**E**
Valid number of data	9	10	8	9	8
Mean	11.89	8.02	12.5	2.04	16.29
Minimum	0	0	0	0	0
Maximum	26.28	14.12	20.37	5.1	41.4
Standard deviation	10.63	5.39	8.71	1.63	15.44

The experimental itaconic acid production data as a function of time for each test condition (whose corresponding graphs are displayed in Figure 1A–E) are listed in Table 2, and were subjected to a kinetic analysis using the modified Gompertz-formula (Equation (1)) to determine the lag/adaptation-phase time, itaconic acid titer potential and maximal titer production rate (Table 4).

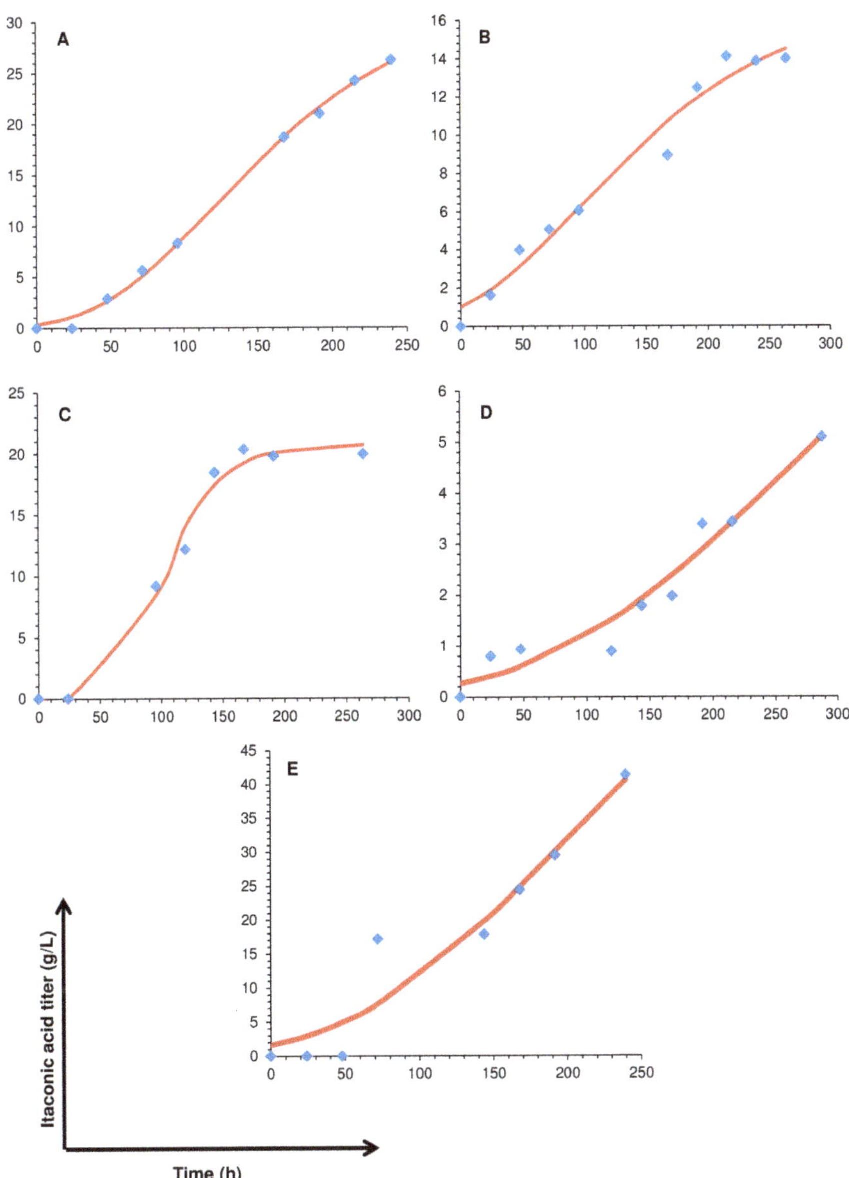

Figure 1. The time profiles of itaconic acid production experiments. Notations (**A**–**E**) are as explained in Table 1. Blue diamonds: Measured data (Table 2); Red lines: Fitted curves derived from the modified Gompertz-model in Equation (1).

Table 4. Results of the kinetic process evaluation based on the data from Table 2.

Kinetic Data	Experimental Setting				
	A	B	C	D	E
P (g/L)	32.70	17.17	20.75	12.98	87.32
R_m (g/L/h)	0.15	0.07	0.26	0.02	0.22
λ (h)	41.61	4.51	63.70	72.44	56.04

Based on these process factors, a ranking was performed, considering that the shorter λ, higher P and R_m are preferred. Bearing this in mind, points (1–5) were assigned in a parameter (λ, P, R_m)-wise manner to demonstrate how they were affected by the 5 different fermentation conditions (Table 5). The final assessment was made by summarizing the given scores. Accordingly, the various pH setting strategies could be ordered, as follows: (E) > (A) > (C) > (B) > (D).

Table 5. Ranking of the various pH setting strategies.

Experimental Setting	Score			Sum of Scores	Final Rank
	P	R_m	λ		
A	4	3	4	11	2
B	2	2	5	9	4
C	3	5	2	10	3
D	1	1	1	3	5
E	5	4	3	12	1

This outcome signifies that the pH-switch strategy (E) was the only one that led to a better itaconic acid formation characteristic than measurement (A), which is the widely-applied approach in the literature and can thus be viewed as the reference setting. In this respect, the findings of Hevekerl et al. [12] are supportive, as it turned out that the best itaconic acid concentration (146 g/L) was attained when the regulation of the pH to 3 began slightly after 2 days of cultivation in the bioreactor. The positive impact was believed to be ascribed to the lower degree of stress on the fungal cells. Actually, this strategy led to a nearly 70% improvement in comparison to the fermentation with the uncontrolled pH [12].

Under the best fermentation condition (E) of this work, the experimental itaconic acid yield—considering the quantity of substrate added—was 0.35 g/g glucose. This seems to relate well with the literature data, where yields in the range of 0.21–0.62 g itaconic acid/g glucose can be found with *A. terreus* strains [22,26,31].

Furthermore, from settings (B) and (C), on can infer that even if the pH is kept constant during the entire biological transformation, the (initial) pH value plays an important role. Accordingly, pH = 2.5 resulted in preferable fermentation kinetic features than pH = 3. This is in agreement with the superiority of setting (E), where the pH was consistently 2.5 from the second day onwards. Besides, it can be concluded that the experimental setting (D) with a pH maintained at 4 was the least attractive by far. Hence, the use of NaOH—recoverable with MED [20–22], as elaborated above—in relatively larger quantities for adjusting the pH to the less acidic range could have an adverse effect. This can make sense in light of the above statement that most studies regarding itaconic acid generation proposed a pH of around 2–3 [8].

4. Conclusions

In this study, itaconic acid fermentation from glucose by *A. terreus* was investigated in relation to the effect of the pH and its regulation strategy. It was found that the initial pH value played a significant role and, additionally, that it did make a difference if the pH was initially adjusted or controlled. Ranking the various pH setting alternatives based on the analysis of fermentation kinetics showed that (under the conditions of the experiments, e.g., bioreactor type, aeration, stirring rate, and substrate concentration), the initial adjustment of the pH to 3 and its adjustment to 2.5 after 2 days was the most promising alternative and should therefore be applied.

Author Contributions: A.K. and G.T. conducted the experiments. K.B.-B. and L.G. supervised the work and contributed to the writing of the manuscript. P.K., N.N. and P.B. evaluated the results and participated in the writing and editing of the manuscript. The authors have equal contribution to this work.

Acknowledgments: This research was supported by the National Research, Development and Innovation Fund project NKFIH K 119940 entitled "Study on the electrochemical effects of bioproduct separation by electrodialysis" and by the financial support of Széchenyi 2020 within project EFOP-3.6.1-16-2016-00015.

Conflicts of Interest: The authors declare no conflict of interest.

References

1. Gonzalez-Garcia, R.-A.; McCubbin, T.; Navone, L.; Stowers, C.; Nielsen, L.-K.; Marcellin, E. Microbial propionic acid production. *Fermentation* **2017**, *3*, 21. [CrossRef]
2. Murali, N.; Srinivas, K.; Ahring, B.-K. Biochemical production and separation of carboxylic acids for biorefinery applications. *Fermentation* **2017**, *3*, 22. [CrossRef]
3. Nghiem, N.-P.; Kleff, S.; Schwegmann, S. Succinic acid: Technology development and commercialization. *Fermentation* **2017**, *3*, 26. [CrossRef]
4. West, T.-P. Microbial production of malic acid from biofuel-related coproducts and biomass. *Fermentation* **2017**, *3*, 14. [CrossRef]
5. Jang, Y.-S.; Kim, B.; Shin, J.-H.; Choi, Y.-J.; Choi, S.; Song, C.-W.; Lee, J.; Park, H.G.; Lee, S.Y. Bio-based production of C2–C6 platform chemicals. *Biotechnol. Bioeng.* **2012**, *109*, 2437–2459. [CrossRef] [PubMed]
6. Yang, L.; Lübeck, M.; Lübeck, P.-S. Aspergillus as a versatile cell factory for organic acid production. *Fungal Biol. Rev.* **2017**, *31*, 33–49. [CrossRef]
7. El-Imam, A.-A.; Du, C. Fermentative itaconic acid production. *J. Biodivers. Bioprospect. Dev.* **2014**, *1*, 119. [CrossRef]
8. Saha, B.-C. Emerging biotechnologies for production of itaconic acid and its applications as a platform chemical. *J. Ind. Microbiol. Biotechnol.* **2017**, *44*, 303–315. [CrossRef] [PubMed]
9. Dwiarti, L.; Otsuka, M.; Miura, S.; Yaguchi, M.; Okabe, M. Itaconic acid production using sago starch hydrolysate by *Aspergillus terreus* TN484-M1. *Bioresour. Technol.* **2007**, *98*, 3329–3337. [CrossRef]
10. Mondala, A.-H. Direct fungal fermentation of lignocellulosic biomass into itaconic, fumaric, and malic acids: Current and future prospects. *J. Ind. Microbiol. Biotechnol.* **2015**, *42*, 487–506. [CrossRef]
11. Pedroso, G.-B.; Montipó, S.; Mario, D.-A.-N.; Alves, S.-H.; Martins, A.-F. Building block itaconic acid from left-over biomass. *Biomass Convers. Biorefin.* **2017**, *7*, 23–35. [CrossRef]
12. Hevekerl, A.; Kuenz, A.; Vorlop, K.-D. Influence of the pH on the itaconic acid production with *Aspergillus terreus*. *Appl. Microbiol. Biotechnol.* **2014**, *98*, 10005–10012. [CrossRef] [PubMed]
13. Karaffa, L.; Díaz, R.; Papp, B.; Fekete, E.; Sándor, B.; Kubicek, C.-P. A deficiency of manganese ions in the presence of high sugar concentrations is the critical parameter for achieving high yields of itaconic acid by *Aspergillus terreus*. *Appl. Microbiol. Biotechnol.* **2015**, *99*, 7937–7944. [CrossRef] [PubMed]
14. Li, A.; Pfelzer, N.; Zuijderwijk, R.; Punt, P. Enhanced itaconic acid production in Aspergillus niger using genetic modification and medium optimization. *BMC Biotechnol.* **2012**, *12*, 57. [CrossRef] [PubMed]
15. Shin, W.-S.; Lee, D.; Kim, S.; Jeong, Y.-S.; Chun, G.-T. Application of scale-up criterion of constant oxygen mass transfer coefficient (kLa) for production of itaconic acid in a 50 L pilot-scale fermentor by fungal cells of Aspergillus terreus. *J. Microbiol. Biotechnol.* **2013**, *23*, 1445–1453. [CrossRef]
16. Cartensen, F.; Klement, T.; Büchs, J.; Melin, T.; Wessling, M. Continuous production and recovery of itaconic acid in a membrane bioreactor. *Bioresour. Technol.* **2013**, *137*, 179–187. [CrossRef]
17. Magalhães, I.-A., Jr.; de Carvalho, J.C.; Medina, J.D.C.; Soccol, C.R. Downstream process development in biotechnological itaconic acid manufacturing. *Appl. Microbiol. Biotechnol.* **2017**, *101*, 1–12. [CrossRef]
18. Varga, V.; Bélafi-Bakó, K.; Vozik, D.; Nemestóthy, N. Recovery of itaconic acid by electrodialysis. *Hung. J. Ind. Chem.* **2018**, *46*, 43–46. [CrossRef]
19. Luo, H.; Cheng, X.; Liu, G.; Zhou, Y.; Lu, Y.; Zhang, R.; Li, X.; Teng, W. Citric acid production using a biological electrodialysis with bipolar membrane. *J. Membr. Sci.* **2017**, *523*, 122–128. [CrossRef]

20. Tongwen, X.; Weihua, Y. Citric acid production by electrodialysis with bipolar membranes. *Chem. Eng. Process.* **2002**, *41*, 519–524. [CrossRef]

21. Pinacci, P.; Radaelli, M. Recovery of citric acid from fermentation broths by electrodialysis with bipolar membranes. *Desalination* **2002**, *148*, 177–179. [CrossRef]

22. Kuenz, A.; Gallenmüller, Y.; Willke, T.; Vorlop, K.-D. Microbial production of itaconic acid: Developing a stable platform for high product concentrations. *Appl. Microbiol. Biotechnol.* **2012**, *96*, 1209–1216. [CrossRef] [PubMed]

23. Wei, Y.; Li, C.; Wang, Y.; Zhang, X.; Li, Q.; Xu, T. Regenerating sodium hydroxide from the spent caustic by bipolar membrane electrodialysis (BMED). *Sep. Purif. Technol.* **2012**, *86*, 49–54. [CrossRef]

24. Chen, M.; Huang, X.; Zhong, C.; Li, J.; Lu, X. Identification of an itaconic acid degrading pathway in itaconic acid producing *Aspergillus terreus*. *Appl. Microbiol. Biotechnol.* **2016**, *100*, 7541–7548. [CrossRef]

25. Klement, T.; Büchs, J. Itaconic acid—A biotechnological process in change. *Bioresour. Technol.* **2013**, *135*, 422–431. [CrossRef] [PubMed]

26. Riscaldati, E.; Moresi, M.; Federici, F.; Petruccioli, M. Effect of pH and stirring rate on itaconate production by *Aspergillus terreus*. *J. Biotechnol.* **2000**, *83*, 219–230. [CrossRef]

27. Vassilev, N.; Kautola, H.; Linko, Y.-Y. Immobilized *Aspergillus terreus* in itaconic acid production from glucose. *Biotechnol. Lett.* **1992**, *14*, 201–206. [CrossRef]

28. Gao, Q.; Liu, J.; Liu, L. Relationship between morphology and itaconic acid production by *Aspergillus terreus*. *J. Microbiol. Biotechnol.* **2014**, *24*, 168–176. [CrossRef]

29. Meena, V.; Sumanjali, A.; Dwarka, K.; Subburathinam, K.-M.; Sambasiva Rao, K.-R.-S. Production of itaconic acid through submerged fermentation employing different species of *Aspergillus*. *Rasayan J. Chem.* **2010**, *3*, 100–109.

30. Nemestóthy, N.; Bakonyi, P.; Rózsenberszki, T.; Kumar, G.; Koók, L.; Kelemen, G.; Kim, S.-H.; Bélafi-Bakó, K. Assessment via the modified Gompertz-model reveals new insights concerning the effects of ionic liquids on biohydrogen production. *Int. J. Hydrogen Energy* **2018**, *43*, 18918–18924. [CrossRef]

31. Kautola, H.; Vahvaselka, M.; Linko, Y.-Y.; Linko, P. Itaconic acid production by immobilized *Aspergillus terreus* from xylose and glucose. *Biotechnol. Lett.* **1985**, *7*, 167–172. [CrossRef]

 *fermentation*

Article

Development of A Low-Alcoholic Fermented Beverage Employing Cashew Apple Juice and Non-Conventional Yeasts

Amparo Gamero [1,2,*], Xiao Ren [3], Yendouban Lamboni [3,4,5], Catrienus de Jong [1], Eddy J. Smid [3] and Anita R. Linnemann [4]

[1] NIZO, P.O. Box 20, 6710 BA Ede, The Netherlands
[2] Food Technology Area, Faculty of Pharmacy, Universitat de València, Avda. Vicent Andrés Estellés s/n, Burjassot, 46100 Valencia, Spain
[3] Laboratory of Food Microbiology, Wageningen University and Research, P.O. Box 17, 6700AA Wageningen, The Netherlands
[4] Food Quality and Design, Wageningen University and Research, P.O. Box 17, 6700AA Wageningen, The Netherlands
[5] International Institute of Tropical Agriculture (IITA), Department of Agriculture and Health, 08 BP 0932 Tri Postal Cotonou, Benin
* Correspondence: amparo.gamero@uv.es; Tel.: +34-96-354-49-59

Received: 27 June 2019; Accepted: 1 August 2019; Published: 3 August 2019

Abstract: Cashew apples are by-products in the production of cashew nuts, which are mostly left to rot in the fields. Cashew apple juice (CAJ), a highly nutritious beverage, can be produced from them. It is rich in sugars and ascorbic acid, but its high polyphenol content makes it bitter and astringent, and therefore difficult to commercialize. The kingdom of fungi contains more than 2000 yeast species, of which only a few species have been studied in relation to their potential to produce aroma compounds. The aim of this research was to develop a new low-alcoholic fermented beverage to valorize cashew apples. For this purpose, a screening was carried out employing non-conventional yeast species and some species of the genus *Saccharomyces* for comparison, followed by a more detailed study with four selected strains cultured at different conditions. The production of volatile aroma compounds as a function of the presence of oxygen, temperature, and yeast species was investigated. The results showed that the more diverse aroma profiles appeared at 25 °C under anaerobic cultivation conditions, where *Saccharomyces cerevisiae* WUR 102 and *Hanseniaspora guilliermondii* CBS 2567 excelled in the synthesis of certain aroma compounds, such as β-phenylethanol and its acetate ester (rose aroma). Further studies are needed to test consumer acceptance of these new products.

Keywords: cashew apple juice; non-conventional yeasts; alcoholic beverages; aroma profile; *Hanseniaspora guilliermondii*; *Torulaspora microellipsoides*; *Saccharomyces cerevisiae*

---

1. Introduction

Cashew (*Anacardium occidentale* L.) is a native crop from tropical America, widely available in several countries of Asia, Africa, and Central America [1]. The most important product from *Anacardium occidentale* L. is the cashew nut. However, the tree also yields a pseudo-fruit called cashew apple to which the nut is attached, which can be either yellow, orange, or bright red [1]. It is a fleshy, fibrous and highly juicy fruit with a soft peel [2], which can be consumed raw or in the form of fresh juice, jam, syrups, candied fruit, jelly, pectin, soft drinks, or other beverages. However, the cashew apple is very bitter and astringent due to its high polyphenol content and therefore, not as palatable as other fruits. For this reason, the cashew apple is almost neglected in commercial terms as compared to the nut. Its industrialization represents not even 10% of the annual Brazilian production [2]. According to official data, 90% of the annual production of about two million tons of cashew apples in Northeast Brazil are lost or underutilized [3]. In 2006, the world production was estimated to be 30 million tons [4,5]. Being considered as an agriculture residue, it leads to a large amount of waste. New products and processes are needed to reduce this high wastage.

Peeling of cashew apples and clarification of cashew apple juice (CAJ) are strategies that can be used to reduce polyphenol content and therefore the bitterness and astringency of the product. The skin of cashew apple contains much higher tannin concentration (516–802 mg/100 g) than the flesh (149–155 mg/100 g) [5] and among several clarification methods, the use of gelatin showed the highest rate of success, being cheap and easily available in Benin [6,7]. However, other innovative strategies, such as the use of microorganisms to decrease polyphenols, could be considered. For instance, yeasts have been demonstrated to be able to metabolize these compounds [8–10].

The most attractive property of cashew apple is its extremely high ascorbic acid content, which is about three to six times that of orange juice and about 10-fold the content in pineapple juice [1]. It also contains thiamine, niacin, and riboflavin in addition to significant concentrations of minerals, such as copper, zinc, sodium, potassium, calcium, iron, phosphorous, and magnesium [1]. Furthermore, its capabilityin reducing sugars (glucose and fructose) makes the substrate appropriate for alcoholic fermentation [11]. In this way, cashew apple would be an excellent, highly available, and low-cost substrate for the production of fermented beverages. In addition, cashew apple juice (CAJ) is believed to have several beneficial properties, such as antibacterial, antifungal, antitumoral, antioxidant, and antimutagenic actions [12–14].

CAJ has been previously used to produce metabolites of interest such as lactic acid, oxalic acid, dextran, mannitol, oligosaccharides, and a biosurfactant in fermentations mediated by bacteria of the genera *Lactobacillus*, *Leuconostoc*, *Pseudomonas* or *Bacillus* [3,11,15–18]. Moreover, yeast species of the genera *Saccharomyces* and *Hanseniaspora* have been demonstrated to have the capacity of producing bioethanol when growing in this substrate [19–21]. Finally, CAJ has been used to produce probiotic beverages employing lactic acid bacteria [22–26] and alcoholic beverages using the yeasts *Saccharomyces cerevisiae* and *S. bayanus* [2,27,28].

Non-conventional yeasts are those that are not used commonly in industrial processes, although some of them present interesting properties, thereby making them an untapped potential for food applications [29]. There is evidence that these species are able to carry out fermentations of sugary substrates producing very diverse aroma profiles [30–33].

The main objective of the current study is to develop a locally feasible process for the exploitation and valorization of cashew apples employing *Saccharomyces* and non-conventional yeasts to develop an innovative low-alcoholic fermented beverage rich in vitamin C.

2. Materials and Methods

2.1. Production and Characterization of Cashew Apple Juice

2.1.1. Plant Materials

Twenty kilograms of yellow cashew apples and twenty kilograms of red cashew apples were harvested at the mature stage in Benin, immediately frozen, and sent to the laboratory in Wageningen, The Netherlands. Ten frozen cashew apples of each batch were randomly chosen and rapidly peeled by grating the epidermis with a razor blade. Next, the frozen flesh was quickly cut into bits, immediately dipped in liquid nitrogen to prevent oxidation, and homogenized to a very fine powder with a blender. Part of the flesh powder was freeze-dried at −20 °C, while the rest was kept at room temperature for juice extraction.

2.1.2. Juice Extraction and Clarification

CAJ was obtained through a mechanical process and clarified by adding gelatin to remove polyphenols and suspended solids [3]. The extraction process was modified according to a local method used in Benin, which is based on direct pressing of the fruits [34]. Peeled cashew apples were cut and ground by a mixer. Then, 1% (*w/v*) gelatin was added to CAJ, stirred for 15 min, and left to settle for 10–15 min at 4 °C. Next, CAJ was filtered through a cheese cloth.

The clarified CAJ was physicochemically characterized and stored frozen (−20 °C) for further analytical studies. Dimethyl dicarbonate (DMDC) was added to CAJ for sterilization and to prevent vitamin loss.

2.1.3. Sugar Determination

Glucose, fructose, and sucrose were determined by HPLC (Thermo Fisher Scientific, Waltham, MA, USA) equipped with a P-2000 pump and ELSD-2100 polymer labs detector. Separations were carried out in a 250 mm × 4.6 mm Alltech prevail carbohydrates column, with an evaporator temperature of 80 °C and nebulizer temperature of 60 °C. The running time was 14 min with a flow rate 1 mL/min on isocratic 75%/25% acetonitrile/water. In brief, 1 mL of sample was mixed with distilled water of about 80 °C, after which the solution was incubated at 80 °C in a water bath for 5 min, then homogenized with an Ultra Turrax T20B for 1 min and centrifuged (ALC PK131R) for 5 min at 2255× g at 20 °C. The extract was diluted up to eighty times with distilled water. For the external standard, sucrose, glucose, and fructose with the range 45–680 μg/mL were used. Subsequently, 2 mL of samples or standards were filtered through 0.45 μm filter and used for HPLC analysis.

2.1.4. Total Polyphenol Determination

Total phenolic content was determined using the Folin–Ciocalteu method as described by Georgé et al. [35], with some modifications. A volume of 0.25 mL clarified CAJ was diluted 1:4 with distilled water. A calibration curve was made, employing tannic acid with a concentration range from 0.03125 mg/mL to 1 mg/mL. The reaction mixture was composed of 1 mL of the sample or standard, 30 mL of distilled water, and 1 mL of Folin–Ciocalteu reagent. After 15 min, the intensity of the blue color that had developed was measured spectrophotometrically (Cary 50-UV visible, Varian, Palo Alto, CA, USA) at 725 nm. The total phenolic content was expressed as Gallic acid equivalents (GAE) per ml of sample.

2.1.5. Ascorbic Acid Determination

The method used for the determination of ascorbic acid (AA) content was described by Hernández et al. [36], with some modifications. An amount of 2.5 mL of sample was mixed with 2.5 mL of the extracting solution containing 3% MPA (metaphosphoric acid) and 1 mM TBHQ (tert-butylhydroquinone) in 10 mL tubes. After homogenizing, the mixture was centrifuged for 5 min at 2255× g at 4 °C. The extract was diluted up to 8 times with distilled water. All extractions were carried out employing ice and under reduced light. For the calibration, commercial L-ascorbic acid was prepared with a concentration range of 1.56–200 μg/mL. Subsequently, 2 mL of the sample or standard were filtered through 0.45 μm filter paper and used for HPLC analysis.

The HPLC system was from Thermo Fisher Scientific P-2000 (USA), equipped with a binary gradient pump and UV 2000 detector. Separations were carried out on a 150 mm × 4.6 mm Varian Polaris C18-A column, with 5.5 min running time and 20 μL injection volumes using an autosampler. The mobile phase employed was 0.2% orthophosphoric acid in distilled water. The flow rate of the mobile phase was 1 mL/min. A UVdetector at a wavelength of 245 nm was employed. The AA peak was identified by comparing its UV-visible spectral characteristics and retention time with the commercial standard of AA.

2.2. Fermentation of Cashew Apple Juice

2.2.1. Yeast Strains

The yeast strains employed in this study are listed in Table 1. The three *Saccharomyces cerevisiae* strains were from the culture collection of the Laboratory of Food Microbiology of Wageningen University, whereas one *S. bayanus* and 21 non-*saccharomyces* yeasts were supplied by Westerdijk Fungal Biodiversity Institute-CBS-KNAW, Utrecht, The Netherlands.

Table 1. Yeast strains employed in this study.

Strain	Genus	Isolation Source/Origin
CBS 77	*Dekkera anomala*	Stout beer, England
CBS 772.71	*Galactomyces geotrichum*	Soil, Puerto Rico
CBS 1545	*Saccharomyces bayanus*	Beer, The Netherlands
CBS 1671	*Zygosaccharomyces rouxii*	Urine, The Netherlands
CBS 1711	*Wickerhamomyces subpelliculosus*	Fermenting cucumber brine, USA
CBS 2499	*Dekkera bruxellensis*	Wine, France
CBS 2567	*Hanseniaspora guilliermondii*	Grape must, Israel
CBS 2568	*Hanseniaspora vineae*	Fruit fly
CBS 2734	*Torulaspora microellipsoides*	Black currants, Denmark
CBS 2796	*Dekkera bruxellensis*	Sparkling Mosselle wine, Germany
CBS 4806	*Brettanomyces custersianus*	Bantu-beer brewery, South Africa
CBS 5552	*Wickerhamomyces subpelliculosus*	Molasses
CBS 5681	*Zygosaccharomyces bailii* var. *bailii*	Moselle wine
CBS 6619	*Hanseniaspora guilliermondii*	Unknown
CBS 6625	*Zygosaccharomyces bailii* var. *bailii*	*Myoporum* sp., Japan
CBS 6641	*Torulaspora microellipsoides*	Sandalwood tree, Hawaii (USA)
CBS 7692	*Starmera caribaea*	Pricklypear cactus, Bahamas
CBS 8344	*Wickerhamomyces subpelliculosus*	Unknown
CBS 8849	*Zygosaccharomyces kombuchaensis*	Kombucha tea, Russia
CBS 8860	*Barnettozyma californica*	Berries, Russia
CBS 9716	*Zygosaccharomyces rouxii*	Honey pot, Germany
CBS 10396	*Kazachstania zonata*	Japan
CBS 10399	*Kazachstania zonata*	Japan
CBS 10400	*Kazachstania gamospora*	Japan
CBS 10404	*Kazachstania gamospora*	Japan
WUR 102	*Saccharomyces cerevisiae*	Masau fruits, Muzarabani (Zimbabwe)
WUR 131	*Saccharomyces cerevisiae*	Masau fruits, Muzarabani (Zimbabwe)
WUR 153	*Saccharomyces cerevisiae*	Masau fruits, Muzarabani (Zimbabwe)

2.2.2. Yeast Screening

Prior to fermentation, CAJ was treated with dimethyl dicarbonate (DMDC) (1ml/L) and stored overnight at 4 °C to inhibit the growth of undesirable microorganisms. Then, juice was inoculated with the chosen strains that had been precultured overnight in GPY medium (2% glucose, 1% bacteriological peptone, 0.5% yeast extract) at optical density at 600 nm ($OD_{600\,nm}$) of 0.1. The 25 yeast cultures were inoculated in 5 mL of yellow CAJ employing microplates. The fermentations were carried out at 25 °C during 7 days in microaerophilic conditions. Out of the 25, four strains were selected to carry out further research. The selection criteria were the following: Growth kinetics and fermentation performance (ethanol content). The fermentations were repeated with the four selected strains, however not only at 25 °C but also at 30 °C and 37 °C under strictly anaerobic conditions employing anaerobic jars. In addition, experiments were carried out at 25 °C under aerobic conditions shaking the microplates.

The first screening was carried out without taking care of creating an anaerobic environment for the fermentations in order to simplify and make a fast selection. The fermentations with the selected yeasts were carried out employing two extreme conditions (strictly anaerobic and aerobic) in order to compare the effect of the oxygen in fermentation performance and aroma profile.

2.2.3. Growth Kinetics

A spectrophotometer (Bio-screen, Los Angeles, CA, USA) was used to determine the growth kinetics by automatically measuring the $OD_{600\,nm}$ at 25 °C during 70 h at 15 min intervals. The data generated were then converted into Microsoft Excel format (Microsoft, Redmond, WA, USA) and processed into growth curves to calculate the specific growth rates of the yeast cultures.

2.2.4. Ethanol Determination

The ethanol content in CAJ was determined by HPLC. Samples were deproteinized using Carrez reagents, which precipitated proteins and colloidal compounds. After 5 min centrifugation, the clear supernatant was diluted 1:1 and transferred to an HPLC vial. The HPLC system used was a Thermo Fisher Scientific (Waltham, MA, USA) equipped with Ultimate 3000 (Dionex), using a 300 mm × 7.8 mm Aminex HPX-87H column (Bio-rad, Hercules, CA, USA) with pre-column. A 5 mM H_2SO_4 was used as eluent at a flow rate of 0.6 mL/min at 40 °C. Detection was by refractive index (Shodex RI 101). Ethanol in the range 1%–10% was used as external standard. Subsequently about 200 µL of samples or standards were used for HPLC analysis. The ethanol peaks were identified and quantified by comparing retention times with those of the external standards.

2.2.5. Aroma Analysis

Aroma compounds were determined by headspace solid-phase dynamic extraction gas chromatography–mass spectrometry (HS-SPDE-GC-MS) employing a 2.5 mL HS syringe with a polydimethylsiloxane active charcoal (PDMS/AC) coated needle (Chromtech, Bad Camberg, Germany). The incubation of the samples was at 60 °C during 15 min at 500 rpm. Afterwards, the sampling was carried out by taking 1 mL of the headspace at 200 µL/s five times. The needle was then desorbed and the headspace injected into the GC column at 25 µL/s. GC–MS employed was a Finnigan Trace GC ultra (Thermo Fisher Scientific, Waltham, MA, USA) equipped with a 20 cm pre-column (CP-Sil 5CB 0.53 mm; df = 1 µm), a VF-1ms (30 m × 0.25 mm; df = 1 µm) capillary column (Varian) and a Combi PAL autosampler (CTC Analytics, Zwingen, Switzerland) in combination with a split injector in splitless mode (1 min) at 250 °C. The carrier gas was helium at a constant flow rate of 1.5 mL/min. The GC oven was initially set at 40 °C for 2 min, raised to 250 °C (10 °C/min) and then kept at 250 °C for 5 min. The total runtime was 28 min. Mass spectral data werecollected over a range of m/z 35–300 in full scan mode (scan time 0.25 s).

2.2.6. Other Analytical Determinations

Residual sugars, total polyphenols, and ascorbic acid were determined in the fermented products with the four selected strains following the same methodology as described in Sections 2.1.3–2.1.5, respectively.

2.3. Statistical Analysis

One-way ANOVA at a 95% confidence level and a Tukey test were employed to compare the composition of the fermented products, whereas multifactorial ANOVA was carried out to determine which factors were significantly affecting the composition of fermented products and the aroma profiles. Finally, principal component analysis (PCA) was used to group the different CAJ and fermented products at the different conditions. All tests were done using XLSTAT (Microsoft, Redmond, WA, USA).

3. Results and Discussion

3.1. Composition of Clarified Cashew Apple Juice

The yellow juice presented several advantages over the red juice, namely a lower polyphenol content and higher ascorbic acid and sugar concentration (Table 2). Consequently, yellow juice was selected to perform the fermentations.

Table 2. Composition of cashew apple juice (CAJ) from yellow and red varieties of cashew apples.

Compound	Yellow CAJ	Red CAJ
Glucose (g/L)	60 ± 3	52 ± 2.5
Fructose (g/L)	52 ± 3	48 ± 1
Total polyphenols (mg/mL)	1.6 ± 0.13	2.3 ± 0.25
Ascorbic acid (mg/mL)	1.1 ± 0.4	0.45 ± 0.2

3.1.1. Sugar Content

The major sugars in the CAJ were glucose and fructose. The obtained quantities were slightly higher than the values reported by Azevedo and Rodrigues [37], who found around 40 g/L for each of these sugars. Usually, fruits such as orange, apple, and pineapple contain more sucrose, which leads to a sweeter taste. This fruit only contains a small amount of sucrose (around 0.8 g/L), resulting in less masking of bitterness. On the other hand, the fact that monosaccharides are the major sugars in CAJ is an advantage for the fermentation since yeasts can directly consume them.

3.1.2. Total Polyphenol

The total phenolic content of fresh yellow CAJ was 1.6 mg/mL, whereas red CAJ contained 2.3 mg/mL (Table 2). These values were in the range of 1.7 to 2.4 mg/mL reported by Adou et al. [34], who also found that the yellow juice had a lower polyphenol concentration than the red one.

3.1.3. Ascorbic Acid

There was a large difference in the concentration of AA between red and yellow CAJ. The ascorbic acid concentration in yellow juice was 1.1 mg/mL whereas red juice only contained 0.45 mg/mL (Table 2). Assunçao et al. [38] reported slightly higher AA levels in yellow CAJ when compared to red CAJ. However, our results showed large differences. According to the same authors, several environmental factors can account for AA levels, such as geographic location, solar intensity, temperature, and soil type.

3.2. Yeast Screening

Twenty-five yeast strains (Table 1) were used to ferment yellow CAJ at 25 °C during seven days under microaerophilic conditions. Four of them were selected for further studies on the basis of their growth kinetics and sugar conversion to ethanol. The selected strains were *Hanseniaspora guilliermondii* CBS 2567, *Torulaspora microellipsoides* CBS 2734 and CBS 6641, and *Saccharomyces cerevisiae* WUR 102.

3.2.1. Growth Kinetics

The growth kinetics of cultures of all strains were tracked by OD measurements over time at 25 °C for 70 h. The strains could be separated into two groups based on their maximum OD values. The strains of the group with the highest maximum OD values (over 1.2) were considered as good growers whereas those with lower maximum OD values apparently could not make proper use of the carbon source in CAJ and were therefore not selected for further research. The specific growth rate was another important criterion for selection. Cultures of strains with higher specific growth rates reproduce faster, which could save time in the fermentation process. The specific growth rates of non-*Saccharomyces* strains ranged from 0.05 to 0.13 h^{-1}, whereas *Saccharomyces* strains presented values of at least 0.18 h^{-1}. Specific growth rates of all cultures are presented in Figure 1.

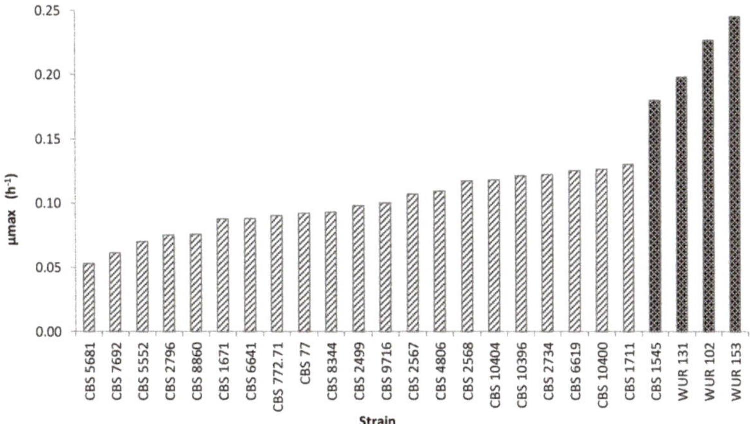

Figure 1. Growth rates of the strains employed in the screening (yellow CAJ, 25°C). Striped bars indicate non-conventional yeasts; dark bars, *Saccharomyces* strains.

Interestingly, some strains showed diauxic growth, such as *Hanseniaspora guilliermondii* CBS 2567. Diauxic growth is observed when an organism is grown in a medium containing two carbon sources and there is a preferential utilization of one carbon source before utilizing the other. This preferred carbon source is consumed first, which leads to rapid growth, followed by a lag phase. After that, the microorganism experiences a slower growth phase, during which the second carbon source is metabolized [39]. The presence of this behavior was taken into account for the yeast selection.

3.2.2. Sugar Conversion to Ethanol

Yellow CAJ contained 112 g/L of sugars (glucose + fructose) prior to fermentation. Most of the strains were able to ferment all the sugars whereas a few strains were not (CBS 772.71, CBS 7692, CBS 8344, and CBS 8860). All of the latter strains belonged to the group with low maximum OD values and low specific growth rates. The conversion of sugars to ethanol reflects the alcoholic fermentation performance of the strains. The strains could be divided into three groups based on their ethanol yield: High, intermediate, and low producers (Figure 2). However, there was an interesting strain, *Torulaspora microellipsoides* CBS 6641, in the low ethanol producing group. Unlike the other strains in this group, CBS 6641 did grow well and consumed all the sugar in the juice but produced alow ethanol amount during fermentation. There might be two reasons. Firstly, because the conditions were not strictly anaerobic, oxygen might have affected the ethanol production by some yeasts while others produced ethanol even with oxygen in the medium due to the "Crabtree effect" [40]. Secondly, it might be due to the redirection of the metabolic flux towards the production of other metabolites, such as organic acids (acetic acid) or glycerol [41–43]. Consequently, CBS 6641 was considered an interesting strain for further research.

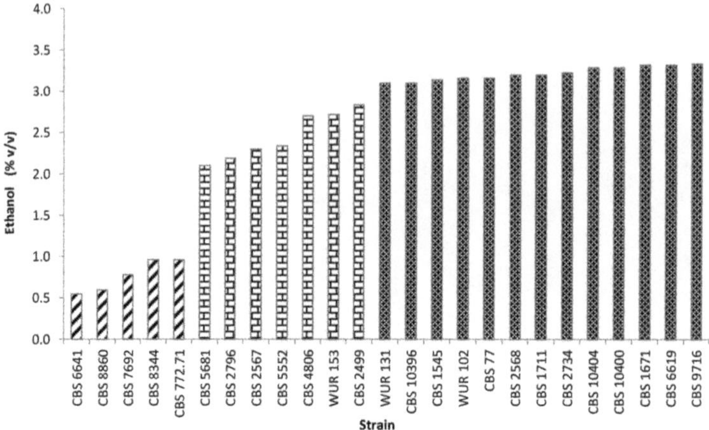

Figure 2. Production during yeast fermentation in yellow cashew apple juice at 25 °C during 7 days. Striped bars depict low ethanol producers; bricked bars, intermediate ethanol producers; dark bars, high ethanol producers.

3.2.3. Yeast Selection

Based on the combined results, four strains were selected for further research. The selected strains were three non-conventional strains and one *Saccharomyces* strain. Table 3 shows the criteria taken into account for the selection and Figure 3 depicts their growth curves. Among the non-conventional yeast strains, the selected yeasts were *Torulaspora microellipsoides* CBS 2734 and CBS 6641 and *Hanseniaspora guilliermondii* CBS 2567, which showed good growth and high, low, and medium ethanol production, respectively. In addition, CBS 2567 was selected for its diauxic growth. Finally, the selected *Saccharomyces cerevisiae* WUR 102 showed a high growth and fermentation performance, thus representing the high ethanol producers.

Table 3. Yeast strains from the screening.

Strain	Species	Ethanol (%)	Optical Density (OD_{max})	μ_{max}	Diauxic Growth
CBS 2567	*Hanseniaspora guilliermondii*	2.3	1.1	0.11	yes
CBS 2734	*Torulaspora microellipsoides*	3.2	1.3	0.12	no
CBS 6641	*Torulaspora microellipsoides*	0.6	1.1	0.09	no
WUR 102	*Saccharomyces cerevisiae*	3.2	1.7	0.23	no

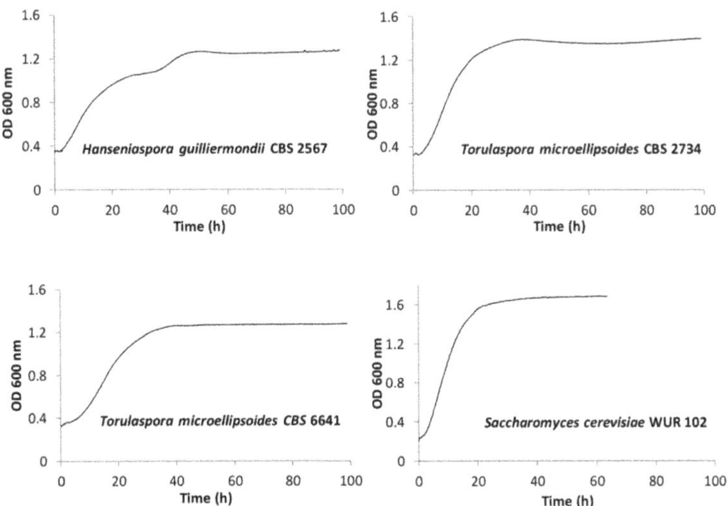

Figure 3. Kinetics of the selected strains at 25 °C employing yellow cashew apple juice.

3.3. Cashew Apple Juice Fermentation with the Selected Strains

Fermentation of yellow CAJ as substrate using the four selected strains was carried out at different temperatures, namely 25 °C, 30 °C, and 37 °C, under strictly anaerobic conditions. In addition, the four strains were grown in yellow CAJ at 25 °C under aerobic conditions.

3.3.1. Sugar Conversion to Ethanol

Analysis of the conversion of sugar into ethanol revealed remarkable differences among the strains (Figure 4). According to multifactorial ANOVA, all three factors studied (i.e., strain, oxygen, and temperature) were significantly affecting ethanol yield. Furthermore, we observed an interaction between the "strain" and "temperature" factors.

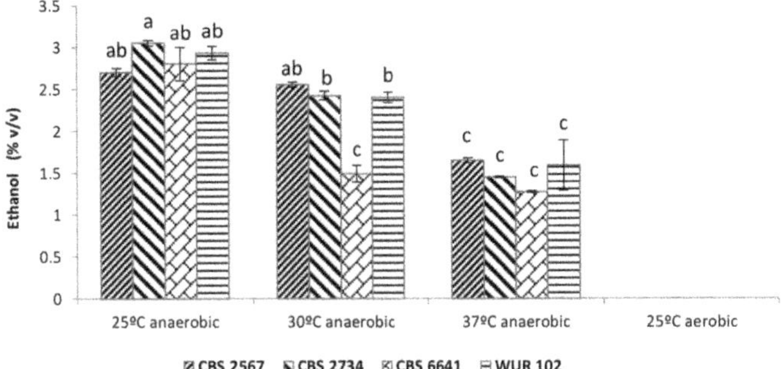

Figure 4. Ethanol production in yellow cashew apple juice during incubation at diferent temperatures employing the selected strains CBS 2567, CBS 2734, CBS 6641, and WUR 102. Letters above bars indicate grouping according to ANOVA.

On the other hand, one-way ANOVA showed three clear groups among the different conditions: (1) Significantly higher ethanol production at 25 °C and 30 °C in anaerobic conditions; (2) significantly lower ethanol production at 37 °C in anaerobic conditions as well as for the strain CBS 6641 at 30 °C in anaerobic conditions; (3) no ethanol production at 25 °C in aerobic conditions. Consequently, in anaerobic conditions, higher incubation temperatures lead to lower ethanol concentrations. This might be due to a reduction in yeast growth and a lower fermentation performance, which is supported by the presence of residual sugars in the juice after fermentation. All four strains produced 2.7%–3% of ethanol at 25 °C. However, ethanol production was reduced to around 2.5% when the temperature reached 30 °C, except in the case of CBS 6641, whose ethanol yield was reduced to only 1.5%, suggesting a poor fermentation performance of this strain at 30 °C. Finally, at 37 °C, the ethanol production was reduced up to 1.5%–1.7%, except in the case of CBS 6641, where ethanol synthesis was only 1.3%. Therefore, strain CBS 6641 proved to be the least robust strain, not being able to ferment efficiently above 25 °C.

Furthermore, CAJ incubated aerobically showed no ethanol formation despite the fact that no residual sugars could be detected at the end of the incubation period. This observation might be explained by either redirection of the metabolic flux towards organic acids instead of ethanol, pointing to the Crabtree negative nature of those strains or by full oxidation of the produced ethanol into CO_2 and H_2O by aerobic respiration.

It is worth mentioning that strain CBS 6641 produced much less ethanol at 25 °C during the screening. This could be explained by the presence of residual oxygen that allowed this strain to produce other compounds instead of ethanol, such as organic acids or glycerol [41–43]. This would mean that this strain is Crabtree negative, since Crabtree positive strains are able to produce ethanol even in the presence of oxygen and when at the same time high levels of glucose are present in the medium as previously commented [40]. During the fermentations of the selected strains, strictly anaerobic conditions were provided, so CBS 6641 and the other selected strains were forced to ferment the sugars.

3.3.2. Ascorbic Acid Content in Fermented Cashew Apple Juice

Multifactorial ANOVA showed that temperature and strain were significant factors determining final ascorbic acid content. In the case of the factor "strain", there was an interaction with the oxygen levels. For instance, under anaerobic cultivation conditions, a reduction in ascorbic acid was observed, which seems to be linked to an increase in temperature (Figure 5).

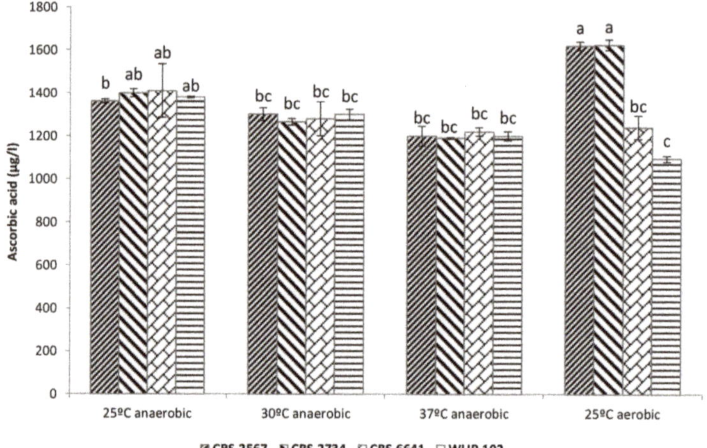

Figure 5. Ascorbic acid content in fermented cashew apple juice employing the selected strains. Letters above bars indicate grouping according to ANOVA.

Furthermore, the factor "strain" significantly affected the ascorbic acid content in aerobic conditions, since strains CBS 2567 and CBS 2734 were even able to increase the levels of this vitamin with respect to the initial CAJ content (1100 ± 350 µg/L). We hypothesize that these strains are capable of synthesizing ascorbic acid, since it has been demonstrated that some yeasts are able to produce this compound under certain circumstances [44]. This production could compensate for ascorbic acid losses due to oxidation.

3.3.3. Total Polyphenol Content in the Fermented Cashew Apple Juice

ANOVA showed no significant differences among the strains regarding the total polyphenol content in the fermented products. In this way, polyphenols were not significantly affected by fermentation with any of the selected strains, temperatures, or oxygen levels tested (Figure 6). However, polyphenol levels in the fermented products were, in most cases, lower than the initial content in yellow CAJ (1.6 mg/L), indicating a possible degradation or bioconversion of polyphenols by the tested strains, which could decrease bitterness and astringency of the final products. The changes in polyphenol content caused by yeasts have been already reported by several authors [8–10]. Another hypothesis to explain the drop in polyphenols could be that part of them were adsorbed to yeast lees and therefore removed prior analysis, as already seen in other alcoholic fermentations [45,46].

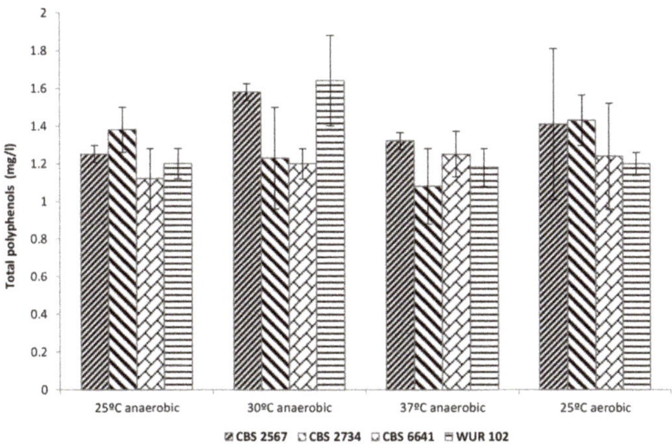

Figure 6. Total polyphenol content in fermented cashew apple juice employing the selected strains.

3.3.4. Aroma Profiles of the Fermented Cashew Apple Juice

The concentrations of the major aroma compounds in fermented CAJ samples were compared. These compounds and their odor descriptions appear in Table 4 and can be divided into five groups: (1) Aldehydes: Acetaldehyde, benzeneacetaldehyde; 3-hydroxybutanal, 3-methylbutanal; (2) alcohols: 2-butanol, isobutanol, 2-methyl-1-butanol, 3-methyl-1-butanol, 1-pentanol, 4-methyl-1-pentanol, 3-methyl-1-pentanol, β-phenylethyl alcohol; (3) acids: Acetic acid; (4) acetate esters: Isoamyl acetate, n-amyl acetate, β-phenylethyl acetate; and (5) ethyl esters: Ethyl butyrate, ethyl lactate, ethyl caproate. According to Garruti et al. [2], the major compounds of fermented CAJ were 3-methyl butanol, isoamyl acetate, isobutanol, and several ethyl esters—ethyl hexanoate, ethyl decanoate, ethyl 3-methyl butyrate, ethyl lactate, and ethyl butyrate. Most of them were found in our study and they are, among others, related to sweet and fruity aromas [2].

Table 4. Aroma compounds found in fermented cashew apple juice.

	Compound	Odour Description
A	Acetaldehyde	Pungent, fruity
B	2-Butanol	Sweet, cashew, fermented, oily, wine, alcoholic
C	3-Hydroxybutanal	Pungent
D	Isobutanol	Sweet, sweaty-chemical, whiskey, fermented, stinky
E	3-Methylbutanal	Fruity, almond, toasted, malty
F	Acetic acid	Vinegar, fermented fruit
G	3-Methyl-1-butanol	Smoky, overripe cashew
H	2-Methyl-1-butanol	Sour
I	1-Pentanol	Alcoholic, burnt
J	Ethyl butyrate	Ethereal, fruity, buttery, ripe fruit
K	Ethyl lactate	Ethereal, rum-buttery, milky, acid, plastic
L	4-Methyl-1-pentanol	Oily green-fruity, herbaceous, yeasty-fermented
M	3-Methyl-1-pentanol	Mild alcoholic
N	Isoamyl acetate	Sweet, fruity, banana, pear
O	n-Amyl acetate	Banana, ethereal, fruity
P	Ethyl caproate	Fruity, wine, apple, banana, pineapple
Q	Benzeneacetaldehyde	Bitter almonds, wild cherry, vanilla
R	Phenylethyl alcohol	Sweet, dried fruit, tea, tobacco
S	β-Phenylethyl acetate	Sweet, fresh, floral, rose, hyacinth

In our study, CAJ was fermented by four different strains (CBS 2567, CBS 2734, CBS 6641, and WUR 102) at three temperatures (25 °C, 30 °C and 37 °C) and employing two levels of oxygen (anaerobic and aerobic) in the case of the fermentations at 25 °C. In order to investigate the differences in the resulting aroma profiles, principal component analysis (PCA) was carried out (Figure 7). The first observation that stood out, was that the condition that yielded less aromatic products was incubation at 25 °C in the presence of oxygen. This was specifically the case of the unfermented CAJ and of the juice fermented with strain CBS 6641. At 25 °C under aerobic conditions, just a few aroma compounds appeared at high concentrations (i.e., benzeneacetaldehyde, phenylethyl alcohol, 3-methylbutanal and 2-butanol). On the contrary, the fermentation condition that yielded more diverse aroma profiles was anaerobiosis at 25 °C, especially for the fermentations carried out by CBS 2567 and WUR 102, followed by the other strains, the unfermented CAJ, and the fermentations at 30 °C. Unfermented CAJ at 30 °C and 37 °C and the fermentations at 37 °C presented intermediate aroma profiles in terms of compound diversity.

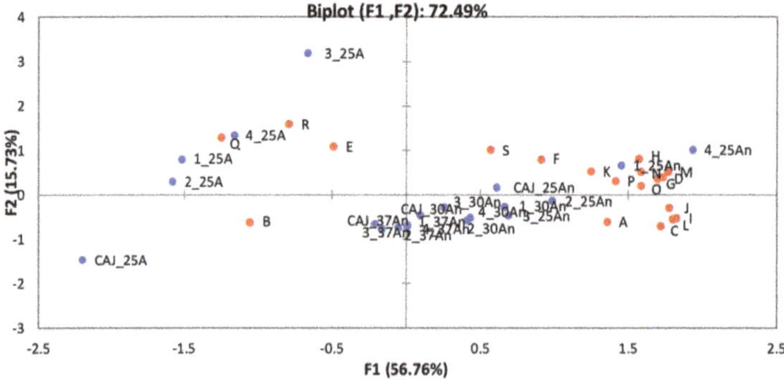

Figure 7. Principal Component Analysis of the aroma profiles of the different fermentations employing the four selected strains. Blue dots: CAJ: cashew apple juice; 1: CBS 2567; 2: CBS 2734; 3: CBS 6641; 4: WUR 102; 25A: 25 °C, aerobic; 25An: 25 °C anaerobic; 30An: 30 °C, anaerobic; 37An: 37 °C, anaerobic; Red dots: Capital letters from A to S indicate the aroma compounds according to the notation in Table 4.

Moreover, multifactorial ANOVA was carried out in order to assess which factors (oxygen, temperature and/or strain) were affecting the aroma profiles. In the case of unfermented CAJ, just two factors (i.e., oxygen and temperature) were taken into account. The results show that both factors affected most of the aroma compounds studied. Exceptions were: 3-methylbutanal, 2-butanol, β-phenylethyl alcohol, acetic acid, and ethyl lactate. In case of the fermented CAJ, the three factors were investigated. The results indicate that oxygen affected the synthesis of almost all the aromas. In addition, a significant interaction between oxygen and strain factors was observed in β-phenylethyl acetate and acetic acid production. The second factor studied was temperature, which affected the production of approximately half of the aroma compounds. Finally, strain type significantly influenced the synthesis of all the ethyl esters, acetic acid, and almost all the aldehydes, with the exception of 3-hydroxybutanal. In addition, it was possible to find interaction among factors. An interaction between strain and temperature in case of 4-methyl pentanol and several esters, ethyl caproate, n-amyl acetate, and β-phenylethyl acetate; and an interaction between strain and oxygen in case of all the ethyl esters, almost all aldehydes (except acetaldehyde), half of alcohols, acetic acid, and almost all the acetate esters (except isoamyl acetate).

Certain aroma compounds were only synthesized under anaerobic conditions, such as ethyl butyrate, 4-methyl-1-pentanol, and n-amyl acetate, whereas the production of 2-butanol by CBS 2567 and CBS 2734 was exclusively under aerobic conditions. In fact, under aerobic conditions, most of the qualitative differences in the aroma profiles were observed. In particular, strain CBS 2567, which was not able to produce 3-hydroxybutanal and ethyl lactate as the other strains, was the only one able to synthesize 3-methyl-1-pentanol. In the anaerobic group, alcohols and esters were the major components. Acetate esters were derived from the condensation of higher alcohols and acetyl CoA, whereas ethyl esters came from condensation of ethanol and acyl-CoA. Temperature, the secondary factor, imposed restrictions on most aromas, especially at 37 °C, except for 3-methylbutanal and acetaldehyde, which appeared at higher levels at this temperature. Regarding the effect of the strains on the aroma profiles in anaerobic fermentations, a result that stands out is the high production of certain aroma compounds by strains WUR 102 and CBS 2567. Strain WUR 102 synthesized higher quantities of benzeneacetaldehyde, 3-methylbutanol, β-phenylethyl alcohol, amyl acetate, ethyl butyrate, and ethyl caproate at 25 °C. In addition, this strain synthesized high levels of phenylethyl acetate at 37 °C. The levels of all these aroma compounds were on average two to four-fold higher than those found in juices fermented by the other strains. On the other hand, the non-conventional yeast CBS 2567 excelled in the production of β-phenylethyl alcohol at 25 °C and its corresponding acetate ester (phenylethyl acetate) at all three temperatures, showing a huge production of up to 35-fold the level found for the other strains. Finally, the non-conventional yeast CBS 2734 yielded high β-phenylethyl alcohol as well. The so-called "fusel or higher alcohols", such as β-phenylethyl alcohol, are either derived from the conversion of amino acids via the Ehrlich pathway or synthesized from sugars [27,47]. They contribute positively to the aroma profile by themselves and because they serve as precursors for the formation of acetate esters [27]. CAJ is rich in amino acids such as alanine, serine, leucine, phenylalanine, proline, glutamic acid, tyrosine, and aspartic acid. The aromatic β-phenylethyl alcohol is derived from the amino acid phenylalanine, whereas phenylethyl acetate is synthesized from β-phenylethyl alcohol. The presence of both compounds is remarkable for the global aroma as they are known to impart herbaceous and rose nuances in the fermented beverages [27].

The odour descriptions were obtained from Garruti et al. [2] and Leffingwell and Associates, Flavor-base 10th edition [48].

4. Conclusions

The aim of this study was to investigate the use of CAJ as a low-cost substrate for developing a fermentation process for the production of a novel alcoholic beverage. Although the clarification step of the CAJ did not help to reduce the levels of total polyphenols in the juice, it cleared the juice without affecting other properties, such as the concentration of ascorbic acid.

Yellow CAJ proved to have better properties than the red juice, such as higher ascorbic acid and sugar contents and lower total polyphenols, making it the preferred substrate for the production of a fermented beverage. However, initial quality control of the raw materials should be included as the composition of cashew apples can vary due to the type of soil, climatic conditions, stage of maturation and seasonality besides type of cultivar [1].

Non-conventional yeasts showed, in general, a poor fermentation performance when compared to *Saccharomyces* strains. However, some of them were able to grow and ferment CAJ efficiently, especially at the lowest temperatures tested (25 °C and 30 °C).

The selected yeast strains were evaluated at different temperature and oxygen conditions. The best condition in terms of aroma production was 25 °C under anaerobic conditions, where the aroma profiles showed the highest diversity. *S. cerevisiae* WUR 102 and *H. guilliermondii* CBS 2567 excelled in the synthesis of certain aroma compounds, especially in this condition, giving a high production of interesting aroma compounds such as β-phenylethanol and its corresponding acetate ester (rose aroma). In addition, the strains showed the potential to metabolize polyphenols while maintaining high ascorbic acid levels.

In conclusion, the optimal fermentation conditions for the production of a fermented alcoholic beverage from CAJ proved to be 25 °C in the absence of oxygen. The final products were slightly yellowish, acidic in taste, low in alcohol (around 3% *v/v*), contained relatively lower total polyphenols, and maintained high levels of ascorbic acid. Further studies are needed to test consumer acceptance of these new products.

Author Contributions: Conceptualization, A.G., Y.L., E.J.S., and A.R.L.; formal analysis, A.G. and X.R.; investigation, X.R.; resources, C.d.J., E.J., and A.R.L.; supervision, A.G., E.J.S., and A.R.L.; visualization, A.G., X.R., and Y.L.; writing—original draft, A.G. and X.R.; writing—review and editing, A.G., Y.L., C.d.J., E.J.S., and A.R.L.

Funding: This research was founded by the European Comission-Marie Curie Initial Training Network CORNUCOPIA (FP7-PEOPLE-2010-ITN, grant agreement nr. 264717).

Acknowledgments: The authors thank the European Comission for the funding; the International Institute of Tropical Agriculture, Benin station (Africa) for providing the cashew apples and Westerdijk Fungal Biodiversity Institute-CBS-KNAW for providing most of the strains.

Conflicts of Interest: The authors declare no conflict of interest.

References

1. Das, I.; Arora, A. Post-harvest processing technology for cashew apple—A review. *J. Food Eng.* **2017**, *194*, 87–98. [CrossRef]

2. Garruti, D.S.; Franco, M.R.B.; da Silva, M.A.A.P.; Janzantti, N.S.; Alves, G.L. Assessment of aroma impact compounds in a cashew apple-based alcoholic beverage by GC-MS and GC-olfactometry. *LWT Food Sci. Technol.* **2006**, *39*, 373–378. [CrossRef]

3. Lopes, T.; Rabelo, M.C.; Barros, L.R.; Saavedra, G.A.; Rodrigues, S. Fermentation of cashew apple juice to produce high added value products. *World J. Microbiol. Biotechnol.* **2007**, *23*, 1409–1415.

4. Michodjehoun-Mestres, L.; Souquet, J.M.; Fulcrand, H.; Bouchut, C.; Reynes, M.; Brillouet, J.M. Monomeric phenols of cashew apple (*Anacardium occidentale* L.). *Food Chem.* **2009**, *112*, 851–857. [CrossRef]

5. Michodjehoun-Mestres, L.; Souquet, J.M.; Fulcrand, H.; Bouchut, C.; Reynes, M.; Brillouet, J.M. Characterisation of highly polymerised prodelphinidins from skin and flesh of four cashew apple (*Anacardium occidentale* L.) genotypes. *Food Chem.* **2009**, *114*, 989–995. [CrossRef]

6. Cormier, R. *Clarification of Cashew Apple Juice and Commercial Applications*; Oxfam: Quebec, Benin, 2008; pp. 1–9.

7. Jayalekshmy, V.G.; John, P.S. 'Sago'—A natural product for cashew apple juice clarification. *J. Trop. Agric.* **2004**, *42*, 67–68.

8. Jayabalan, R.; Marimuthu, S.; Swaminathan, K. Changes in content of organic acids and tea polyphenols during kombucha tea fermentation. *Food Chem.* **2007**, *102*, 392–398. [CrossRef]

9. Hainal, A.R.; Ignat, I.; Volf, I.; Popa, V.I. Transformation of polyphenols from biomass by some yeast species. *Cellul. Chem. Technol.* **2011**, *45*, 211–219.

10. Ettayebi, K.; Errachidi, F.; Jamai, L.; Tahri-Jouti, M.A.; Sendide, K.; Ettayebi, M. Biodegradation of polyphenols with immobilized *Candida tropicalis* under metabolic induction. *FEMS Microbiol. Lett.* **2003**, *223*, 215–219. [CrossRef]

11. Silveira, M.S.; Fontes, C.P.M.L.; Guilherme, A.A.; Fernandes, F.A.N.; Rodrigues, S. Cashew apple juice as substrate for lactic acid production. *Food Bioproc.* **2012**, *5*, 947–953. [CrossRef]

12. Kubo, I.; Ochi, M.; Vieira, P.C.; Komatsu, S. Antitumor agents from the cashew (*Anacardium occidentale*) apple juice. *Food Chem.* **1993**, *41*, 1012–1015. [CrossRef]

13. Melo, A.A.; Rubensam, G.; Picada, J.N.; Gomes, E.; Fonseca, J.C.; Henriques, J.A. Mutagenicity, antioxidant potential, and antimutagenic activity against hydrogen peroxide of cashew (*Anacardium occidentale*) apple juice and cajuina. *Environ. Mol. Mutagen.* **2003**, *41*, 360–369. [CrossRef] [PubMed]

14. Melo, A.A.; Rübensam, G.; Erdtmann, B.; Brendel, M.; Henriques, J.A.P. Cashew (*Anacardium occidentale*) apple juice lowers mutagenicity of aflatoxin B1 in *S. typhimurium* TA102. *Genet. Mol. Biol.* **2005**, *28*, 328–333.

15. Nogueira, A.K.; Martins, J.J.L.; Lima, J.G.; Giroa, M.E.A.; Franco, K.; Maciel, V.M.; Loiola, O.D.; Valderez, M.; Barros, L.R.; de Santiago, R.S. Purification and characterization of a biosurfactant produced by *Bacillus subtilis* in cashew apple juice and its application in the remediation of oil-contaminated soil. *Colloids Surf. B Biointerfaces* **2019**, *175*, 256–263. [CrossRef] [PubMed]

16. Betiku, E.; Emeko, H.A.; Solomon, B.O. Fermentation parameter optimization of microbial oxalic acid production from cashew apple juice. *Heliyon* **2016**, *2*, e00082. [CrossRef] [PubMed]

17. Chagas, C.M.; Honorato, T.L.; Pinto, G.A.; Maia, G.A.; Rodrigues, S. Dextran sucrase production using cashew apple juice as substrate: Effect of phosphate and yeast extract addition. *Bioprocess Biosyst. Eng.* **2007**, *30*, 207–215. [CrossRef] [PubMed]

18. Rocha, M.V.; Souza, M.C.; Benedicto, S.C.; Bezerra, M.S.; Macedo, G.R.; Pinto, G.A.; Gonçalves, L.R. Production of biosurfactant by *Pseudomonas aeruginosa* grown on cashew apple juice. *Appl. Biochem. Biotechnol.* **2007**, *137–140*, 185–194.

19. Barros, E.M.; Rodrigues, T.H.S.; Pinheiro, A.D.T.; Angelim, A.L.; Melo, V.M.M.; Rocha, M.V.P.; Gonçalves, L.R.B. A yeast isolated from cashew apple juice and its ability to produce first- and second-generation ethanol. *Appl. Biochem. Biotechnol.* **2014**, *174*, 2762–2776. [CrossRef]

20. Talasila, U.; Vechalapu, R.R. Optimization of medium constituents for the production of bioethanol from cashew apple juice using Doehlert experimental design International. *J. Fruit Sci.* **2015**, *15*, 161–172. [CrossRef]

21. Teles, A.D.; da Silva, A.; Meneses, E.; Ceccato, S.R.; Marques, S.J.; Valderez, M.; Rocha, L. Mathematical modeling of the ethanol fermentation of cashew apple juice by a flocculent yeast: The effect of initial substrate concentration and temperature. *Bioprocess Biosyst. Eng.* **2017**, *40*, 1221–1235.

22. Kaprasob, R.; Kerdchoechuen, O.; Laohakunjit, N.; Thumthanaruk, B.; Shetty, K. Changes in physico-chemical, astringency, volatile compounds and antioxidant activity of fresh and concentrated cashew apple juice fermented with *Lactobacillus plantarum*. *J. Food Sci. Technol.* **2018**, *55*, 3979–3990. [CrossRef] [PubMed]

23. Kaprasob, R.; Kerdchoechuen, O.; Laohakunjit, N.; Somboonpanyakul, P.B. Vitamins and prebiotic fructooligosaccharides of cashew apple fermented with probiotic strains *Lactobacillus* spp., *Leuconostoc mesenteroides* and *Bifidobacterium longum*. *Process. Biochem.* **2018**, *70*, 9–19. [CrossRef]

24. Kaprasob, R.; Kerdchoechuen, O.; Laohakunjit, N.; Sarkar, D.; Shetty, K. Fermentation-based biotransformation of bioactive phenolics and volatile compounds from cashew apple juice by select lactic acid bacteria. *Process. Biochem.* **2017**, *59*, 141–149. [CrossRef]

25. Dantas, F.N.; Bezerra, J.; da Costa, M.; dos Santos, M.; Bertoldo, M.T.; Estevez, M.M.; de Souza, J.; Leite, E. Potential prebiotic properties of cashew apple (*Anacardium occidentale* L.) agro-industrial by product on *Lactobacillus* species. *J. Sci. Food Agric.* **2017**, *97*, 3712–3719.

26. de Godoy, E.; Soares, T.H.; Narciso, F.A.; Fernandes, A.L.; Narain, N.; Sousa, E.; Rodrigues, S. Chemometric evaluation of the volatile profile of probiotic melon and probiotic cashew juice. *Food Res. Int.* **2017**, *99*, 461–468. [CrossRef] [PubMed]

27. Araújo, S.M.; Silva, C.F.; Moreira, J.J.S.; Narain, N.; Souza, R.R. Biotechnological process for obtaining new fermented products from cashew apple fruit by *Saccharomyces cerevisiae* strains. *J. Ind. Microbiol. Biotechnol.* **2011**, *38*, 1161–1169. [CrossRef] [PubMed]

28. Mohanty, S.; Ray, P.; Swain, M.R.; Ray, R.C. Fermentation of cashew (*Anacardium occidentale* L.) 'apple' into wine. *J. Food Process. Pres.* **2005**, *30*, 314–322. [CrossRef]

29. Gamero, A.; Quintilla, R.; Groenewald, M.; Alkema, W.; Boekhout, T.; Hazelwood, L. High-throughput screening of a large collection of non-conventional yeasts reveals their potential for aroma formation in food fermentation. *Food Microbiol.* **2016**, *60*, 147–159. [CrossRef] [PubMed]

30. Dashko, S.; Zhou, N.; Tinta, T.; Sivilotti, P.; Lemut, M.S.; Trost, K.; Gamero, A.; Boekhout, T.; Butinar, L.; Vhrovsek, U.; et al. Use of non-conventional yeast improves the wine aroma profile of Ribolla Gialla. *J. Ind. Microbiol. Biotechnol.* **2015**, *42*, 997–1010. [CrossRef]

31. Zhou, N.; Schifferdecker, A.J.; Gamero, A.; Compagno, C.; Boekhout, T.; Piskur, J.; Knecht, W. *Kazachstania gamospora* and *Wickerhamomyces subpelliculosus*: Two alternative baker´s yeasts in the modern bakery. *Int. J. Food Microbiol.* **2017**, *250*, 45–58. [CrossRef]

32. Gutiérrez, A.; Boekhout, T.; Gojkovic, Z.; Katz, M. Evaluation of non-*Saccharomyces*yeasts in the fermentation of wine, beer and cider for the development of new beverages. *J. Inst. Brew.* **2018**, *124*, 389–402. [CrossRef]

33. Van Rijswijck, I.M.H.; Wolkers-Rooijackers, J.C.M.; Abee, T.; Smid, E.J. Performance of non-conventional yeasts in co-culture with brewers' yeast for steering ethanol and aroma production. *Microb. Biotechnol.* **2017**, *10*, 1591–1602. [CrossRef] [PubMed]

34. Adou, M.; Kouassi, D.A.; Tetchi, F.A.; Amani, G. Phenolic profile of cashew apple juice (*Anacardium occidentale* L.) from Yamoussoukro and Korhogo (Côte d'Ivoire). *J. Appl. Biosci.* **2012**, *49*, 3331–3338.

35. Georgé, S.; Brat, P.; Alter, P.; Amiot, M.J. Rapid determination of polyphenols and vitamin C in plant-derived products. *J. Agric. Food Chem.* **2005**, *53*, 1370–1373. [CrossRef] [PubMed]

36. Hernández, Y.; Lobo, M.G.; González, M. Determination of vitamin C in tropical fruits: A comparative evaluation of methods. *Food Chem.* **2006**, *96*, 654–664. [CrossRef]

37. Azevedo, D.C.S.; Rodrigues, A. Obtainment of high-fructose solutions from cashew (*Anacardium occidentale*) apple juice by simulated moving-bed chromatography. *Sep. Sci. Technol.* **2000**, *35*, 2561–2581. [CrossRef]

38. Assunção, R.B.; Mercadante, A.Z. Carotenoids and ascorbic acid from cashew apple (*Anacardium occidentale* L.): Variety and geographic effects. *Food Chem.* **2003**, *81*, 495–502. [CrossRef]

39. George, S.E.; Costenbader, C.J.; Melton, T. Diauxic growth in *Azotobacter vinelandii*. *J. Bacteriol.* **1985**, *164*, 866–871.

40. De Deken, R.H. The Crabtree effect: A regulatory system in yeast. *J. Gen. Microbiol.* **1966**, *44*, 149–156. [CrossRef]

41. Bely, M.; Stoeckle, P.; Masneuf-Pomarède, I.; Dubourdieu, D. Impact of mixed *Torulaspora delbrueckii*-*Saccharomyces cerevisiae* culture on high-sugar fermentation. *Int. J. Food Microbiol.* **2008**, *122*, 312–320. [CrossRef]

42. Erasmus, D.J. Impact of yeast strain on the production of acetic acid, glycerol, and the sensory attributes of icewine. *Am. J. Enol. Vitic.* **2004**, *4*, 371–378.

43. Finogenova, T.V.; Morgunov, I.G.; Kamzolova, S.V.; Chernyavskaya, O.G. Organic acid production by the yeast *Yarrowia lipolytica*: A review of prospects. *Appl. Biochem. Microbiol.* **2005**, *41*, 418–425. [CrossRef]

44. Petrescu, S.; Hulea, S.A.; Stan, R.; Avram, D.; Herlea, V. A yeast strain that uses D-galacturonic acid as a substrate for L-ascorbic acid biosynthesis. *Biotechnol. Lett.* **1992**, *14*, 6. [CrossRef]

45. Mazauric, J.P.; Salmon, J.M. Interactions between yeast lees and wine polyphenols during simulation of wine aging: II. Analysis of desorbed polyphenol compounds from yeast lees. *J. Agric. Food Chem.* **2006**, *54*, 3876–3881. [CrossRef] [PubMed]

46. Salmon, J.M.; Fornairon-Bonnefond, C.; Mazauric, J.P. Interactions between wine lees and polyphenols: Influence on oxygen consumption capacity during simulation of wine aging. *J. Food Sci.* **2002**, *67*, 1604–1609. [CrossRef]

47. Nisbet, M.A.; Tobias, H.J.; Brenna, J.T.; Sacks, G.L.; Mansfield, A.K. Quantifying the contribution of grape hexoses to wine volatiles by high-precision [U^{13}C]-glucose tracer studies. *J. Agric. Food Chem.* **2014**, *62*, 6820–6827. [CrossRef] [PubMed]

48. Leffingwell & Associates. Flavor-base 10th Edition. Available online: https://www.leffingwell.com/flavbase.htm (accessed on 27 June 2019).

MDPI

St. Alban-Anlage 66

4052 Basel

Switzerland

Tel. +41 61 683 77 34

Fax +41 61 302 89 18

www.mdpi.com

Fermentation Editorial Office

E-mail: fermentation@mdpi.com

www.mdpi.com/journal/fermentation